CAMBRIDGE LIBRARY COLLECTION

Books of enduring scholarly value

Earth Sciences

In the nineteenth century, geology emerged as a distinct academic discipline. It pointed the way towards the theory of evolution, as scientists including Gideon Mantell, Adam Sedgwick, Charles Lyell and Roderick Murchison began to use the evidence of minerals, rock formations and fossils to demonstrate that the earth was older by millions of years than the conventional, Bible-based wisdom had supposed. They argued convincingly that the climate, flora and fauna of the distant past could be deduced from geological evidence. Volcanic activity, the formation of mountains, and the action of glaciers and rivers, tides and ocean currents also became better understood. This series includes landmark publications by pioneers of the modern earth sciences, who advanced the scientific understanding of our planet and the processes by which it is constantly re-shaped.

A Description of Active and Extinct Volcanos

First published in 1826, at a time when the earth sciences were in a state of confusion and controversy, this pioneering study of volcanic action by Charles Daubeny (1795–1867) was significant in promoting the scientific method and the science of geology, at the same time establishing the author's international reputation. Having studied medicine, Daubeny increasingly turned his attention to chemistry, volcanos and earthquakes. Elected a fellow of the Royal Society, he also sought to elevate the status of science in Britain. He presents evidence here, gathered from his travels across Europe, in a methodical fashion, developing contemporary ideas regarding the processes at work beneath the surface of the earth. This reissued first edition provides an opportunity to examine Daubeny's reasoning prior to the revisions of the 1848 edition (also reissued in the Cambridge Library Collection), which was updated to take account of the work of Charles Darwin.

Cambridge University Press has long been a pioneer in the reissuing of out-of-print titles from its own backlist, producing digital reprints of books that are still sought after by scholars and students but could not be reprinted economically using traditional technology. The Cambridge Library Collection extends this activity to a wider range of books which are still of importance to researchers and professionals, either for the source material they contain, or as landmarks in the history of their academic discipline.

Drawing from the world-renowned collections in the Cambridge University Library and other partner libraries, and guided by the advice of experts in each subject area, Cambridge University Press is using state-of-the-art scanning machines in its own Printing House to capture the content of each book selected for inclusion. The files are processed to give a consistently clear, crisp image, and the books finished to the high quality standard for which the Press is recognised around the world. The latest print-on-demand technology ensures that the books will remain available indefinitely, and that orders for single or multiple copies can quickly be supplied.

The Cambridge Library Collection brings back to life books of enduring scholarly value (including out-of-copyright works originally issued by other publishers) across a wide range of disciplines in the humanities and social sciences and in science and technology.

A Description of
Active and Extinct Volcanos

With Remarks on their Origin,
their Chemical Phaenomena,
and the Character of their Products

CHARLES DAUBENY

CAMBRIDGE
UNIVERSITY PRESS

CAMBRIDGE
UNIVERSITY PRESS

University Printing House, Cambridge, CB2 8BS, United Kingdom

Cambridge University Press is part of the University of Cambridge.

It furthers the University's mission by disseminating knowledge in the pursuit of
education, learning and research at the highest international levels of excellence.

www.cambridge.org
Information on this title: www.cambridge.org/9781108075114

© in this compilation Cambridge University Press 2014

This edition first published 1826
This digitally printed version 2014

ISBN 978-1-108-07511-4 Paperback

This book reproduces the text of the original edition. The content and language reflect
the beliefs, practices and terminology of their time, and have not been updated.

Cambridge University Press wishes to make clear that the book, unless originally published
by Cambridge, is not being republished by, in association or collaboration with,
or with the endorsement or approval of, the original publisher or its successors in title.

VOLCANO OF JORULLO.

G.Hollis. fc.

Extentam tumefecit humum, ceu spiritus oris
Tendere vesicam solet, aut direpta bicornis
Terga capri. Tumor ille loci permansit et alti

A DESCRIPTION

OF

ACTIVE AND EXTINCT

V O L C A N O S;

WITH REMARKS ON

THEIR ORIGIN, THEIR CHEMICAL PHÆNOMENA, AND THE CHARACTER OF THEIR PRODUCTS,

AS DETERMINED BY THE CONDITION OF THE EARTH DURING THE PERIOD
OF THEIR FORMATION.

BEING THE SUBSTANCE OF SOME LECTURES DELIVERED
BEFORE THE UNIVERSITY OF OXFORD,

WITH MUCH ADDITIONAL MATTER.

———————

By CHARLES DAUBENY, M.D. F.R.S.

FELLOW OF THE GEOLOGICAL SOCIETY, AND OF THE COLLEGE OF PHYSICIANS
IN LONDON;
HONORARY MEMBER OF THE BRISTOL PHILOSOPHICAL AND LITERARY
INSTITUTION; CORRESPONDING ASSOCIATE OF THE GIOENEAN
SOCIETY OF NATURAL HISTORY AT CATANIA;
PROFESSOR OF CHEMISTRY, AND FELLOW OF MAGDALEN COLLEGE, OXFORD.

πολλα δ'ενερθ' υδεος πυρα καιεται.

London:

PRINTED AND PUBLISHED BY W. PHILLIPS, GEORGE-YARD, LOMBARD-STREET;
AND BY JOSEPH PARKER, OXFORD.

1826.

TO

JOHN KIDD, M.D. F.R.S.

Regius Professor of Medicine at Oxford,

WHO, BY HIS LECTURES ON GEOLOGY, FIRST CALLED THE ATTENTION OF THE
UNIVERSITY TO THIS IMPORTANT BRANCH OF NATURAL KNOWLEDGE,

AND TO

The Rev. WILLIAM BUCKLAND, D.D. F.R.S.

&c. &c.

WHO HAS SINCE SO LARGELY CONTRIBUTED BY HIS EXERTIONS, BOTH AS A MAN
OF SCIENCE, AND AS A LECTURER,

TO OPEN A WIDER FIELD, AND AWAKEN AN INCREASED INTEREST,
IN THE SAME STUDY,

THE FOLLOWING REMARKS

ON

𝕿𝖍𝖊 𝕮𝖍𝖊𝖒𝖎𝖈𝖆𝖑 𝖆𝖓𝖉 𝕲𝖊𝖔𝖑𝖔𝖌𝖎𝖈𝖆𝖑 𝕻𝖍𝖆𝖊𝖓𝖔𝖒𝖊𝖓𝖆 𝖔𝖋 𝖁𝖔𝖑𝖈𝖆𝖓𝖔𝖘,

ARE INSCRIBED

BY ONE WHO IS SENSIBLE HOW MUCH HE IS INDEBTED TO THEIR JOINT
INSTRUCTIONS;

HAVING RECEIVED HIS FIRST LESSONS IN CHEMISTRY FROM THE FORMER,
AND IN GEOLOGY FROM THE LATTER.

In the present state of geological science, a mineralogist could hardly employ himself better, than in traversing those ambiguous countries where so much has been ascribed to the antient operation of volcanic fire, and marking out what belongs clearly to the erupted or unerupted lavas, and what parts are of doubtful formation, containing no mark by which they may be referred to the one of these, more than to the other. Such a work would contribute very materially to illustrate the Natural History of the Earth.

Illustrations of the Huttonian Theory.

PREFACE.

—◆◆—

THE circumstance of a work like the present proceeding from a Professor of Chemistry seems to call for some explanation; for notwithstanding the near connection that subsists between the latter Science and every Department of Geological inquiry, yet it must be confessed, that the Study of Volcanos embraces in itself a field of such extent, that it ought to be entered upon as a principal, rather than as a subordinate occupation.

It is fair therefore to myself to mention, that the subject was first taken up at a time, when there appeared a reasonable prospect of my obtaining an appointment, which would have entailed the necessity of a five years absence from my native country.

The appointment in question I indeed ultimately lost, owing, as it was understood, to certain doubts that had been started with regard to my eligibility as a Candidate; but, as I had already formed the plan, and in some degree advanced in the details of the enquiry, I continued to prosecute it at intervals, not only for several years after my hopes of the situation alluded to had been frustrated, but even at a time when the office I afterwards obtained in the University of Oxford might have rendered a somewhat different line of pursuits more appropriate.

a 2

I have been obliged however in consquence to curtail in a considerable degree the scheme I had formed, which comprehended originally an examination of the Volcanos in the New, as well as of those in the Old World; and am under the necessity of now bringing forwards as a compilation, many parts of the work in which I had intended to introduce nothing but original matter.

It is satisfactory for me however to reflect, that I have visited most of the principal localities in Europe, noticed in my two first Lectures as the seats of volcanic action, so that with respect to them, even where facts are stated which did not fall within the compass of my own observation, I have been able to ascertain by going over the same ground, what degree of credit is due to the individuals on whose authority they are given.

In treating of the other Quarters of the Globe in which Volcanos occur, I have spared no pains in availing myself, to the best of my ability, of those resources, which a proximity to extensive public libraries has placed at my disposal, and therefore hope that this part of the work at least may be of use to future travellers; not merely by putting before them what is already ascertained, but likewise by directing their attention to those points which still require investigation.

I venture therefore to offer these Lectures as supplying in some degree, even in their imperfect state, a deficiency long felt in the geological literature of Great Britain; no treatise on the subject of Volcanos having appeared in this language since that of the Abbè Ordinaire, except indeed the recent publication of Mr. Poulett Scrope, which, though containing many ingenious views on the theoretical parts of the subject, is not calculated to supersede the demand for another work, expressly designed to convey a detailed statement of facts, with regard to the characters and situation of the rocks which owe their origin to subterraneous fire.

I have only further to add, that the remarks, made at the commencement of the First Lecture, with respect to the little attention that has been paid in Great Britain to the Department of Geology, which forms the subject of this work, must be understood as applying solely to that portion of it, which relates to Products confessedly Volcanic; for in no country has more important light been thrown on the nature of Trap and Basaltic Districts, than by the labours of Dr. Macculloch and of other English Geologists;* to some of whom I feel personally indebted, either for much of the information which forms the ground work of such an enquiry, or for the friendly assistance afforded me in the prosecution of it.

---

* See particularly the Memoir on the Coast of Antrim, by my friends the Rev. W. Buckland and W. Conybeare, in the Third Volume of the Geological Transactions, Supplementary to Dr. Berger's paper on the same district.

CONTENTS.

LECTURE I. & II.

DESCRIPTION OF THE VOLCANOS EXISTING CHIEFLY IN COUNTRIES VISITED BY THE AUTHOR.

LECTURE I.

ON THE VOLCANOS OF FRANCE AND GERMANY.

LECTURE II.

ON THE VOLCANIC COUNTRIES VISITED BY THE AUTHOR, IN 1823—24.

LECTURE III.

DESCRIPTION OF VOLCANOS IN COUNTRIES NOT VISITED BY THE AUTHOR.

AFRICA.

ASIA.

Asia Minor.

Contents.   xvii

LECTURE IV.

GENERAL INFERENCES RESPECTING VOLCANIC PHÆNOMENA.

LECTURES ON VOLCANOS.

LECTURE I.

ON THE VOLCANOS OF FRANCE AND GERMANY.

Introductory remarks.—Volcanos when to be considered active—when extinct—Characters by which the latter may be known—Classification of Volcanic products according to their Mineralogical constitution.

Extinct Volcanos of France—In Auvergne two classes— Post-diluvial and Ante-diluvial—the former class considered—their antiquity —Problematical Rock of the Puy de Dome, &c.—Mode of its formation discussed.— Second Class of Volcanic Rocks—1st, near Clermont, where they consist of Tuffs alternating with Freshwater Limestones.—2d, at the Mont Dor—its Basaltic and Trachytic Formation considered—3d, in Cantal.—Volcanic Rocks of the Puy en Velay—of the Vivarais— their age.--Volcanic Rocks in other parts of France briefly considered.

Volcanic Rocks of Germany—those near the Rhine—1st class, Post-diluvial.—Volcanos of the Eyfel—Crater of Laach—Lava of Niedermennig, Gerolstein, Mosenburg, Bertrich.—Trass of the Rhine considered—Whether the Post-diluvial Volcanos were in action since the existence

A

*of Historical Records.—2d Class, Ante-diluvial con-
sidered.—Of the Siebengebirge—Of the Westerwald—
Of the Vogelsgebirge—Basalts near Eisenach, Budingen, Hanau, Frankfort.—Volcanic Rocks of the Brisgau,
near Constance and in Wirtemburg, in Silesia, Bohemia,
Moravia, &c. briefly considered.*

It has often been a subject of dispute amongst Geologists, whether the processes, to which the earth is supposed
to owe its actual condition, were the same with any that are
taking place at present, differing only in magnitude, extent,
and duration; or whether they must be explained by assuming a totally distinct system of causes, which, since the
commencement of the present order of things, have ceased
to exist.

The latter is the opinion expressed by Dr. Kidd in the
close of his Geological Essay; and it is favoured more particularly by the appearances presented by the rolled masses
met with every where at the bottom of vallies, which are
now attributed by almost universal consent, to a body of
water differing both in its cause and mode of action from
our present rivers.

But this remark, however applicable it may be to the
ether forces that are now in operation, does not seem to
extend either to earthquakes or volcanos, from both which
agents effects have resulted even within the narrow limits of
our own observation, which, although inferior in point of
magnitude to some of those produced at former periods,
seem nevertheless analogous in kind. It is therefore rather
remarkable, that the progress of other departments of geology in this country should have tended so little to advance
our knowledge of these particular subjects, and that the
inquiries instituted with regard to the changes which the
earth has undergone, should not have produced more frequently an appeal fto a class of phænomena so capable of
illustrating them.

This is the more surprising when we consider that until lately the very questions that engrossed a more than ordinary share of attention, were precisely those which bore the nearest relation to the phænomena of volcanos; such, for instance, as the discussions relative to the origin of rocks of the trap family.

It is true that the resemblance existing between the products of actual volcanos, and the last mentioned class of formations, was often urged in proof of the general operation of heat; but the discussions to which this hypothesis gave rise, do not appear to have occasioned the same minute inquiry into the structure of volcanic districts, which they were the principal means of exciting with regard to whinstone and basalt.

This neglect may perhaps be ascribed in the first instance to the circumstances of the times, which were such during the period at which these questions were most warmly agitated, as rendered the parts of Europe wherein active volcanos occur but little accessible; and at a later period perhaps the same effect may have been produced by the preponderance of the arguments in favour of the igneous theory, which might seem to render it superfluous to hunt for proofs in distant countries, while facts were every day accumulating from sources nearer home.

Those however who have been withheld by the latter consideration from attending to these phænomena, ought to recollect, that it cannot be reckoned sufficient to have established the bare position that trap rocks are produced by heat, since many subordinate inquiries still remain with regard to the particular structure and relations of these rocks, which call for a more minute examination, and render an appeal to the phænomena of existing volcanos still of importance.

Nor will it be viewed as an unprofitable undertaking, if we are able to shew by an extensive induction of particulars, that the differences between the products of antient and modern volcanos in all these respects are such, as we

might deduce à priori from considering the condition of
the earth's surface at the several periods at which they were
in action.

I recollect so long ago as the year 1816, when I was pursuing my studies at Edinburgh, being led by something
like this train of thought, to meditate the excursions that I
have since accomplished, in the hope of supplying in some
measure this gap in our geological knowledge. It is true,
that at the time I made this resolution, I was far from viewing the question as I do at present, or from being persuaded,
as I now am, that volcanic and trap rocks are, for the most
part at least, analogous formations, calculated mutually to
reflect light upon each other; on the contrary, I was then
rather a convert to the views of Professor Jameson, whose
opinions on all subjects connected with geology, were received among his pupils with that respect, to which his
acknowledged accuracy and extent of practical information,
justly entitled them.

Still, with all my deference for the Professor's judgment,
I never rose from the enquiry without a conviction that
something was yet wanting to compleat the chain of his
proofs, and that in order to determine whether trap rocks
were really of igneous origin or not, the most effectual method would be to compare them in all their details with
products universally acknowledged to be volcanic.

I felt that for this purpose a mere examination of hand
specimens was not sufficient, the very spots themselves
should be visited, and the circumstances of geological
position, as well as the nature of the rocks associated, care-
fully compared with what we see in the trap districts, which
have excited so much attention and dispute.

I could not help wondering that this inquiry, intimately
connected as it is with the basaltic question, should never
have been taken up by any of the zealous supporters of
either system, that the volcanos of Auvergne for instance,
should be known to us chiefly through a French work of

rather an old date,* or a short German tract of Von
Buch's,† and that of the reputed volcanos of Hungary, we
should possess absolutely no authentic account, since one
author represented the whole country as of aqueous origin,‡
whilst another described the very craters from whence the
lava was ejected.‖

Our information with regard to the volcanic rocks of the
Vicentin, of Sicily, and even of the country round Naples,
was at that time still more imperfect, and although I am far
from wishing to conceal that since the period to which I
allude,§ many additions have been made to our knowledge
of these and other volcanic districts, yet I believe it must be
at the same time admitted, that even up to the present date
the information communicated, at least in the English language, has been too scanty to enable us to determine the
relation of these rocks to the trap formations of this and
other parts of Europe; whilst in elucidating the nature of
existing volcanos, the only direct fruit of the Huttonian

---

* Montlosier sur les Volcans d'Auvergne, 1802.

† Mineralogische Briefe aus Auvergne, in the 2d volume of his Geognostiche Beobachtungen, Berlin. 1809.

‡ Esmarck Kurze Beschreibung Einer Reise durch Ungarn, 1799.

‖ Fichtel Mineralogische Bemerkungen von den Karpathen, Wien, 1791.

§ I allude to the winter of 1816—17, which I spent at Edinburgh. In 1819
I visited Auvergne, and published a short account of my observations in
Jameson's Journal, vol. 3 and 4. In the same year the first notice of
Beudant's researches in Hungary appeared in Daubuisson's Traite de Geog-
nosie. They were three years afterwards more fully detailed in his own
work, from which I have drawn a great part of my account of the structure of
the volcanic rocks of that country. In 1820 Professor Buckland examined the
Vicentin, and satisfied himself with respect to the analogy which the strata in
that country seen alternating with volcanic products bear to the beds above
the chalk in England. He also visited Auvergne and part of Hungary. Mr.
Bakewell likewise has noticed Auvergne in his "Travels in the Tarentaise,
London, 1823." The foreign contributions to the knowledge of volcanos will
be seen by reference to my general list of works on the subject. It is to be
regretted that the death of Professor Playfair so soon after his return from
Italy should have prevented the public from being benefited by his researches
in that country.

controversy was the work of Sir G. Mackenzie on the rocks of Iceland.

I have therefore been since induced to devote such a portion of my time as could be spared from other occupations, to an examination of various volcanic districts, and as the facts collected in the course of several journies on the Continent, undertaken chiefly with that object, however inadequate they may be to supply the deficiency complained of, promise at least to contribute something to the existing mass of information, I propose to embody them in the following sketch of the general structure of volcanos in various parts of the world.

The most obvious and practical division of volcanos appears to be that into active and extinct, the former class comprehending those which at any period since the existence of authentic records have been in a state of eruption; the latter such, as, though incontestably of the same nature, have never been remembered to exhibit signs of activity.

By this definition we exclude from the immediate consideration all rocks, which, whether of igneous origin or not, are so constituted as to evince that they have been formed in a manner different from those of existing volcanos, and confine ourselves to such as consist in part at least of products, which not only are known to result from fire, but seem never to have occurred without its intervention.

Thus independently of the circumstances connected with the figure of the mountain, the direction of its strata, and the existence of a crater, by which a volcano is usually characterized, there are certain circumstances in the aspect of the individual masses which appear to afford decided indications of a similar origin.

When for instance we observe a mountain constituted of materials possessing even in part a vitreous aspect and fracture, together with a cellular structure, especially if these cells are elongated in the same direction, if they are in general unoccupied by crystalline matter, and have a glazed internal appearance, we need not hesitate in pronouncing the whole mass as volcanic, although all vestiges of a crater may be lost, and the form possess no analogy to that which belongs in general to mountains of the same class.

Now if we examine the rocks, which, from their possessing the above mechanical structure, and from other circumstances, we are led to consider as incontestably volcanic, it will be found, wherever their mineralogical characters are not wholly obliterated, that they appear to belong to one of two substances distinguished by the nature of the simple minerals of which they consist.

The 1st, which, from the harsh and earthy feel it often possesses, has been denominated trachyte, is essentially composed of crystals of glassy felspar, often cracked, which are imbedded in a basis generally considered as being itself a modification of compact felspar. To this are sometimes superadded crystals of hornblende, mica, iron pyrites, specular iron, and more rarely augite, and magnetic or titaniferous iron ore.

The 2d substance resulting from volcanic operations, appears to be some modification of basalt, consisting essentially of augite, felspar, and titaniferous, or magnetic iron ore, generally accompanied with olivine, and occasionally with hornblende. In many cases indeed the ingredients are too intimately mixed to allow of our ascertaining their nature, except perhaps by the ingenious mechanical method of Monsieur Cordier ;* but it is always easy to distinguish this kind of volcanic product from trachyte, which even when it has the colour of basalt, melts before the blowpipe into a white enamel, whilst the latter retains its original colour after being fused.

As the only essential ingredient therefore of trachyte appears to be felspar, whereas basalt always contains augite, I shall, in speaking of volcanic products, use the liberty of

---

* See Cordier sur les substances Minerales dites en Masse. Journ. de Phys. Vol. 82—83.

employing occasionally the term felspathic lava as synonymous to the former, and that of augitic lava to the latter.

In order however not to prejudge the question with reference to trap rocks, I intend to confine myself to the consideration of such trachytes, and such basalts, as possess, at least in part, that mechanical structure which I have laid down as characteristic of indisputable lavas; and hence, as it will appear in the sequel, my attention will be limited to those formations which were produced at a comparatively recent epoch in the history of our planet.

Whatever conclusions indeed we may be disposed to draw with respect to the origin of certain of the older rocks, as for instance of the basalts and porphyries of the county of Antrim, it is clear that they ought to be distinguished from those forming the subject matter of these Lectures, which, from their close resemblance to the products of existing volcanos, are at once recognized as analogous formations; since the nature of the inference in these two cases differs not in degree, but in kind, the one being a deduction of reason—the other an immediate result of observation.

On the Volcanos of France.

The volcanic country to which my attention was originally directed, in the hope of obtaining some additional insight into the nature and origin of basalt, was that which occupies a considerable tract in the centre of France, known under the general term of Auvergne. This, and some other parts of the same country, also of a volcanic nature, were visited by me in the year 1819, and the result of my observations was in part inserted in the Edinburgh Journal of Science for the ensuing year. In the course of the summer of 1823, I also spent a few days in the same country, and cleared up two or three points which in my former visit I had left undecided.

In my letters to Professor Jameson above referred to, I distinguished these rocks into two classes, the one formed *before*, the latter *since* the vallies were excavated; and in order to keep constantly in view this important difference, proposed to apply the term *post-diluvial* to the one, and *ante-diluvial* to the other. The adoption of such a nomenclature must not be supposed to imply the expression of any opinion on my part with respect to the much agitated question as to the identity of the particular deluge recorded in the Mosaic history, with the cause to which the excavations of the vallies and the formation of beds of gravel are to be referred.* All that is intended to be conveyed is a statement as to the *relative* antiquity of the rocks referred to these respective classes; and the terms *ante-diluvial* and *post-diluvial* seem preferable to *ancient* and *modern*, because the latter imply only their relation to each other; the former also their connection to the rocks with which they are associated.

Now the post-diluvial volcanic products in Auvergne are distinguished, as we shall see, from the remaining class in their external characters, as well as their position. They are more cellular, have in general a harsher feel, and more of a vitreous aspect, their surface presenting a series of minute elevations and depressions, and the scanty portion of soil which covers them affording but little pasturage, and that generally of the worst description.

The mountains referred to this division constitute a chain which rises considerably above the elevated granitic platform on whieh they rest, and extend at intervals over a space of above eight leagues from north to south; from whence the rocks which compose them may be often traced a considerable way into the valleys contiguous. Above

---

* See the note on that subject attached to my paper entitled " Sketch of the Geology of Sicily." Edinburgh Philos. Journal for October 1825 *sub fine.*

sixty of these eminencies might, I believe, be enumerated
within the boundary marked out; but as their number renders selection necessary, I shall simply notice such as are
most remarkable, beginning with that of Volvic near Riom,
the lava of which furnishes a considerable part of the build-
ing-stone used in that neighbourhood, and in spite of its
porous character, is exceedingly durable.

The fact of its having descended in a liquid form from
the mountain above, and that at a period subsequent to any
of the great revolutions which have changed the face of our
planet, is demonstrated by the exactness with which the
stream has modelled its course to the slope of the valley;
and that its fluidity was owing to heat, is evident enough
from its porous texture and semi-vitreous aspect; so that its
connection with volcanos now in activity seems sufficiently
apparent. On the summit of the Puy de Nugere, is a bason-
shaped cavity of an oblong form, broken away on the side
down which the lava has taken its course, and, notwithstanding the changes which time has effected in its form,
still retaining marks of having been once the crater, from
whence the lava of Volvic was ejected.

It is interesting to remark, that the stream in its descent
appears to have been arrested by a sort of knoll of granite,
which probably rose considerably above the general level,
and by the obstacle it opposed to its progress, caused it to
divide into two branches, between which this little granitic
eminence is seen protruding, a solitary vestige of the rock
which formerly existed on the surface, but is now overspread with lava. The two branches of the main stream
appear to have become reunited below; and having descended the slope of the hill, to have spread themselves over
the valley of Volvic, extending to within a mile of the town
of Riom.

It has been remarked by Von Buch that the lava of Volvic
seems chiefly to have consisted of felspar, and to have resulted from the fusion of some trachytic rock; and it is
certain that it is of a lighter colour than most other lavas,

and does not contain augite. It is also distinguished by exerting no action upon the needle, containing little or no magnetic iron ore; but on the other hand, it is quite full of specular iron, which occurs in the crevices of the rock, as well as disseminated through its porous structure in minute plates of a bright metallic lustre. This mineral, which is a common product of volcanos now in activity, from which it is supposed to have been sublimed, is met with in various other parts of Auvergne, as at the Puy de la Vache, the Puy de Dôme, and among the trachytes of Mount Dor.

The lava of the Puy de Côme, a mountain a few miles to the south-west of Clermont, is described by M. Montlozier, the author of a little essay on the Theory of the Volcanos of Auvergne, as equally interesting with that of Volvic. The lava that has flowed from the hill divides, he says, into two branches, one of which flows directly into the bed of the river Sioule, whilst the other takes the direction of a place called Tournebise, reaches the village of Mont Gibaud, and terminates like the other, by flowing into the bed of the river, about three miles lower down.

A torrent of this description might naturally be expected to effect singular changes in the face of the country which it traverses. Accordingly we shall find that it has blocked up a little valley which formerly seems to have had a drainage to the west, on the side of Chambois and Massanges, and has converted it into a sort of swamp, known by the name of the Lac de Côme.

Lower down the same lava has occasioned still greater changes. The rivers Sioule and Monges formerly ran parallel, in a direction from south to north, and entered the plain of Mont Gibaud by two defiles separated by the intervention of two hills. But one branch of the lava of Côme has so obstructed the course of the river Monges, that its waters have been turned aside to the left, where they have worked themselves a passage through an argillaceous hill, made immense excavations in it, and in this manner have reached the bed of the river Sioule, a league and a half higher up than they would naturally have done.

Compelled however to flow in a direction contrary to the slope of the country, a large portion of the waters constantly stagnates in its channel, and has formed a swamp which goes by the name of the " Etang de Fung."*

A somewhat similar circumstance has happened in the case of the lake Aidat, which seems to have been originally formed by a stream of lava now stretching across it. In this case, however, a still greater impediment existing to the escape of the waters by any other outlet, they have in process of time succeeded in cutting themselves channels through the parapet of lava thrown across them, the projecting portions of which appear like islands in the midst.

The stream of lava that has occasioned this impediment appears to have been furnished by one of three mountains, all of which have given out *coulées* flowing in the same direction, and therefore intermixed one with the other. The most considerable of these mountains is called the Puy de la Vache, the whole of which is composed of scoriaceous lava very different from that of Volvic, as it contains much iron in the state of magnetic, as well as in that of specular iron ore, the oxydation of which imparts a general redness to the rock, and likewise occasional crystals of augite and olivine. There would seem to have been formerly a crater on the summit, three sides of which are now standing, whilst the fourth was perhaps broken away by the stream of lava which descended from that quarter. The *Coulée* is easily followed with the eye along the valley as far as the lake, in consequence of the irregularities of its surface, and the ridge which it forms above the level plain.

The most perfect crater, however, which exists in Auvergne is that of the Puy Pariou, north of the town of Clermont. It is perfectly round, and according to M. Ramond, more than 250 feet in depth. Its structure is simple enough, consisting wholly of loose masses of slaggy

* This is very well laid down in Desmarest's Map of Auvergne. Paris 1823.

lava, sufficiently decomposed to allow of the growth of turf. It has given off a stream of lava which may be traced southward to the place called " les Barraques," where it divides into two branches, descending the slope of the gra- nitic hills between that spot and Clermont.

Of the modern Volcanos, however, in this neighbourhood, there is probably none more interesting than the Puy Graveneire. This mountain, which lies within two miles of Clermont, seems, as we approach its summit, to consist of an entire mass of cinders, so that we may in some degree comprehend the origin of a ludicrous opinion ascribed to a professor of the Academy of Clermont, when the volcanic nature of the rocks of Auvergne was first asserted, and maintained by an appeal to the structure of this particular mountain, who, it is said, accounted for the scoriæ found on its surface, by gravely remarking that he had heard of iron-founderies having formerly been established on the spot. Notwithstanding such strong indications of its having been in a state of ignition at a comparatively recent era, no trace of its crater can be detected, nor has it that abrupt and conical form characteristic of volcanic hills, being rather a long, round-backed eminence, rising abruptly indeed on two of its sides, but to the north connected with the chain of the Puy de Dôme, and to the south reaching into the plain of Limagne. In spite of the absence of a crater, two streams of lava appear to have pierced the sides of this mountain, and to have descended into the valley, one on the side of the village of Royat, the other on that of the Puy Montaudoux. These coulées display a singular intermixture of compact and cellular lava, the former gene- rally occupying the centre, and surrounded by the latter variety, but without any marked line of demarcation be- tween the two. The compact rock is a basalt, remarkable for its large distinct crystals of augite and olivine, and its being seen in connection with a lava of so cellular and vitreous an aspect affords, in common with the facts I shall detail, with respect to the German volcanos, a sufficient proof, that pressure is not always necessary for the formation of such products.

I shall not stop to particularize any larger number of the more recent class of volcanos, as they are much the same in their characters with those already enumerated, and differ very little from such as are at present in activity in other parts of the world. Indeed even the streams of lava which they have given out, are often so little decomposed, so partially covered with vegetation, that we not only readily admit their post-diluvial origin, but even imagine they must have been formed within the limits of authentic history. The records nevertheless of their eruption are no where to be found, and the evidence we are in quest of can only, it would seem, be collected from the volume of nature, which in this instance speaks a language so intelligible ; for with regard to the popular names of certain of the mountains and vallies,* to which some have referred as indications of a remote tradition, it seems more probable that they were applied to the places they designate, in consequence of the ideas which their appearances were calculated to suggest to the minds of their first inhabitants, than from the latter having been themselves eye-witnesses of the events which occasioned them.

The high antiquity of the most modern of these volcanos is indeed sufficiently obvious. Had any of them been in a state of activity in the age of Julius Cæsar, that general, who encamped upon the plains of Auvergne, and laid siege to its principal city,† could hardly have failed to notice them. Had there been even any record of their existence in the time of Pliny or Sidonius Apollinaris, the one would scarcely have omitted to make mention of it in his Natural History, nor the other to introduce some allusion to it among his descriptions of this his native province.

---

* Such are Montbrul, Vallée d'Enfer ; and perhaps the very name of the province may be derived from certain appearances that might have reminded its first settlers of the lake *Avernus* near Naples.

† Gergovia, near Clermont.

The case is even stronger, when we recollect that the poet's residence was on the borders of the Lake Aidat, which owed its very existence to one of the most modern volcanos; and that he was aware of the nature of such phænomena, appears from a letter extant of his addressed to the Bishop of Vienne,* in which, under the apprehension of an attack from the Goths, he informs him that he is going to enjoin public prayers, similar to those which the bishop had established, at the time when *earthquakes demolished the walls of Vienne, when the mountains opened and vomited forth torrents of inflamed materials, and the wild beasts, driven from the woods by fire and terror, retired into the towns, where they made great ravages.*

A passage of this kind, though it may be brought forward as an argument in favour of the modern date of some of the volcanos in the neighbouring province of Vivarais, affords I think strong negative evidence of the antiquity of those in Auvergne, and disposes us to assign to them an æra as remote, as is consistent with the fact of their posteriority to the formation of the vallies.

Let us now proceed to the consideration of another description of rocks found in the same neighbourhood, the nature and origin of which appear to be somewhat more problematical, and less in harmony with the phænomena of volcanos at present in activity.

The department of which Clermont is the capital, has received its name from a mountain, which as the highest in the province, and occurring in some degree detached from the rest, has acquired more importance than it might in other situations have obtained, although indeed its height is considerable, exceeding 4000 feet. The Puy de Dôme, the hill to which I allude, is of a conical form, and remarkable for the distinctness of its outline, rising abruptly from the midst of a sort of amphitheatre of volcanic rocks, which it considerably overtops, but which by a little stretch of the

* See Sidon. Apoll. Lib. 7. Epist. 1. ad Mamertum.

imagination might be supposed to have constituted the crater from whence this great central mass was projected.

However this may be, the mineralogical characters of the mountain are such, as differ entirely from those of the hills on either side of it. The Puy de Dôme seems to consist almost entirely of a rock with a felspar base, allied as it would seem to trachyte, but of a more earthy character, and containing more rarely crystals of glassy felspar. These however do occur even in the most pulverulent part of the rock, and are common in the more compact portions, where indeed the resemblance to trachyte is often so perfect, as to leave us in little doubt with respect to the real nature of the rock in general. The term Domite therefore, which was originally assigned to it from its occurring in the Puy de Dôme, must be considered as expressive merely of a variety or subspecies of trachyte, marked by the earthy character of its basis, and by its whitish or greyish colour. It contains numerous plates of mica disseminated, as well as of specular iron, which also forms a thin superficial coating on the stone between its crevices.

It also contains occasionally quartz, grains of which are sometimes so disseminated as to give an arenaceous character to the rock.

The most remarkable circumstance relating to this substance, is that it is confined to this hill, and two or three in its immediate vicinity; which, though they all present some modifications of aspect, still possess sufficient of a common character to be referable to the same class.

They are all conical, all detached, and have surrounding them hills of a volcanic nature, which bear not the slightest analogy to them in appearance. I shall refer to M. Montlozier, and the other writers who have described them, for an account of the Grand and Petit Cliersou and the Sarcouy, and shall confine myself to some remarks on the Puy Chopine, the most extraordinary certainly for the assemblage of rocks of which it is made up.

This mountain, which is situated to the N. W. of Clermont, about half way between that city and the village of Pont Gibaud, has long puzzled geologists, from the singular confusion and anomalous structure of the rocks which compose it. Owing indeed to the quantity of debris which every where covers its sides, where not concealed by vegetion, it is difficult to determine with precision the position they occupy, or the relations they bear to each other. On climbing to its summit, I found, *in situ,* a rock analogous to domite, unaltered granite, and a conglomerate with a granitic base, rocks which seem to be related to each other. Lower down I observed a granular hornblende rock, which appeared to pass into the granite; and these four substances make up, so far as my observations extend, the higher portions of the mountain. Lower down, we have lavas, both compact and vesicular, none of which, so far as I observed, occupy the summit, although Mons. Montlozier, who examined the spot doubtless with more attention, says he saw one small portion extending thus high. It should be remembered that the Puy Chopine, even more distinctly than the Puy de Dôme, is encircled by an amphitheatre of hills, which are comprehended under the names of the Puy Chaumont, and the Montagne des Gouttes. I examined these hills and found them all to be volcanic, consisting chiefly of a tuff containing portions of scoriæ, and lavas of various denominations, all cemented together by an ochreous paste.

Such, so far as I observed, appears to be the constitution of the Puy Chopine; and the singular assemblage of rocks which it comprises, whilst it serves to explain its own formation, may perhaps furnish us with a clue to the theory of the Puy de Dôme, and the other mountains similarly constituted. Encompassed on all sides by volcanic rocks, and bearing in themselves evidences of the agency of fire, the igneous origin of these latter mountains will scarcely be disputed; but the precise manner in which they have been affected by this agent still admits of a question. It may be said, for instance, either that the rocks now consisting of domite

always occupied the same position, but have been altered
in character by volcanic agency; or it may be supposed
that they are altogether new formations, thrown up from a
great depth by the same agent to which they owe their
mineralogical characters. Again it may be asked, whether
admitting the latter view of the case to be the correct one,
we are to suppose each of the domite hills to have been
ejected separately, or to have formed parts of one continuous stream of lava, the intervening portions of which have
since been swept away.

With regard to the first hypothesis which assumes an
alteration of character to have taken place in this substance,
without any change of position, it is sufficient to observe
that domite differs so completely from any other known
rock, that we cannot imagine the change to have taken
place without a fusion of the ingredients, which renders it
impossible that the rocks could at that time have stood at
their present elevation. Discarding therefore this hypothesis, let us confine ourselves to the inquiry, whether on
the assumption that these rocks are foreign to the situation
they occupy, we are to consider them as separately thrown
up like the volcano of Jurullo in Mexico, which will be
afterwards considered,* or as the relics of one continuous
stratum of felspathic lava. The latter opinion has found an
advocate in Mons. Daubuisson, who seems to consider the
Puy de Dôme, and its accompanying hills as outliers of the
great trachytic formation which extends over the Mont
Dor. Von Buch, on the contrary, whose opinion on such
subjects is entitled to great weight, imagines that the mountains composed of domite have been thrown up from below,
elaborated from the materials of the fundamental granite,
altered partly by the effect of heat, and partly by elastic
vapours.

The difficulty of supposing so complete a destruction of a
stratum as is implied by Mons. Daubuisson's hypothesis,

* See my Third Lecture.

leads me to prefer that part of Von Buch's theory relating to the separate elevation of the domite mountains, which is also favoured by their conical form, and by their being placed in the midst of a volcanic country.

From what materials this singular rock can have been produced, seems still more problematical; if, as it is most probable, from the subjacent granite, what is become of the greater part of the quartz, which forms so essential and abundant an ingredient in the latter rock? why has a heat, capable of dissipating so large a portion of this refractory material, and of reducing the felspar to a glassy, and often to a pulverulent state, left the crystals of hornblende untouched, and apparently effected no alteration in the mica? These are questions, which in the present state of our knowledge it may be difficult to answer, although, it should be remarked, that in estimating the amount of the quartz dissipated, we should be wrong to calculate it from the difference between the quantity of that mineral existing in the granite and in the domite; but must deduce it from a comparison of the quantity of silex that would be contained in the quartz we suppose to have disappeared, and that existing in the felspar substituted. Vauquelin has shewn that the domite of Sarcouy contains 92 per cent. of silica, a proportion I should conceive quite equal to that existing in most granites. With regard to the crystals found in the domite, these probably are rather the results of the igneous action, than the unaltered relics of the original stratum.

On the other hand, it is impossible to return from viewing the Puy Chopine, without feeling a persuasion that the granites and domites there seen associated, are in a certain degree connected with each other, and that in all probability the latter has been formed by the agency of heat, modified by peculiar circumstances, out of the materials of the former.

Adopting this theory therefore as the most probable that has been offered, we may account for the intermixture of hornblende rock, by supposing that it formed beds in the granite which was thrown up, whilst the unequal operation of heat may explain the occurrence of the latter substance as well as of the former, unaltered in the midst of domite. With regard to the volcanic trap or lavas which occupy the lower portions of the mountain, we may consider these as detached from the rocks that occur in the Montagne de Giouttes and Puy Chaumont contiguous, elevated as it would appear by the same process which thrust up the granite and domite through the midst of them.

The geologist, who adopts this view of the subject, will regard the modifications of appearance, observable in the rocks which have been referred to the general head of domite, as arising from some difference, either in the intensity of the heat to which they were severally subjected, or in the mode of its application, rather than in the material from which it was produced.

Thus, M. Montlozier has observed that the Puy de Monchar, a mountain to the N. W. of the Puy de Dôme on the road to Aurillac, seems merely to have been forced up, without having experienced any material alteration in structure, for though partly composed of scoriæ, and other volcanic products, yet it is also made up of masses of unaltered granite, unaccompanied, as at the Puy Chopine, with domite, but in such disorder, as plainly demonstrates, that they do not exist in their natural position.

The second stage of alteration is seen in the case of the Puy Chopine, where the granite is not only raised by some expansive force from the spot it originally occupied, but also partially converted into the state of domite, whilst a portion still unchanged, remains as an evidence of the materials from which the former was produced.

Lastly, in the case of the two Cliersous, the Grand Sarcouy, and the Puy de Dôme, the change from granite into domite is complete throughout, and the whole reduced into a spongy pulverulent mass, as is particularly seen in the Puy Sarcouy. It is stated that the latter rock still exhales the odour of muriatic acid, and that the presence of that substance has been ascertained in it by Vauquelin; but

whether this was one of those elastic vapours which according to some theorists assisted in altering the characters of
the granite, must be left for future investigation.

Neither have we any certain data on which to found a
conjecture respecting the age of this rock; it is probable
indeed that its date is not the same as that of the trachytes
of Mont Dor, which we shall next consider, since we rarely,
if ever, find amongst that extensive formation a rock exactly
corresponding to that of the Puy de Dôme, and the analogy
of the neighbouring volcanic rocks would lead us to infer
that it was more modern; yet it must be confessed that we
want in this instance that direct evidence which is afforded
us by those volcanos that have given forth currents of lava.

On the Ante-diluvial Volcanic Rocks near Clermont.

W E are not to suppose, because the neighbourhood of Clermont is the principal seat of the more recent volcanos, that no
older rocks, referable to the same class, are therefore found
about it. The basalt of Montaudoux, which Dr. Boué has
remarked as identical with the rock of Calder, between Glas-
gow and Edinburgh, evidently belongs to an era much more
remote than that of the scoriaceous lava of Graveneire, to
which it is so near. The mountain Gergovia, situated a
little farther to the south, is capped with a basalt, partly
compact, and partly amygdaloidal, containing minute crystals of mesotype, pretty abundantly disseminated, which,
of course, must be attributed to a period of time, anterior
to the excavation of the valley which it overlooks. The
Puy Charade, on the other side of Clermont, is also capped
with a volcanic rock, whilst the sides and base of the mountain are composed entirely of granite. These instances are
sufficient to establish the ante-diluvial date of many of the

trap rocks near Clermont, and that they are of volcanic origin will be admitted even by those who contend for the aqueous theory in general, from their association with beds of tuff containing scoriæ and other cellular products.

Thus at the Puy Marmont near Veyre, about three leagues south of Clermont, on the road to Brioude, the rock is capped with basalt, like that of Gergovia, underneath which is a calcareous rock, identical both in its external characters and imbedded petrifactions with that which occurs in the latter locality. This is followed by a thick stratum, composed of a sort of tuff containing imbedded portions, not only of basalt and other trap-rocks, but even of limestone. The paste by which these ingredients are held together, partakes in some places of the characters of wacke, but where it approaches the bed underneath it, the cement itself becomes calcareous, in which case the only difference between the strata consists in the presence or absence of the imbedded fragments. In like manner the limestone bed is at bottom interspersed with fragments of the volcanic matter, which becoming by degrees more and more frequent, give it at length the characters of a tuff. Basalt also seems to occur interstratified with the rocks above noticed, and the whole series, which, from its present highly inclined position, seems to have undergone some change since the period of its original formation, rests finally upon the limestone of the plain of Limagne.

Whether the tuff in this instance contains any scoriform lava, I am not prepared to say, but its origin is clearly the same with that of the other rocks consisting of the same material, which occur in various places resting upon the same limestone in the plain of Limagne.

Two of these rocks occur a mile or two south of Clermont, near the road of Brioude, one of them constituting a little knoll, hardly perceptible until we are close to it; the other attaining a considerable height, and remarkable for the abruptness with which it rises out of the midst of so

level a plain, reminding us of the Roche de St. Michael, and
the other eminences which will be noticed as occurring
near the Puy en Velay.

The first of these rocks is called the Puy de la Pege or
Puy de la Poix; the other or larger one, the Puy Crouelle.
The Puy de la Pege consists entirely of a kind of tuff,
strongly impregnated with bitumen, which covers the external surface with a kind of varnish, and fills all the crevices in the rock. This tuff often contains fragments of
vesicular as well as of compact lava, the former connecting
its origin with modern volcanic products, the latter with the
tuffs of the surrounding country. In some cases the brecciated or conglomerated character of the rock was lost,
and the prevailing substance seemed to be a species of trap-
rock, which from the unequal manner in which it has decomposed, exhibits a number of light spots disseminated
through a darker ground.

The Puy Crouelle, which is about half a mile distant,
is composed of the same tuff equally penetrated with asphaltum. Unaltered portions of the limestone occur in the
midst of it, and the same substance constitutes the stratum
which immediately supports the tuff. The line of junction
between the two formations is as might be expected irregular, so that the limestone seems in some places to send up
wedge-shaped processes into the incumbent rock. The
commencement of the volcanic stratum forms a sort of
natural boundary, beyond which no vineyards are to be
seen, whether it is that the greater abruptness of this rock
prevented their growth, or that the nature of the material
itself was hostile to vegetation.

In my letter to Professor Jameson, I stated it as my
opinion, that these isolated hillocks of tuff were raised by
some volcanic agency from beneath, through the limestone
on which they seem to repose, an hypothesis which I see is
adopted by a German geologist, M. Steininger, who has
written still more lately on the subject. Nevertheless on
reconsidering the question, and comparing the tuff of

Limagne with that near the Puy en Velay, I am at present more inclined to believe that these rocks, with others of the same description that lie scattered in the different parts of this district, are relics of a more extensive stratum, the intervening portions of which have been removed.

It may be difficult perhaps to suppose so complete a destruction of a rock as is implied in the latter hypothesis, but it seems still more so to reconcile the general phænomena of the tuff, its alternation with Neptunian products, the occasional presence of shells, and of unaltered portions of extraneous rocks, &c. to igneous ejection.

It would be premature however to enter at present into the question relative to the origin of volcanic tuff, since an inquiry of this nature would oblige me to appeal to the volcanic rocks in other parts of Europe which have not yet come before us. I shall therefore only observe that vestiges of a similar tufaceous rock holding the same relation to the limestone, are seen in various parts of the plain of Limagne, especially at the town of Pont du Chateau, where the surface of the stone is coated with those fine Chalcedonies so highly prized by collectors.

I must not forget a formation somewhat different from the above which occurs a little to the west of the Puy Marmont at the hill of Mouton. It bears some resemblance to the trass of the Rhine volcanos, containing imbedded portions of pumice as well as of scoriæ and compact lava, but, like the tuff of the contiguous hill, the paste by which the fragments are held together, seems to be often calcareous.

It should be noticed, as one of the few localities in Auvergne where pumice is to be found, which seems the more remarkable, as this substance is a common product of that class of volcanos, which consists of trachyte,

Now besides the problematical rocks near Clermont already noticed, as consisting of a variety of that substance, an extensive trachytic formation occupies the principal portion of the neighbouring chain of Mont Dor, and extends over the greater part of the adjoining department of Cantal.

It is connected with the ante-diluvial rocks of volcanic origin near Clermont, by its age, as well as other circumstances, being seen in Cantal resting upon tertiary freshwater limestone, and covered by rocks which recall to our recollection the basalts of the plain of Limagne.

The table land known under the name of Mont Dor, embraces a circumference of about 12 leagues between the towns of Rochefort, La Tour d'Auvergne, and Besse, and the highest point in the range attains the elevation of 5829 feet above the level of the sea.

Though its shape at a distance corresponds more to what we observe in many transition districts, yet it appears to be entirely of volcanic origin, and two distinct classes of rocks, both referable to this cause, may be distinguished.

On the summit we generally observe a basaltic formation, associated with a sort of trap tuff or breccia, and a very cellular description of lava.

Underneath this is that porphyritic felspar rock generally known under the name of trachyte, with which are associated a conglomerate made up of fragments of the same rock, as well as of basalt and cellular lava, cemented by a felspathic paste, and a rock apparently homogeneous, but in reality analogous to trass, consisting of finely comminuted portions of volcanic matter. Underneath these, but at so low a level, as hardly to justify its being considered as making a part of the Mont Dor range, is seen the granite, which seems to constitute the original substratum throughout the whole of this Province; the volcanic and even the freshwater formation, hardly deserving to be viewed as parts of the crust of the earth, but being rather in some sense extraneous to its composition.

I shall consider these rocks in succession, beginning with the highest in the series :—

1st. The basaltic formation of Mont Dor comprises several rocks, differing from each other much in appearance and external characters.

The rock which has induced me to give this name to the whole, is a compact and sonorous basalt, containing occasionally crystals of olivine, and more commonly some of augite and hornblende, the latter having frequently an acicular form.

Associated with it is a vesicular rock, to which the name of lava may not improperly be given, since the cavities which it contains possess no appearance of having been once *filled,* like those in the basaltic amygdaloids of secondary countries, and its general aspect obliges us to refer it to the same class with the undisputed products of volcanos now in activity. We meet also with scoriæ, either in detached fragments, or with portions of compact trap, the whole cemented together by iron-clay, so as to constitute a species of volcanic tuff.

I found the order of superposition on a hill which I examined near Lake Gery, a few miles from the village called " Les Bains de Mont Dor," to be as follows :

On the summit a thin bed of scoriæ. Underneath, a tuff, containing fragments of the more compact united with the vesicular variety of lava, but in some places in a state of such extreme division, that the whole might be mistaken at a distance for red sandstone. Beneath all was a compact and crystalline basalt, made up of a confused assemblage of these acicular crystals of hornblende, together with augite and felspar.

It would seem that this is the general order of superposition; and the pressure of the vesicular matter above may possibly have contributed to give to the rock underneath the hardness and consistency which it possesses, although it must be confessed at the same time, that basalt exists amongst modern lavas in situations, where it would seem impossible to assume the existence of any kind of pressure, at least at the time the lava was ejected.

From the description we have given, it will be easily understood, that the subjacent rock, the trachyte, is chiefly seen exposed on the sides and in the bottom of the vallies; the great and elevated table land, which composes the range of Mont Dor, and extends with little interruption into Cantal, having its upper strata generally composed of basalt, and the other rocks associated with it.

Yet this arrangement does not appear to be universal, for the basaltic platform has not extended over the whole surface of the table land, and I observed in particular that the summit of the highest peak called the Pic de Sancy, consists of trachyte; nor is it confined to the more elevated situations, but in many places, especially in the flanks of the Mont Dor, is seen at a lower level than the greater part of the porphyritic stratum.

The latter case occurs at the " Cascade du Quereuil," near the Baths, where the basalt is seen in fine columns incumbent upon a rock, which, from its analogies to one found among the Rhine volcanos, I shall call trass, and covered with volcanic tuff. Now as the former rock rests immediately upon granite, without the intervention of any trachyte, a recent German writer, M. Steininger, has concluded that the latter formation probably lies above this basaltic rock, and consequently that the Mont Dor range contains two sets of basaltic lavas, the one superior to the trachyte, the other occurring in the midst of it. But as this Geologist does not appear to have adduced any decided instance, in which the trachyte is seen superimposed on the basalt, it may be better to imitate the caution of Monsieur Ramond, whose experience in the geology of Auvergne is probably greater than that of any other observer, and who, without altogether denying the fact of an alternation between the two formations, confesses that the general aspect of the phænomena leads him to conjecture, that the irregular disposition of the basalt may be explained by supposing it to have descended from the highest points of the chain, and filled up the hollows and crevices in the porphyritic mountains beneath.

At all events the parallel which Steininger has attempted to draw between his two formations of basalt in the Mont Dor, and the antient and modern basaltic lavas near Clermont, fails in a very essential point, since the most modern of the volcanic rocks of Mont Dor have been produced antecedently to the formation of the vallies, whilst those near Clermont are, as we have seen, post-diluvial.

There is indeed near a little lake, called the Lake of Servieres, which stands on the summit of the table land, and in its form resembles a crater, a rock in height not exceeding fifty feet, which, from its conical form, and the analogy which the substances composing it bear to the products of recent volcanos, seem to be more modern than the surrounding country.

This fact however does not affect the question with regard to the general antiquity of these basalts, which supposing them to have been thrown out since the existence of the present order of things, would have been found at the bottom of the vallies by which the country is intersected.

2. The trachytic formation is essentially composed of crystals of glassy felspar, imbedded in a base which seems to be of the same material. Its fracture is more commonly rough and earthy, but is not unfrequently compact. In the latter state it is that mica, hornblende, and other minerals, are found imbedded, whilst it is the former variety which contains the finest and most regular crystals of glassy felspar. It passes sometimes into pitchstone porphyry, as at the Vallée d'Enfer, at others into a kind of hornstone porphyry, both near the village where the baths are situated.

It is frequently coloured red by iron, and sometimes incloses flattened balls of clay ironstone. In its fissures are also found plates of specular iron ore, a substance which I have noticed as occurring among the recent volcanos near Clermont, and at the Puy de Dôme.

Associated with the trachyte, and as I believe interstratified with it, are those singular beds which I have compared with the trass of the Rhine volcanos, consisting of an apparently homogeneous rock, bearing a resemblance to tripoli, possessing a rough earthy feel and slaty fracture, generally grey, but sometimes of an ochreous yellow colour, from the intermixture of oxide of iron.

Whether these beds are to be attributed to the disintegration of the trachyte, and the subsequent agglutination of its finely divided fragments into an uniform mass, or whether they have not rather been the result of ejections of finely pulverized matter, may admit of dispute. The same uncertainty extends to those beds of volcanic breccia likewise found accompanying this formation, in which the fragments are cemented by a paste often resembling the trachyte itself.

To the latter we may probably also refer those fragments of a breccia containing sulphur and alum rock,* found in the Gorge d'Enfer, near the village of the Baths, in the bed of the River Dordogne, which takes its rise in the mountains above.

Of this rock M. Cordier has published in the "Annales des Mines," a description as well as an analysis, from both which he infers that it is analogous to the alum rock of Tolfa, like which it yields, on exposure to heat and moisture, numerous capillary crystals of alum. It has never been met with *in situ*, but it seems probable, that if the middle regions of the Pic de Sancy, above the spot to which it has been brought by the torrents, could be explored, the beds of tuff which there exist might be found to contain it.

* Noggerath, in the German Translation he has lately published of these Letters, remarks, that alum rock and alum stone must be distinguished, the former serving to designate a species of rock, the latter a simple mineral. See also Leonhard's Characteristic der Felsarten, vol. 2. 553.

The following is the result of Cordier's analysis of the alum rock of Mont Dor, compared with that of the alum-rock and the alum stone of Tolfa, Hungary, and Montione.

	Alum Rock of Mont Dor. *Cordier,*	Alum Rock of Tolfa. *Vauquelin.*	Alum Rock of Tolfa. *Klaproth.*	Alumstone of Hungary. *Klaproth.*	Alumstone of Tolfa. *Cordier,*	Alumstone of Montione. *Collet Descotils.*
Sulphuric Acid ..	27,03	35,00	16,5	12,50	35,495	35,6
Alumine	31,80	43.92	19.0	17,50	39,654	40.0
Alkali	5,79	6,08	4.0	1.00	10,021	13,8
Water	3,72	4.00	3,0	5,00	14,830	10,6
Silex..........	28,40	24,00	56,5	62,25	....	....
Oxide of Iron	1,44	....	....	....	a trace	....
Loss	1,82	....	1.0	1,75	....	....
	100.	100.0	100.	100.	100.	100.

In my letter to Professor Jameson, published in the 4th number of his Journal, I noticed some dykes of vesicular lava which traverse the trachytic formation at the waterfall called the Grande Cascade de Mont Dor.

From the absence of any dislocation or hardening of the rock which they traverse, I inferred that they were processes given off from the basaltic lava of the cliff above, which might have insinuated itself in a liquid state into the cracks or fissures of the subjacent rock, rather than the rents caused by the basalt in the act of attaining the position which it now occupies.

I perceive that M. Ramond adopts the same opinion with reference to the place in question,* nor am I aware of any thing in the particular circumstances of the case, that militates against such a conclusion; but I must at the same time remark that my principal ground for adopting it is greatly weakened, from having remarked in the volcanic districts which I have since examined, ·that the dykes that appear to be of the most modern date, have not so generally affected the rock contiguous, as the more antient ones have done.

I should also expect from what I have since seen among the German volcanos, that the basalt which caps the table land of Mont Dor, has been ejected through the medium of *dykes*, rather than of *craters*,† and it is therefore not improbable that those of the Grande Cascade de Mont Dor may be among the number of these vents. I am still however of opinion, that the dykes of volcanic tuff that occur in Cantal, of which several are mentioned by Steininger, and one has been noticed by myself in the communication alluded to, are nothing more than an upfilling of fissures that existed in the subjacent rock, and I am confirmed in this idea, from having seen at the foot of the Siebengebirge, on the Rhine, similar veins of trass filling up the cracks in the rock of the same description which there encircles the trachyte.

In describing the rocks found at Mont Dor, I have said almost all that appears necessary respecting the trachyte of Auvergne, for that of Cantal is distinguished chiefly by its more compact form, and by the rarer occurrence in it of scorified matter.

* Nivellement Barometrique de Mont Dor.

† Steininger in his Tract on Auvergne above cited, notices dykes of trachyte near Murat, and of basalt at Theyzac; both in Cantal. I did not observe them.

This formation is also occasionally capped with basalt, as at the Puy Griou, and the same rock likewise descends, as at Mont Dor, to a comparatively low level, so that Steininger has adopted with respect to it the same hypothesis as to the existence of a basaltic formation interstratified with trachyte, though apparently without adducing any more decided proofs of such an alternation.*

The highest rocks in Cantal are however mostly capped with porphyry slate, a substance only found at Mont Dor, composing two isolated peaks near Rochfort, called Sanadoire and Thuiliere.

The mineralogical characters of the porphyry slate approach so nearly to those of some varieties of trachyte that Daubuisson regards it as a modification of that rock; but in the case before us, there does not appear to be any passage from one rock into the other, and the limits of the two formations are very distinctly marked, especially in the former instance, by the more harsh and rugged outline of that portion of the mountain which is composed of porphyry slate. From the indestructible nature of this rock, the hills in Cantal are usually covered towards their surface with massive fragments of rock, of a greyish colour and great hardness, whereas the trachyte beneath decomposes in a more rapid and uniform manner.

Nor can there be a greater contrast than between the luxuriance of some of the vallies, as that of Theyzac, in which the substratum is of trachyte, and the extreme barrenness of the higher parts, which are composed of clink-stone.

I am therefore inclined to believe this rock to be a formation distinct from the trachyte underneath it, of the same age probably as the basalt of Mont Dor; and when we recollect that in the Siebengebirge near the Rhine, basalt and trachyte occur together without any determinate order, and that in the hills near the Lake of Constance, the former species of

* See Steininger Erloschenen Vulkane in Süd frankreich, p. 202.

rock is found associated with clinkstone in such a manner
as leads one to suppose that both are products of the same
æra, we need not be surprised at seeing the porphyry slate
to be in the place in Cantal of the basalt, which is spread
over the surface of the Mont Dor.

I have already alluded to the tuff which in the latter chain
is found associated with trachyte. A similar rock occurs in
greater abundance throughout Cantal, and is there distinguished by the grotesque appearance which it assumes,
presenting to the eye a range of mural precipices, broken
into a number of fantastic shapes, a circumstance very characteristic of rocks of this description, both here and in the
neighbourhood of the Puy en Velay.

In Cantal, the tuff is best displayed near the village of
Theyzac, on the road from Aurillac to Murat, where it is
placed between two beds of trachyte, being found rather
less than half way up on either side of the hills which bound
the valley, whilst the summit and base alike consist of
trachyte. It dips gradually to the east; so that about half
a league from Theyzac, on the road to Murat, it reaches
the level of the road. Different as the tuff appears from the
trachyte which it accompanies, it will be found on examination, that the fragments of which it consists are
cemented always by a basis of the latter rock, and that a
passage from the one to the other proceeds by imperceptible
gradations. The fragments consist in general of a trachyte
of a more compact character than the paste which cements
them, but we also find basalt and cellular lava intermixed ;
and I remarked beds or veins of the same description of
stone which I have compared to the trass of the Rhine volcanos when speaking of Mont Dor. A little beyond Theyzac, near Vic en Carladez, a mass of this rock occurs
inclined apparently in the midst of the tuff, and with its
layers irregularly incurvated, forming a sort of arch, which,
though on a smaller scale, reminded me of one of clay
porphyry which I had observed in Arran, and which is re-

c

presented in the plates to Dr. Mac Culloch's work on the Western Islands.

The tuff in some places, as at Salers, is composed of minute fragments so highly charged with oxide of iron, that it has much the appearance of a ferruginous sandstone. In this state it sometimes contains impression of leaves and branches of trees, which appear in no respect mineralized, but carbonized and reduced to an impalpable powder by the ordinary process of decay. In other cases, where the tree has wholly disappeared, the hollow which it occupied in the midst of the tuff still remains. This circumstance tends in a still greater degree to identify the tuff of Auvergne with the trass of the Rhine volcanos.

The greater part of the tuff of Cantal, as well as the trachyte, appears to be posterior to the calcareous formation of the vallies; in one instance only, in the valley of Font-anges, have I observed an appearance of alternation between them, in which case a thin calcareous bed of considerable hardness is seen resting upon a tuff in which the fragments are held together by a trachytic base, and is covered by the same material.

Thus the age of these rocks in the Cantal seems to be nearly the same with that of the older volcanos of the plain of Limagne, for the limestone which supports and some-times alternates with them, appears in both cases to be identical.

The shells that are found at Gergovia near Clermont, and at Aurillac in Cantal, both belong to the freshwater forma-tion, and the recent discovery of bones belonging to the Mastodon, and to extinct species of several existing genera of animals, in the volcanic tuff of Mont Perrier, near Issoire, completes the resemblance with the rocks of the Paris basin.*

* This discovery is announced in the Bulletin des Sciences for November 1824, p. 328, in an extract from a memoir read by M. le Comte Laizer at the annual meeting of the Philosophical Society of Clermont in Auvergne. Between Champeix and Issoire, an elevated platform of basalt and tuff occurs, the latter

The same remark likewise applies to the volcanic rocks that are met with about fifty miles farther to the south in the neighbourhood of the town of Puy en Velay, where an extensive deposit of tuff occurs, resting on strata of tertiary formation, but covered by diluvial detritus.

The tuff is composed of fragments of scoriform lava and basalt with the debris of various other rocks, all cemented together by sand and wacke. It appears at one time to have overspread the valley in which the town of Puy is situated, but since to have been in great measure swept away by diluvial action and other causes. Owing however to the unequal degree of consistence possessed by different portions of this formation, it has been affected by these agents in a very irregular manner, and hence it has happened, that in the midst of the valley caused by the destruction of the tuff, several detached hillocks appear, which, from their singularly abrupt and almost pyramidal form, look at a distance rather like artificial constructions, than the result of natural causes.

Such is the hill on which the cathedral and part of the town of Puy is situated; that near the village of Expailly, celebrated for the crystals of zircon and hyacinth which it contains, and still more remarkably the rock of St. Michael,

composed of fragments of pumice and trachyte cemented by the usual argillaceous paste. In this aggregate are the bones of no less than twenty extinct species of Mammalia, several of which have been pronounced by Cuvier to be new.

Among the Pachydermata are, the Mastodon, Elephant, Rhinoceros, Hippotamus, Tapir.

Ruminantia—two species of Ox, like the Auroch, two species of Stag,—all four extinct.

Rodentia—a Beaver.

Carnivora—two new species of Bear, three species of the G. Felis like the Panther, one species of the Hyæna, one species of Fox, one species of Otter,— all of them new.

Besides the above, occur bones of Birds, and impressions of Fish.

Drawings of these bones are announced as about to be published by subscription.

c 2

the height of which, according to Faujas, is two hundred,
and its diameter only one hundred and seventy feet.*

It is curious that the same tapering figure which we view
with surprise in the Alps, and consider as characteristic of
rocks of the most compact texture, should in this instance
be found belonging to a stratum consisting of loose materials,
and of so modern a date.†

The tuff about the Puy is also associated with masses of
compact basalt, which seem in general to rest upon, but in
some instances have the appearance of being intermixed
with it. Thus near the tufaceous rock of Expailly, just
noticed, is another isolated knoll, which consists of columnar basalt, and in like manner at St. Pierre Eynac, a village
at a short distance from the Puy, the basalt seems to alternate in beds with the tuff.

It appears to me, that a key to the true explanation of
these phænomena may be obtained by considering the structure and position of an isolated mass of basalt near the Puy,
called the *Rocher Rouge.* This rock is superimposed on
the slope of a granitic hill, from which it rises to the height
of more than one hundred feet, and has all the appearance
of an enormous dyke, both from the shattered condition of
the granite round it, and from the manner in which the
latter is seen on one side to lean against the protruding mass.

Where the granite has been by accident removed, the
basalt is seen rising as it were from beneath it, and from the
principal mass are seen to spring two dykes, which penetrate

---

* Une espece de grande obelisque, faconné des mains de la nature. Faujas,
Volc. du Viv. p. 342.

† Mr. Herschell has however given the Geological Society an account of a
still more abrupt figure belonging to hills made up of even softer materials. It
occurs in the Tyrol, where masses of diluvial matter, consisting of pebbles
loosely cemented by sand, are seen to form a succession of pyramidal hills even
more precipitous than those of the Puy. In this instance the effect is explained
by observing that the portions of the diluvial matter which rise in this abrupt
manner above the rest, have been protected from the action of the rain, for we
may observe on the top of every one of these a large stone which shielded the
parts immediately underneath it.

the former rock horizontally to a considerable distance. Fragments of granite are also to be met with imbedded in the substance of the basalt.

Yet notwithstanding these unequivocal proofs of igneous ejection, it seems impossible that the basalt should have originally stood in the position it now occupies, unless it had been at that time supported by some surrounding stratum ; and it therefore seems most simple to suppose that the granite in this situation was once covered with tuff, which the action of the waters has since swept away.

By the occurrence of similar dykes thrust up in other places through the midst of the tuff, and sometimes penetrating it horizontally, I should account for the basalt seen intermixed with, as well as covering this volcanic breccia.

I cannot pretend however to have studied the geology of this neighbourhood with the attention requisite for determining the relation of all the rocks included under this series to each other, and must therefore refer you to M. Bertrand Roux's excellent description of the environs of Puy, for further particulars.

From his statement it would appear, that the basaltic rocks of this neighbourhood are of very different ages, though I cannot admit that we are justified in estimating their relative antiquity by comparing together the depth to which the several parts of this formation have been worn away. M. Bertrand Roux himself furnishes us in my opinion with a convincing proof that the effect has not been dependant on the longer or shorter continuance of causes now in action, when he mentions that the rock on either side of the old Roman roads, none of which can be less than 1300 years old, has undergone since that period scarcely any sensible decay. Instead therefore of considering with M. Roux the amount of the destruction that has taken place in different parts of the formation, a sort of chronometer to assist us in determining their relative age, I should rather adopt the *converse* of the proposition, and argue that the time required would, according to his own shewing, have been so immense, that we are in a manner driven to suppose the effect to have been brought about by causes differing in their mode of action from those at present in operation.

The conclusion arrived at by either process of reasoning corresponds, however, in assigning to the volcanic products alluded to a very remote antiquity; for whilst M. Bertrand Roux is bound to suppose them as much older than the Roman roads, as the whole amount of the degradation they have experienced exceeds that which has taken place since the date of the latter; my conclusion leads me to place their formation at an epoch at least somewhat more remote than that of the last general revolution which has affected the face of our planet.

A limit on the other hand is set to the age that can be assigned to this volcanic breccia, by the circumstance of its being superposed on strata, containing freshwater shells, and bones of mammalia* similar to those of the basin of Paris. Hence the eruptions to which the materials of this tuff owe their existence, though anterior to the period at which the vallies were excavated, must date from one subsequent to the formation of the tertiary rocks found in that neighbourhood.

The same remark will apply to that extensive volcanic formation near the Puy, which ranges from south-east to south-west, forming the elevated ridge which separates the Velay from the neighbouring province of the Vivarais.

The principal rocks in this district are trachyte and porphyry slate, the latter generally superimposed; on the flanks however are occasionally seen detached patches of basalt, which seem to belong to the formation seen covering the tuff on the hills about the Puy.

---

* Cuvier has ascertained that they belong to the genus Paleotherium and Anthracotherium; the former contained iu a gypseous deposit similar to that of Montmartre; the latter in a calcareous rock, in which were found freshwater shells. The same bed inclosed bones of other Mammalia, and portions of the shell of the Turtle.

It would nevertheless appear that the whole of the trachytic formation was of later date than great part of the
tuff, for if this were not the case, we ought to find fragments
of trachyte and of clinkstone generally distributed throughout this aggregate, whereas in reality they are only met with
in the beds of alluvial formation which alternate with the
upper members of the series.

It would therefore seem that the process which occasioned
the latter deposit was going on at the very time at which the
trachyte was formed, so that this latter must be attributed
to the same epoch in the history of our planet, namely,
to that immediately preceding the commencement of the
existing order of things.

The most elevated point in the trachytic and clinkstone
formation of this part of France is the Mont Mezen, which
rises to the height of 5900 feet above the level of the sea :
and the next in point of height is the singular conical hill
called Gerbier de Jones, which is composed entirely of
clinkstone porphyry, so fissile as to be used for a roofing slate.

Near the Mount Mezen I observed a small lake, that of
St. Front, having somewhat the appearance of an extinct
crater, but there is little in the character of the surrounding
rocks to countenance such an opinion.

The whole indeed of this elevated table-land appears o
have been formed not only at a very ancient period, but
even under different circumstances from those of existing
volcanos ; and even supposing craters to have ever occurred
in the midst of it, we may conclude that they would have
been obliterated by the same agents which excavated the
vallies seen every where intersecting them.

It is interesting to remark, that the same tuff occurs mantling round this ridge on its eastern as on its western slope ;
and here also its date is probably earlier, as no fragments of
trachyte are found imbedded ; so that it seems difficult to
account for the relative position of the two, unless we suppose the trachyte of the Mezen to have been thrown up
through it by the agency of subterranean heat.

Thus during the period immediately antecedent to that at which man and other existing species of Mammalia first came into being, at a time when the lower parts of the country was still under water, but the higher had become peopled with various tribes of land animals, the neighbourhood of the Puy appears to have been agitated by volcanos, which overspread the country with their ejected materials, caused the destruction of the animals that existed there, and, according to M. Roux, obstructed the drainage of the district, and consequently raised the waters to a still higher level than before. The ejected materials, intermixed with fragments of older rocks washed down at the same time from the neighbouring high ground, were deposited at the bottom of the water, forming the immense masses of ,tuff which now cover the valley of Puy, and during the latter part of the period occupied by this process, the same volcanic forces elevated from the midst of the then existing lake the trachytic rocks which constitute the ridge of Mont Mezen.

But besides these traces of volcanic action at a period antecedent to the formation of the valleys, the neighbourhood of the Puy, no less than the province of the Vivarais which bounds it on the south-east, exhibits also decided evidence of post-diluvial eruptions having taken place.

West of the town of Puy is a series of little volcanos, amounting according to M. Bertrand Roux to more than a hundred, the two most remarkable of which are the Lake de Bouchet and the Crater of Bar. The former, which is situated near the villages of Cayre and Bouchet, is of an elliptical form, and without any outlet. Its depth is about 90 feet, and its greatest diameter 2300. The character of the rocks in its neighbourhood corresponds very well with the idea of its volcanic origin. The crater of Bar is placed on an isolated mountain in the midst of granite, forming a truncated cone about 20,000 feet in circumference at its base, and 830 in height. It is composed entirely of *lapilli* and scoriform lava, and on its summit is the crater, which is almost perfect, 1660 feet in diameter, and 130 in depth. It appears that a lake once existed there, but it is now nearly dried up.*

Among the volcanos that occur to the east of the basin of the Puy among the mountains of the Vivarais, one of the most remarkable is that called the Coupe au Colet d'Aisa, near the town of Entraigues. It is a conical hill, having on its slope a stream of basaltic lava, which according to Faujas St. Fond may be traced on the one hand from a crater-shaped cavity on its summit, and may be followed on the other to the very bottom of the valley at its foot.

It is interesting to remark that a stream, the course of which was at one time obstructed by the lava which has flowed into the valley, has cut itself a new channel through the midst of it, and that the portion of the rock which forms the bed of the river, or lies immediately above, alone exhibits a columnar arrangement.

The same remark applies to that line of basaltic columns seen extending along the borders of the river Volant, between Vals and Entraigues, which, though they doubtless extend above the level attained by the stream at present, are nevertheless confined to the lower portions of the rock.

Similar basaltic colonnades occur in many of the other vallies that intersect the mountains of the Vivarais, and I believe their position in every instance corresponds with that already pointed out.

In the case of the one which extends along the borders of the river Auliere, near the village of Colombier, the following section is exposed, which serves very well to illustrate the age and mode of formation which I have assigned to this rock.

1st. The base of the hill, on the borders of the river, is a bed of rolled pebbles, consisting of the granite and older volcanic rocks of the neighbouring province.

2d. Resting on this is a bed of quartzy sand mixed with particles of black lava

* See M. Bertrand Roux sur les env. de Puy.

3d. Masses of black porous lava covered by the prismatic basalt, which again serves as a support for masses of porous lava extending upwards for 200 feet towards the conical portion of the hill of Montpezat, from which it has probably descended, although all vestiges of a crater are obliterated on its summit.*

I shall not detain you by enumerating the other post-diluvial volcanic products which I noticed in the Vivarais, as my examination was of a very hasty nature, and limited to ascertaining their general relations. There is indeed so much of a common character in the post-diluvial lavas of that country, that a detailed description would be of more interest with reference to the topography of that part of France, than to the natural history of volcanos in general. Compared however with the products of Etna and Vesuvius, they present one important difference, namely, in their more compact and basaltic appearance; connected with which is the general occurrence of a columnar arrangement, not derived like that met with in the latter localities, from the contraction, and consequent splitting of the mass, but arising probably out of a tendency to form a series of globular concretions, which the pressure exerted by the parts upon each other has reduced to prisms more or less regular according to the circumstances of the case.†

The great similarity that exists between the volcanic formations of Auvergne and of the Vivarais, in age as well as in character, is sufficiently apparent from the facts already stated.

In both countries we have proofs of volcanic eruptions which, though not noticed by history or tradition, must have been posterior to the formation of the vallies, and in both instances we observe a more extensive formation of trachyte and of tuff, indicating the operation of the same causes at a period which, though antecedent to the latter

---

* See Faujas St. Fond, Volcans du Vivarais.
† See this more fully explained in my fourth Lecture.

epoch, were either subsequent to, or coincident with, that of the deposition of the latest class of rock formations.

In either case the general tenor of the phænomena leads us alike to the conclusion, that the period, at which these latter mentioned eruptions must have occurred, was that, at which the great mass of the ocean had retired, having left however in the hollows those lakes, to which we attribute the formation of the calcareous and gypseous deposits containing freshwater shells, seen alike in the vallies of Cantal, of the Limagne, and of the Puy.

These eruptions seem more generally to have given rise to a formation of volcanic tuff or breccia, but associated with the latter are those trachytic lavas, which seem referable to the same or to a somewhat more-recent epoch. The trachyte of Cantal, for example, seems to be contemporaneous with the tuff which alternates with the freshwater limestone at Salers, but its greater compactness, and the absence of cellular products, lead us to imagine it more ancient than the analogous formation which I have described as existing in the Mont Dor. This corresponds well with what M. Bertrand Roux has inferred with respect to the age of the trachyte and porphyry slate of the Mezen, which he concludes to be more modern than the greater part of the tuff that surrounds it, from the occurrence of fragments of these rocks only in its upper strata.

Evidences of volcanic agency are not exclusively confined to the districts of France we have been considering, they occur likewise still farther south among the Cevennes, and near the shores of the Mediterranean in the neighbourhood of Marseilles, and of Montpellier.

Of these the rock which has been most noticed by geologlists is that of Beaulieu, near Aix in Provence, described by Saussure, Faujas St. Fond, and still more lately by Menard de Groye.* It is stated by the latter as about 1200

---

* See Journ de Phys. vol. 82, 83.

fathoms in length, six or seven hundred in breadth, and
rising to about two hundred fathoms above the level of the
sea. It is composed of basalt which, as we trace it downwards, is seen to pass into a very crystalline greenstone, and
is covered by an amygdaloidal wacke, the cells of which are
empty near the surface, probably from the decay of the
rock, but in the interior are filled with calcareous matter.
The latter is sometimes so diffused through the substance of
the rock, that it forms with it a kind of breccia, and even
swells out into nests or geodes of considerable size, imbedded
in the midst of the tuff. Shells are also contained in this
formation, and serve to connect it still more closely with
the limestone covering it, the recent origin of which may
be inferred from the existence in it of bones of ruminating
animals. The limestone is compact, and passes into a siliceous kind of rock, probably a chert, which is called by
Haüy *quartz agate calcifere.*

Upon the whole I suspect that the basalt of Beaulieu, like
the rocks of Auvergne and the Vivarais, belongs to the epoch
at which the tertiary class of rocks were formed, though M.
Menard de Groyé's account is not such as to be completely
decisive.

Neither is there any thing in the structure of the rock of
Beaulieu, so far as we can judge from his description, which
can be viewed as establishing its volcanic origin more fully
than might be done in the case of almost every other trap
rock, for the analogies he has pointed out between the products of existing volcanoes and those of which he has given
us an account, would hold good equally in every other case,
and the evidence of igneous injection derived from the presence of dykes is in this instance wanting.

A league to the south-east of Agde, a town placed near
the sea at a distance of about twelve miles west of Cette in
Languedoc, is the hill of St. Loup, which belongs to the
most recent order of volcanos.*

* See Journal des Mines, No. 141, p. 231.

Its crater, which is considerable, has given rise to two currents of lava, on one of which the town of Agde is erected, whilst the other, having taken the direction of the sea, has formed a neck of land called Cape Agde, and a little island at a short distance from the shore.

M. Marcel de Serres has also described two other rocks to which he assigns the same origin and date; the one that of St. Thibery about four miles north of Agde, the other that of Montferrier near Montpellier. The latter I visited some years ago, and found to consist entirely of compact trap, so that I should be disposed to view it as considerably more ancient.

On the Volcanos of Germany.

After this general description of the Volcanos of France, I shall proceed to lay before you a short sketch of those which occur distributed over various parts of Germany.

Although no active volcanos are found in any part of that extensive country, and the recognition of those which are extinct dates only from the last century, yet those who have visited the spots themselves will feel no more doubt as to their having once existed, than an American who had witnessed the burning mountains of his own hemisphere, but had never heard of those in Europe, would entertain with respect to the real nature of Vesuvius, if landed at its foot when it chanced to be in a tranquil state.

This remark applies to no case more completely than to that of the rocks which occur in a district commonly known by the name of the Eyfel, situated between the Rhine and the present frontier of the Netherlands.*

* This account of the Rhine volcanos is principally drawn (where the reverse is not stated) from observations made by myself during a tour in that country in the Summer of 1825.

This country is bounded on the south-east by the Moselle, on the north-east by the Rhine, on the west by the Ardennes and the other mountains round Spa and Malmedi, and on the south by the level country about Cologne.

The fundamental rock which comes to view is clay-slate, associated with greywacké, and with a saccharoid magnesian limestone containing trilobites and other petrifactions, which stamp it as belonging to the transition series.

These rocks in a few places support horizontal beds of what appears to be the second or variegated sandstone formation. Scattered however over the greater part of the district alluded to, are a number of little conical eminences, often with craters, the bottoms of which are usually sunk much below the present level, and have thereby in many cases received the drainage of the surrounding country, thus forming a series of lakes, known by the name of " Maars," which are remarkably distinguished from those elsewhere seen by their circular form, and by the absence of any apparent outlet for their waters.

Steininger, * a geologist of Treves, who has published the most circumstantial account of this district that has yet appeared, distinguishes these craters into three classes.

The first includes those properly speaking known by the name of " Maars,"—volcanos, which have ejected nothing but loose fragments of rock with sand and balls of scoriform lava.   In this class are :

1. The Lake of Laach.
2. The Maar of Ulmen.
3. Three Maars at Daun.
4. Two at Gillenfield.
5. One at Bettenfield.
6. One at Dochweiler.
7. One at Walsdorf.
8. One at Masburck.

No. 6 and 7, however, have fallen in.

* See, for an enumeration of his works, the Appendix.

The second class is distinguished from the preceding in consisting of those which have ejected fragments of slag, sometimes loose, and sometimes cemented together into a paste. Of this denomination are :

1. Three Craters at Gillenfield.
2. Two at Bettenfield.
3. One at Gerolstein.
4. One at Steffler.
5. Two at Boos.
6. One at Rolandseck.

The third class includes such volcanos as have given out streams of lava as well as ejections of loose substances. Of these latter we may mention :

2. Two at Bertrich (one very small).
3. One at Bettenfield (the Mosenburg).
4. One at Ittersdorf.
5. One at Gerolstein.
6. One at Ettringen.

Thus the whole number of craters in the Eyfel district, including those of the same date that are scattered along and near the left bank of the Rhine within the limits marked out, appears to be not less than thirty.

The sides of these craters, wherever their structure was discernable, appeared to me to be made up of alternating strata of volcanic sand and fragments of scoriform lava, dipping in all directions away from the centre at a considerable angle, and the same kind of material has in many instances so accumulated round the cones, as to obliterate in great measure the hollow between them, and to raise the level of the country nearly up to the brim of the craters.

The formation of these cones seems likewise to have been in some instances followed by an ejection of substances of a pumiceous character, and the same kind of material (whether derived from these or from some antecedent eruptions, will

be afterwards considered), is spread widely over the country
bordering on the Rhine, either in loose strata alternating
with beds of a loamy earth, derived probably from substances in a minute state of division thrown out by the
same volcano, and mixed up into a paste with water ; or
else forming masses of considerable thickness, in which the
fragments of pumice are intermixed with the latter substance, and constitute together with it a coherent mass
known by the name of Trass.

The volcanos of the Eyfel are also, as above noticed,
accompanied by streams of lava, but these have not, in my
opinion, like the generality of those seen elsewhere, been
satisfactorily traced to the craters, but seem rather to have
flowed from the sides or base of the mountains with which
they are respectively connected.

These Coulées, like the volcanic cones themselves, are
sometimes almost buried under heaps of matter subsequently
ejected, so that in the lava of Niedermennig, the quarry,
from whence the millstones are obtained, is worked at a
depth of eighty feet from the present surface. They are in
some cases analogous to the ejections of existing volcanos,
but at others they possess more of a basaltic character, being
freer from cells than true lavas generally are, although it
can be demonstrated that they too are (geologically speaking) of modern formation, inasmuch as they follow the inclination of the vallies, and must therefore have flowed since
the latter were excavated.

The above remarks may suffice for a general description
of the Eyfel volcanos. I shall now therefore proceed to par-
ticularise two or three of the more important.

Beginning with the country nearest the Rhine, which is
known by the name of the Maifeld, let us first consider the
remarkable crater called the Lake of Laach, which occurs
near Andernach, a few miles west of the Rhine.

The lake of Laach is perfectly round, and embraces a
circumference of two miles. Its sides are overgrown with
wood from the level of the water up to the brim of the

crater, which is reached externally by a gentle ascent not at all in proportion to the depth of the internal cavity.

The thickness of the vegetation renders it difficult to discover the nature of the subjacent rock, but it appears to consist of a black cellular lava full of augite. Besides this however, the sides of the crater present numerous loose masses, which appear to have been ejected, and consist of glassy felspar, ice-spar, sodalite, hauyne, spinellane, and leucite. It will be seen, when I come to speak of the neighbourhood of Naples, how remarkable is the resemblance between these products and the ejected masses found on the slope of Vesuvius.

Not much above a mile from this spot is the rock of Niedermennig, so extensively quarried for millstones; but though this has all the appearance of a stream of lava, no one has as yet succeeded in tracing it either to Laach or to any other neighbouring volcano.

The lava is divided by vertical fissures into irregular columnar masses, some twenty feet in height, and these columns cut horizontally, and having their angles rounded off, are fashioned into millstones,* for which they are well adapted from the unevenness of their fracture, derived from the infinite number of minute cells distributed through the substance of the rock.

It is only the middle portion however of this bed which can be so employed, for in the upper not only are the pores too considerable, but the concretions, being smaller than they are below, scarcely afford masses of sufficient magnitude for the purpose intended.

This difference arises from the greater size of the fissures

* The same use appears to have been made of the Lavas of Etna by the ancients. See Corn. Sev. Etna.

Quin etiam vario quædam sub nomine saxa
Toto monte liquant; illis custodia flammæ
Vera tenaxque data est; *sed maxima causa molaris*
Illius incendi lapis, is sibi vindicat Ætnam.

D

which extend vertically through the substance of the rock;
these, of considerable width at top, contract gradually as
they descend, until they at length disappear altogether, and
in consequence impart to the separate columns of rock a
tapering form, gradually enlarging in bulk from above downwards, until the whole is at bottom confounded into one
solid and entire mass.

Hence the rock of Niedermennig is distinguished by the
workmen into three portions, which they compare to the
top-branches, the trunk, and the root of a tree. The upper
or top-branches * is that in which the columnar concretions
are too small to be worth working; the lower or root, that
in which the concretionary structure is lost by the final dis-
appearance of the fissures; the central or trunk, that in
which the columns are of a size adapted for millstones of
the usual dimensions.

The lower part of the rock having never yet been penetrated, it is impossible to state the exact thickness of the
whole, but the upper portion not worked is stated at seven,
and the central from 15 to 40 feet in thickness.

The lava of Niedermennig is interesting for the variety of
extraneous substances imbedded in it. In the midst of the
volcanic mass we observe, not only hauyne, magnetic iron
ore, and other crystallized minerals, but even portions of
limestone, of quartz, or of an intermixture of quartz and
felspar in the state of kaolin, which looks like the altered
fragments of some granitic rock. The whole is buried at
a depth of more than sixty feet from the present surface
under a succession of strata consisting either of loam, the
origin of which will be discussed afterwards, or of a congeries of rolled masses of pumice and scoriform lava with
those of all the different rocks found in the neighbourhood,

* The upper portion is called *kopfe, glochen* or *aeste;* the middle *schienen* or
stamme; the lower *sohlgestein* or *dielstein.*   For these and other particulars see
Nöggerath über fossile Baumstamme, Bonn 1819, P. 60 *et seq.*

the whole loosely bound together by an earthy, loose, loamy sand.

From the lake of Laach to the little town of Gerolstein, about thirty miles westward, a continued succession of volcanic cones, with craters, presents itself.

Of these none is more curious than that of Gerolstein, which derives moreover an additional interest, as the spot which Von Buch has selected for one of his proofs of the singular theory he has lately advanced with respect to the conversion of common limestone into dolomite by volcanic agency.

The town of Gerolstein is built in a narrow valley between two ridges of limestone, each side presenting a precipitous escarpment, owing to the projection of occasional masses of the rock beyond the soil which covers in general the surface of the hill. This limestone is highly crystalline, and has all the characters of dolomite; it contains numerous corallines, trilobites, and other petrifactions belonging to the transition series.

The slope of the hill fronting Gerolstein, like that at the back of the town, is wholly calcareous, but on reaching its summit we soon discover traces of volcanic operations.

If we commence our examination with the western extremity of the ridge, which lies about half a mile beyond Gerolstein, we observe a conical mass of slaggy lava thrown up between the limestone strata, which are distinctly seen both north and south of it.

Connected as it would appear with this, is a mass of lava which occupies the centre of the western slope of the ridge, and may be traced to a considerable distance over the plain at its foot, divided however into two branches by a projecting mass of the calcareous rock which it meets with on its descent.

The cellular and even scoriform appearance of this lava, coupled with the manner in which it has accommodated it-

GEROLSTEIN.

self to the present slope of the mountain, leads one to consider it a modern *coulée ;* but I could not satisfy myself that it had proceeded, as is usually the case, from a crater. There is indeed on the summit of the ridge from whence the lava proceeds, a little beyond the conical eminence above noticed, a crater-shaped cavity surrounded by volcanic matter, but with this the stream of lava seen on the slope of the mountain does not appear to have any connection.

Other patches of the same kind of rock, chiefly of a slaggy character, occur along the summit of this ridge east of the crater, which perhaps have been separately thrown up by a succession of small volcanic ejections unaccompanied by streams of lava. At the eastern extremity however of the ridge we are somewhat surprised to discover a small but abrupt knoll of compact basalt called the Casselburg, the site of a ruined castle, placed apparently under the same circumstances with the cones of scoriform lava before mentioned. It covers a bed of variegated sandstone which rests upon the dolomite, but the line of junction is so concealed with wood, that it is impossible to ascertain whether the basalt lies upon the contiguous rock, or, as is most probable, has been forced through it.

Von Buch, who has persuaded himself from some facts observed in the Tyrol, that dolomite limestone derives its magnesia from the augite of lavas, appeals to its occurrence at Gerolstein in connection with so much volcanic matter as confirming the truth of his hypothesis.* But it seems difficult to reconcile this opinion with the age which we are compelled to assign to the volcanic operations here as well as in other parts of the Eyfel. As it is evident that no foreign ingredient could penetrate the substance of the rock in its present hardened condition, so as to unite with the

---

* Von Buch Ueber das Vorkommen des Dolomite in die nahe der vulkanischen Gebilde der Eifel, in the 3rd vol. of Nöggerath's Gebirge in Rhinland Westphalen.

other constituents, and diffuse itself uniformly throughout the mass, it seems necessary for Von Buch's hypothesis to suppose the limestone to have previously been at least softened by the heat, which occasioned the sublimation of the magnesia. Hence we should be obliged to fix the period at which this process took place as antecedent to the formation of the vallies, for these would be necessarily obliterated by any softening of the limestone which now overhangs them.

Indeed it would be necessary to carry back this supposed softening of the calcareous rock to some period antecedent to the retirement of the ocean, when sufficient pressure might be exerted to prevent the carbonic acid from being driven off from the limestone when exposed to the heat required for softening it.

But all this is completely contradicted by the phænomena of the volcanic products in question, the cellular appearance of which plainly indicates the absence of pressure, and which even seem from the existence in them of craters, and by the manner in which they have accommodated themselves to the present slope of the vallies, to have been formed since the commencement of the present order of things.

The following section may serve to give a clearer notion of the relative situations of the rocks.

Section of the rock on either side of Gerolstein from North to South.

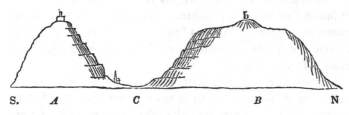

S.　　*A*　　　　　*C*　　　　　*B*　　　　N

A and *B* are two hills composed of dolomite, and in the centre of B is the little crater *b*, consisting of slag and scoriæ. *C* the valley on which Gerolstein stands, which would be filled up, if the rocks on either side had been fused or softened.

MOSENBERG.

A few miles south of Gerolstein occurs another volcanic hill, called the Mosenberg, the structure of which is also very remarkable.

It constitutes a long round-backed eminence, on the edge of which, and but slightly elevated above the general level, are three craters, two of which are perfect, the third broken away on its southern extremity.

These craters, as well as the whole of the high ridge on which they stand, are composed of slag, but at the basis of the hill, just below the point at which the crater is broken away, is seen a mass of basaltic lava, very slightly cellular, which continues in an easterly direction for about half a mile. It has all the appearance of a coulée, and most writers have regarded it as derived from the crater which is broken away on the side nearest it. With regard to the latter point, I must confess myself not altogether satisfied, having failed in tracing the basalt up to the crater itself, and I am the more confirmed in my scepticism, because the ridge which it forms appeared to me to extend considerably west of the point at which it would have commenced, had it descended from the side of the crater which is broken away.

Among the phænomena connected with this mountain, not one of the least curious is the crater-shaped cavity called the Meerfeld, which lies at its northern extremity. It is perfectly round, and sinks regularly towards the centre, where is a small lake, so that its resemblance to the other "Maars" of the Eyfel district is complete. On examining however its sides, we do not find a single trace of volcanic matter, the whole circumference of the hollow being composed of clay slate.

Viewed in connection with the surrounding country, it seems impossible to attribute it to any other than to volcanic agency, differing only from the other craters in its being caused by a disengagement of elastic vapours, unaccompanied by any ejection of solid matter.

BERTRICH.

The only other volcanic appearances I shall stop to particularize, are those exhibited near the Baths of Bertrich,
not far from the road between Coblentz and Treves.
Bertrich lies in a narrow valley that has been excavated,
in the great table-land of greywacke slate, which occupies
so large a portion of the district we are considering. To
the north, just where the slope begins, we may observe a
small conical hillock, called the Falkenlei, presenting a precipitous front towards the valley, but on the opposite side
sinking more gradually into the table-land. It consists entirely of slaggy or cellular basaltic lava, the former variety
constituting the upper, the latter the lower parts of the
formation. At a little distance to the N. W. occurs another hillock, rising conically from the same table-land near
the commencement of the valley, which differs from the
Falkenlei in enclosing a small but regular crater. Divided
from the above by a round hollow in the clay-slate formation, or more properly a circular enlargement of the valley,
is another more considerable volcanic hill, called the Facher-
höhe, situated likewise on the brim of the hill, but rather
to the S. E. of the other eminences. This is also conical,
and has a crater broken away on its southern side.

Thus these three volcanic knolls, together with two other
less considerable ones, that lie betwixt the Falkenlei and the
Facherhöhe, form a semicircle corresponding nearly with
the figure which the escarpment of the table-land presents
towards the valley of Bertrich. All of them lie very near
the brow of the escarpment, whilst no vestige of any thing
of the same kind occurs in the valley below.

In the latter however we observe a large assemblage of
volcanic products, but of a different character. They are
all basaltic, and contain but few cells or cavities, whilst the
cones above are chiefly made up of scoriæ, and contain
basalt but rarely. The basaltic lava of the valley may be
traced from the village of Bertrich, which lies near its

BERTRICH.

southern extremity, almost continuously along the sides of
the brook called the Issbach for upwards of a mile, but its
thickness is greatest at a point, where this brook is enlarged
by another rivulet which flows into it from the west.

This point lies almost under that part of the escarpment
of the table-land on which the Falhenhöhe is situated, and
it would appear that two streams of basaltic lava meet about
this spot, the one proceeding from the direction of the
Facher-höhe, the other running more from the north-west,
and having the appearance of being derived from one of the
cones lying in that quarter. When however we attempt to
follow the above coulées up the slope of the hills, down
which, on the above supposition, they must have flowed,
we are immediately stopped by the extreme thickness of
the wood, which clothes every where the sides of the valley.
If we take the circuitous path which brings us to the summit of the hills, and examine the two volcanic cones, we discover no signs of a stream of lava having been derived from
either. Nor do I think that geologists are warranted in
inferring from analogy, that the basaltic lava has descended from these eminences, since we have already seen
that an equal degree of uncertainty exists with respect to
the origin of the other lavas in the Eyfel country; and
indeed if the basalt of Bertrich actually came from that
quarter, its situation would probably be marked by a corresponding convexity on the slope of the hill, if not by its
being destitute of the timber which is so abundant elsewhere.

But whatever may be the fact with regard to the origin of
these basaltic coulées, it is evident that they were formed
since the excavation of the valley, inasmuch as they follow
its inclination and are not seen except near its bottom.
They have however been cut through by the little stream of
the Issbach which flows along the valley, and where that is
the case, manifest a columnar structure.

BERTRICH.

At the point where the two basaltic coulées meet, an interesting fact occurs,—a natural grotto is seen in the midst of the lava about six feet high, three broad, and twelve or fifteen long, open at both extremities, and thus making part of a foot-path which overlooks the ravine containing the torrent of the Issbach. The walls of this grotto are composed of basalt slightly cellular, and forming a number of concentric lamellar concretions, piled one upon the other, and in general somewhat compressed, so that the interstices between the balls are filled up. The grotto itself has obtained the name of the Cheese-cellar (Kase-keller), from the resemblance which the configuration of the basalt bears to an assemblage of Dutch cheeses piled one above the other; it beautifully illustrates the origin of the jointed columnar structure which this rock so often assumes, since a little more compression would have reduced these globular concretions into a prismatic form, each ball constituting a separate joint in the basaltic mass.

The most probable way of accounting for the existence of this natural grotto, is to suppose the lava which forms its walls to have cooled near the surface before the mass had ceased to flow in its interior: hence a hollow would be left in which the basalt had room freely to assume the form most natural to it, and the concretions, being but little compressed on account of the cavity within, retain their original globular figure. In further proof of this, it may be remarked that the lava above the grotto consists of irregular prisms, and not of balls, as is the case with that which constitutes its walls.

Descending into the bed of the rivulet we have an opportunity of observing the line of junction between the clay-slate and superimposed lava; on the former no change seems to have been effected, but there intervenes between it and the compact basalt which composes the mass of the coulée, a very thin stratum (perhaps not exceeding an inch in thick-

BERTRICH.

ness) of highly cellular lava, a fact which may perhaps be explained by the extrication of steam from the damp surface over which the lava flowed, in the manner supposed by Sir G. Mackenzie with respect to some of the lavas in Iceland.*

Thus much for the basalt above the village of Bertrich; below it, occurs another isolated basaltic mass, which it would be even more difficult to refer to any of the volcanos on the hills above. It forms part of a low, oblong, round-backed hillock, rising in the midst of the valley, which is here somewhat wider than ordinary, the rest consists of the clay-slate which seems to have been thrown up by the protrusion of the basaltic mass accompanying it.

The formation of this hillock has, if I mistake not, turned the stream of the Issbach aside from its original direction, and obliged it to cut a new channel which winds round the base of this little eminence.

The fact of the valley being larger just before we reach it than it is above, may I think be explained, by supposing, that until the stream had cut itself a new channel, the waters so accumulated as to form a lake, and this may possibly have been the cause of the two other circular enlargements of the valley mentioned as occurring above. I have already had occasion, when speaking of Auvergne, to notice other cases in which lakes have evidently been produced by beds of lava obstructing the course of rivulets, and shall employ this hypothesis to account for the formation of the Dead Sea, as a consequence of the volcanic eruption that destroyed the cities in the vale of Siddim.*

The warm baths of Bertrich, and the carbonic acid which rises in so many places in the Eyfel district,† imparting to the springs through which it passes the briskness of soda water, are phænomena which seem to indicate a continuance of

---

* See my third Lecture:
† See Edinb. Phil. Journal, vol. 13. p. 191.

the volcanic operations formerly so prevalent, but it will be afterwards shewn that these effects are far from being confined to volcanic countries, and occur too generally to lead to any certain conclusion.

It would detain us too long were I to go through the description of the other volcanic craters distributed over this country, though the geological traveller ought not to omit visiting the beautiful Crater of the Pulvermaar, the three Lakes at Daun, or the singular Hill of Rodderburg near the Rhine, fronting the romantic Island of Nonnerwehrt, composed entirely of slag, though placed by the side of the Rolandseck, which consists of prismatic basalt.

There is one species of volcanic product however which requires a short notice, although I have not been able to introduce it under the head of any of the preceding volcanos, as it has never been distinctly traced to any one of them in particular.

I allude to that immense deposit of trass or pumiceous conglomerate which occurs in various places in this district, as at the foot, and mantling round the sides, of the older trachytes and basalts of the Siebengebirge and the Westerwald, and still more extensively throughout the neighbourhood of Andernach, in the low ground bounded by the Eyfel mountains on the one hand, and those round Coblentz on the other. In no situation however does it occur so well displayed as near the village of Brohl, in a glen which meets at right angles the great longitudinal valley in which the waters of the Rhine find a passage.

The valley of Brohl seems to have been originally excavated solely in the greywacke formation, but it has since been partially filled up by an immense deposit of trass, which occupies its lower portion, but does not extend to the summit of the hills which bound it. This substance varies much in point of consistence at top and at bottom; in the latter situation being almost compact enough for a building stone, whereas in the former it is more loose and friable. It contains imbedded fragments not only of pumice, but also of clay-slate, as likewise trunks of trees carbonized, and in one place even land-shells.

It has been observed that pumice in an entire state is most abundant on the right bank of the Rhine above Coblentz, about Engers, Neuwied and Bendorf, which agrees very well with what we should expect, if we suppose the source from whence it proceeded was in the direction of the Eyfel volcanos. The origin however of this substance has given rise of late to much discussion in Germany, and in particular between Professor Nöggerath of Bonn, and M. Steininger,* to whom I have already alluded as the author of certain geological tracts on the country we are considering.

The former regards trass as the result of showers of ashes and pumice, which having fallen into water, became mixed up with the mud which that fluid was in the act of depositing, together with the fragments of the contiguous rocks carried down by torrents into the lowest situations, the latter, objecting to the above hypothesis, suggests that it may be possible to trace its existence to mud eruptions similar to those recorded as having occurred among the volcanos of Equinoxial America. The latter hypothesis however I shall not stop to consider, as I conceive it will obtain few advocates, when we recollect the thickness and extensive distribution of this deposit in the vicinity of Andernach, the want of any connexion with the neighbouring craters, and its analogy to the Puzzolana of Naples, which it would be impossible to refer to any such local origin. As however the rejection of one hypothesis does not imply the admission of another, it will be necessary to consider how far the opinion of Professor Nöggerath be reconcileable with the situation and other phænomena of the trass of the Rhine.

* See the elaborate paper under the title " Gibt Tacitus einen historischen Beweis von volkanischen Eruptionen am Niederrhein," by Professors Nees Von Esenbeck and Noggerath, in the 3rd volume of the work entitled " Das Gebirge in Rheinland Westphalen." And for the arguments on the contrary side, the works of Steininger, especially his " Gebirgkarte Mainz." 1822. p. 33.

I assume as an acknowledged fact that this material is derived in part from comminuted masses of pumice, and similar felspathic substances ejected by volcanic action. When we look therefore at the country in the neighbourhood of the Rhine, especially near Andernach, we discover pumice in two different states, either entering as one of the ingredients into the composition of trass, or constituting a congeries of loose fragments not united by any cement, which alternate in beds with a sort of loamy earth of the very same nature with that deposited at present by lakes, or rivers which flow through a soft stratum.

The first idea that suggests itself in order to account for this difference, is that the mass was formed from pumice deposited below the surface of water, whilst the loose fragments were ejected upon dry land; but the alternation of the latter with beds of loam is sufficient to prove that this distinction is admissible.

That since the epoch of the general deluge, water must have covered the face of this country to a considerable depth, is evident enough, I conceive, from the existence of that extensive deposit of marl which occupies the bottom of the valley of the Rhine from Basle (as it is said) to Cologne, the thickness of which often exceeds 40 feet, although it is the most recent of all the strata, resting often upon the diluvium, and covered by nothing except vegetable mould. The nature of the shells that occur in it, which are the same with those found at present in the country, prove that its origin cannot be very remote, and its apparent identity with the loamy beds which alternate with the congeries of loose fragments consisting chiefly of pumice, that occur near Andernach, tends to shew that at a period subsequent to the excavation of vallies in general, a large expanse of fresh-water occupied the lower situations in this part of Germany. As however the water has not in these cases given rise to the deposition of trass, it seems necessary to refer the latter to some more antient period, where the circumstances of the case were so far varied as to occasion a different result.

To this conclusion I know but of two objections, one, the presence in the trass of Brohl of land shells, similar to those at present existing, the other, its filling up a valley ; but the former, being found only in one place, may have been washed there by rain or torrents at some subsequent period, and the latter may have been coeval with the transition rocks themselves, or at least not be one of those produced by diluvial action.

If these suppositions be objected to as gratuitous, I would remark that they are adopted merely to reconcile the formation of the trass of the Rhine with that of the analogous formations of Hungary and Naples, which I shall afterwards consider ; and to explain the fact of its occurrence in all cases underneath the marl, and never alternating with it, as is the case with those beds of pumice which exist uncemented by any paste. I am therefore upon the whole inclined to refer to a period antecedent to the formation of vallies of denudation the production of trass, but the same inference cannot be extended to the Eyfel volcanos, the lavas from which, having in all cases flowed in conformity to the slope of the vallies, must have burst forth subsequently to their formation.

Whether however the latter, though proved to be geologically speaking of modern date, have been in activity within the limits to which historical documents extend, is quite a distinct question, and one that requires proof of another kind.

Steininger * and others indeed appeal somewhat absurdly to a passage of Tacitus, which they interpret as referring to

---

* The following is the passage to which they allude:—

" Sed civitas Juhonum, socia nobis, malo improviso afflicta est. Nam ignes, terrâ editi, villas, arva, vicos passim corripiebant, ferebanturque in ipsa conditæ nuper coloniæ mœnia. Neque extingui poterant, non si imbres caderent, non fluvialibus aquis, aut quo alio humore : donec inopiâ remedii, et irâ cladis, agrestes quidam eminùs saxa jacere, dein, residentibus flammis, proprius aggressi, ictu fustium, aliisque verberibus, ut feras, absterrebant. Postremo tegmina corpori direpta injiciunt, quanto magis profana, et usu polluta, tantò magis oppressura ignes.—*Tac. Ann.* l. 13. c. 57.

an eruption of one of the Eyfel volcanos at so late a period
as the reign of Tiberius Cæsar.

That historian, it is true, notices in his Annals, a fire that
broke out from the earth, and ravaged the country of the
Juhones, extending even to the walls of Cologne.

This fire it is added could not be put out by water, and
baffled every other expedient, until the inhabitants, finding
that they checked it by volleys of stones, came near, and
at length succeeded in smothering it by throwing heaps of
clothes upon the flames.

This passage has given rise to the very elaborate paper by
Professors Nöggerath and Nees von Esenbeck, published in
the work before quoted, in which it is shewn, first, that there
is no certainty as to the country which the people called by
Tacitus " Juhones," inhabited, or whether indeed that be
the true reading; secondly, that a fire which reached the
walls of Cologne could not be volcanic, for no volcanic appearances exist within twenty miles of that place, the nearest
crater being that of Rodderburg, near Rolandseck; and
thirdly, that the description of Tacitus applies much better
to some artificial combustion, such as what might be caused
by setting fire to the woods or heaths in a dry season, than
to any thing of a volcanic nature. It is indeed quite ridi-
culous to refer to any such cause a fire which was checked
by volleys of stones, and stifled by throwing over it dirty
clothes; and if we conceive ourselves obliged to adhere to
the words of the historian, and believe the flame to have
" emanated from the ground," it seems more likely that it
should have arisen from a disengagement of inflammable
gas, than from the usual concomitants of an eruption.

When however we recollect that at the period to which
we allude, a colony planted at this distance from the seat of
empire bore much the same relation to Italy, which the back
settlements of Canada, or the wilds of Newfoundland do to
Great Britain, it seems scarce an impeachment upon the
general accuracy of the historian, to suppose, that he may
have been misled by popular rumour in the details of an

64   *Antiquity of the Eyfel Volcanos.*

event, at once so remote, and so little affecting the general
interests of the state.

Steininger in a late publication has communicated another
fact which he thinks conclusive as to the modern date of
the trass about Bendorf, namely, the discovery of a coin
of Vespasian imbedded in the substance of the mass. Unfortunately however it turns out, that Roman altars are preserved cut out of this very same rock, so that we must account for the presence of the piece of money, as perhaps it
is not very difficult to do, from the action of rains or torrents
washing down upon the coin a quantity of this trass, which
afterwards became consolidated. We may also remark that
Steininger did not find the coin himself, but received it
from one of the workmen, whose account of its actual position cannot be greatly depended upon.*

It is also certain that the lava of Niedermennig existed in
the time of Augustus, for the pillars of the ancient Roman
bridge at Treves are built of this material.†

* See the Memoir before alluded to in "Nöggerath's Rheinland Westphalen."
† It is curious that Becanus, in his " Origines Antwerpianæ, Lib. i. p. 60,
attributes the fire recorded by Tacitus, to the combustion of some of the bituminous or coaly matter so common near Lieges in the Netherlands. " Ubi
sunt specus illæ subterraneæ, pigra barathra, montes piceo lapide sive Asphal-
tide constantes, et sylvæ obscuræ.

But this hypothesis is altogether irreconcilable with the statement of the
historian as to the fire having reached the walls of Cologne, a city, which from
its distance from Lieges, as well as from the structure of the intervening country, must have been quite beyond the possible influence of such an event.
Becanus likewise supposes, with not much more probability, that this was also
the spot referred to by Claudian as having âfforded Ulysses an entrance into
the infernal regions.

> Est locus, extremum quâ pandit Gallia littus,
> Oceani prætentus aquis, ubi fertur Ulysses
> Sanguine libato populum movisse silentûm.
> Illic umbrarum tenui stridore volantûm
> Flebilis auditur questus: simulachra coloni
> Pallida, defunctasque vident migrare figuras.
>   Lib. i. Carm. 3. 123.

But those who consider the above passage as a sufficient ground for assuming
the existence in Gaul at that time of appearances analogous to those of the
Phlegrean Fields near Naples, will look to the Volcanos of the Eyfel district,
and still more naturally to those of Auvergne, as their probable site, rather
than to the coal country of the Netherlands.

We are therefore under the necessity of attributing to the Eyfel volcanos, a date historically very antient, though geologically speaking modern, since geological research may be said almost to terminate, where history begins: if we adopt the opinions of Professor Buckland respecting the excavation of the vallies, we must suppose these rocks to have been formed, like some of those in Auvergne, subsequently to the deluge recorded by Moses ; or, if limiting ourselves to those views in which all geologists concur, we choose to speak more indefinitely, the date of their eruptions must be pronounced to be posterior to the event which reduced the surface of the globe to its present condition.

It would seem however that these are the only volcanos in this part of Europe to which the same remarks will apply, for the remaining rocks near the Rhine to which we attribute a similar origin, no less than those in Hessia and on the borders of the Thuringerwald, belong evidently to a more remote period. The latter indeed all appear to have been submitted to the same agents that have affected the older formations; their craters, if ever they existed, have been obliterated : the evidences of their destruction are seen in the rolled pebbles at their foot; and they are intersected by deep vallies, which the action of running water could never have occasioned.

Yet it would appear that the Rhine volcanos have been principally formed during the period at which the tertiary beds were deposited, for I know of no instance where they are covered by any thing older than diluvial detritus, and in many cases they are seen to rest on the brown coal and other beds belonging to the plastic clay formation. Such is the case at the Habichwald near Cassel, at the Meisner near Eschwege, in the Westerwald east of Coblentz, and in the Siebengebirge near Bonn.

The latter chain of hills presents in some measure the general features of the whole, and shall therefore be noticed more particularly.

E

Siebengebirge.

On the eastern bank of the Rhine, rising abruptly from the borders of the river, are the mountains which have obtained this appellation from the seven peaks which strike the eye at a distance. To the north they seem to sink more gradually into the plain, but on the south they terminate in the " castled crag cf Drachenfels," which offers on that side a very precipitous front.

This chain of hills consists partly of basalt, partly of trachyte, and the association of these two rocks without any determinate order being discoverable ought to stagger those, who with Beudant consider the two formations as produced by different processes, and at distinct epochs.

Here at least it seems difficult to determine which of the two be the most recent, or whether indeed they have not been ejected at one and the same time.

That it is in this manner that the seven mountains have been formed, appears probable from an examination of them on the side that overlooks the river. We here may trace for some distance up the hill a micaceous slaty sandstone belonging to the greywacke formation of the plain below, containing thin seams of anthracite. This is covered by nearly horizontal beds of a kind of trass, like that near the Lake of Laach, but containing less pumice. On ascending to a greater height we meet at length with the trachyte, which seems therefore to lie under the other rocks, at the same time that it rises above them.

This trachyte is traversed by vertical fissures, like the lava of Niedermennig, and like it is often cellular, and contains imbedded portions of a mixed rock looking like altered granite, but in which the felspar is changed into kaolin. The rock of Drachenfels consists of a somewhat different variety of trachyte, marked by its large and regular crystals of glassy felspar, and the mountain adjoining it on the west, called the Wolkenburg, differs from either, consisting of a number of minute acicular crystals of hornblende. It seems evident, that these rocks have both been forced up through the greywacke, which indeed continues until we reach nearly two-thirds of the entire height of the Wolkenburg. It is covered by the same marl which I have before noticed.

Near Königswinter however we observe a very compact quartzose conglomerate, lying betwixt the greywacke and the trachyte, which Professor Nöggerath has referred to the brown coal formation, seen so extensively on the western side of the Rhine. It is rarely observed so compact as in this instance, and it contains masses of opal in which the fibrous structure of the wood is plainly discernible. Thus the date of the trachyte is at least as recent as that of the oldest tertiary formation.

The bed of trass or trachytic conglomerate in the Seven Mountains is traversed by several dykes of a substance allied to the volcanic formation which it encircles.

One of these is seen on the hills above Königswinter pursuing an oblique direction, and at the same time sending out horizontal branches into the trass. The substance of the dyke is very various in composition, approaching in some cases to trachyte, and in others partaking of the characters of basalt.

Opposite to Bonn, at the northern extremity of the Siebengebirge, are the quarries of Obercassel, which have been described by Professor Nöggerath in the work already so often referred to.

They are composed of basalt, which, though more commonly compact, is occasionally found vesicular, and the vacuities, which are generally oval, appear to be of the same age with the rock, and not to have resulted from decomposition. Some of them are filled with calcareous spar, carbonate of iron, and other minerals, but others are void.

The point however most worthy of notice in this rock, is the concentric arrangement of its parts.

The principal quarry, that of Ruckersberg, situated near

the summit of the hill, displays this structure in the most
satisfactory manner.* We observe here not a cluster of globular concretions, like those of Bertrich, nor yet a line of
prisms parallel to each other as in many parts of the Vivarais, but a succession of concentric coats, which on the
northern side of the rock are broken away, but in that
which is entire, are seen encircling one another in such a
manner as to create an impression, that the whole may have
once formed an immense spheroid, composed of a series of
laminæ wrapped round a common centre.

That the form of this mass was elliptical, having its longest
axis in an horizontal direction, appears from this circumstance, that to the north of this quarry, but exactly parallel
to it, another section of the rock is exposed, in which the
same concretionary structure presents itself, but the laminæ
form a portion of a curve turned exactly in the opposite
direction to that of the former locality.

Thus let A be the quarry of Ruckersberg, and B the
section exposed to the north of it, the following will be the
disposition of the strata, from which it is evident that if
the rock had not been broken away in the interval, an
ellipsis would have been formed.

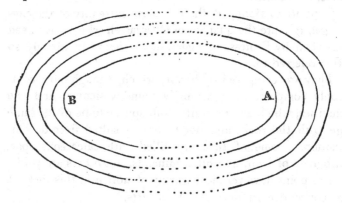

We observe too as we descend, that whenever a section is exposed, the direction of the strata is such as corresponds very well with this idea, the rock being found at a short distance below, dipping towards the same point, but at a slighter angle, and at the bottom becoming at length completely horizontal.

Now this admits of a ready explanation if we suppose the whole mountain to have constituted the same great elliptical mass, since it is evident that agreeably with this structure, the dip of the strata would diminish progressively in proportion to their distance from the supposed centre, so that at length they would appear, when viewed within a limited space, altogether horizontal.

It is necesssary however for this hypothesis to assume, that the upper portion of the mountain has been swept away, since the axis of this immense spheroid is placed not in the present centre of the mountain, but about the site of the quarry of Ruckersburg near its summit. Now as this quarry is about 300 feet above the base of the hill, we must attribute to the latter an original elevation of between 500 and 600 feet.

Besides this division into concentric laminæ, the basalt of Obercassel has likewise a tendency to form columnar concretions, which always range at right angles to the curvature of the strata; hence when the latter are vertically disposed, the prisms are horizontal, and vice versâ; appearing in every intermediate position in accordance with the dip of the laminæ themselves.

The *Seven Mountains*, although they may appear isolated, are in fact a prolongation of the extensive basaltic formation of the Westerwald, which is connected with another considerable volcanic district, north-east of Frankfort, termed the Vogelsgebirge.

From the latter, the isolated basaltic cones of Frankfort

and Hanau on the one side, and of Cassel and Eisenach on the other, seem to be ramifications.

Different as these several chains may be to each other in various particulars, they have the following characters in common, by which, if I mistake not, they are distinguished from the basalts, which occur associated with secondary strata.

Although frequently as compact as the latter, they are always, when viewed on the great scale, found to be accompanied more or less with cellular products, the cavities of which do not appear to have ever been filled with infiltrations of crystalline matter, which is the case with the amygdaloids accompanying secondary basalt. These cellular rocks frequently occur on the highest parts of the chain, as near Rennerod in the Westerwald, where they rest on compact basalt, as though the latter had been the effect of the pressure exerted by the superincumbent stratum. In other places the highest portions of the chain consist of compact basalt, whilst beds of cellular lava occur mantling round it, as if after the formation of the central mass, ejections of cellular matters had succeeded, which ranged themselves round its sides. Such is the case in the Vogelsgebirge.

The volcanos of the Westerwald are also identified with those of the Siebengebirge in consisting partly of trachyte, and likewise in being accompanied with strata of trass, which seem to fill up the bottom of the vallies. Concerning the manner in which the latter has been formed, I shall only remark at present that the existence in it of slag and pumice, and the rare occurrence of fragments of compact lava, like that of the mountains encircling it, seem to shew that it is derived in a greater degree from ejections of pumiceous matters than from the detritus of the surrounding country.

Occasional knolls of basalt lie scattered over the whole space between the Westerwald and Vogelsgebirge, as at Limburg, Wetzlar, &c. which some may consider as the relics of a continuous stratum once covering the country;

whilst others may regard them as produced by distinct eruptions of volcanic matter.

That the latter is the true state of the case appears, I think, from comparing them with the conical basaltic hills near Eisenach, where circumstances have enabled us to ascertain clearly the relative position of the volcanic mass to the contiguous stratum.

As the latter appear to me highly interesting, not only in themselves, but also as affording a key to the structure of the whole country we are considering, I shall proceed at once to describe them, without noticing the rocks intermediate.

These knolls of basalt occur in the second or variegated sandstone formation, and appear on the surface perfectly unconnected one with the other. As they afford a valuable material for the roads, so much of them has in many cases been removed, that the rock which originally rose above the surface of the sandstone, is now worked considerably below the level of that rock. In tracing it downwards the mass is generally found to enlarge, so that its shape appears to resemble that of a wedge.

The quarry in which the relation of the basalt to the contiguous rock is best displayed is the Pflasterkaute near Eisenach on the road to Frankfort.*

* For a knowledge of this locality, and in some measure for the power of examining the geological phænomena therein displayed, I hold myself indebted to a very intelligent road-surveyor, M. Sartorius of Eisenach, who began many years ago taking advantage of his situation, to expose the rock in such a manner as might enable geologists in time to determine, whether the basalt merely lay upon the sandstone, as the Wernerians would suppose, or had been forced up through the midst of it from a considerable depth. His first account of this and other similar spots, was published in 1802, in a small pamphlet entitled " Die Basalte in der Gegend von Eisenach," but as every year's consumption of stone renders the excavations deeper. the fact must at present be exhibited in a more satisfactory manner than it was when he wrote his first pamphlet, as it becomes more and more difficult to explain the position of the basalt by any irregularity of surface in the subjacent sandstone, and less likely that any termination will be found to the dyke as we penetrate downwards.

Sartorius has since published several other Geological tracts, which will be specified in the Appendix.

NEAR EISENACH.

In this instance the excavations are carried to such a depth, that we are enabled distinctly to see the basalt more than 50 feet below the surface of the sandstone. The line of junction is also well displayed, and we observe the sandstone changed from an horizontal to a vertical position, split in all directions, and rendered harder and whiter where the basalt touches it.

The latter is in some places compact, and in others cellular, the cells partly empty, and partly filled with calcareous spar, quartz crystals, and zeolite. The central portion of the mass is always freest from these hollows, and at the surface is generally a kind of tuff made up of fragments of the volcanic rock cemented by clay or sand.

The following may serve to give an idea of the relative position of the several rocks displayed in the quarry.

Vertical Section of the Pflasterkaute near Eisenach.

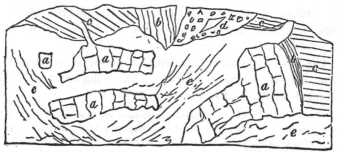

a a a a Basalt. *b b* Sandstone altered. *c c c* Ssndstone unaltered.
d Tuff with fragments of Sandstone. *e e e* Debris.

The Stoffelskuppe and the Kupfergrube, both which lie at no great distance, present similar phænomena, with this additional circumstance, that the masses of sandstone, which occur entangled in the midst of the volcanic rock, are sometimes prismatic. This last fact is however best displayed near the town of Budingen, north of Hanau, in the Wetterau.

NEAR EISENACH.

The prevailing rock in this country is the new red sandstone, but on the slope of the hill south of the town occurs a little eminence called the Wildenstein, consisting of columnar basalt, which seems altogether isolated, being encompassed on all sides by sandstone.

Though we are not enabled, as in the case of the Pflasterkaute, to observe the relation of the trap to the surrounding stratum, yet it can hardly be doubted that it has been forced up in the same manner, and has thus carried with it the portions of sandstone that occur in the very midst of the basalt.

These portions appear to have been curiously altered by the heat to which they were subjected; they are hardened almost to the degree of flinty slate, rendered white and splintery, and in most instances form clusters of little prisms, possessing even greater regularity of form than those of the basalt which encircles them. It is curious to mark the resemblance between the prisms here alluded to, and those produced artificially in several parts of Derbyshire and Yorkshire,* where the soft friable sandstone of the country is rendered serviceable for road-making by exposure to heat, which hardens and causes it to split into small columnar concretions.

Sartorius has described a number of other basaltic eminences in the neighbourhood of Eisenach, all of which he supposes to have been thrown up in a similar manner, and many indeed of which are proved to be so constituted.

He likewise endeavours to shew † that a certain connexion exists between many of these distinct cones, and that they are in some cases grouped round some common centre, forming a system of basaltic eminences, which possess a certain correspondence in position, and, as it should seem, in origin.

In the case to which he appeals, the central point (the Dietrichberg) is a large overlying mass of basalt, the direc-

* As at Rotherham.
† See Sartorius, Geognost, Beobacht. Eisenach, 1821. p. 100.

NEAR EISENACH.

tion of which appears to be conformable with that of the subjacent rock, but in the smaller eminences of the same formation which are distributed round it, the basalt always is placed obliquely, and its inclination is found to vary according to its position with reference to the central mass.

Sartorius from thence concludes that all these detached cones have been thrown up from a common point, and that the focus of the volcanic action lay immediately underneath the Dietrichberg, at a depth which he thinks might be calculated, by considering the distance of any one of these masses from the centre of the system, coupled with the degree of its inclination.*

My principal motive for noticing this statement, which I had no opportunity of verifying, is to lead to an examination of other similarly constituted basaltic districts, with a view to ascertain whether any such arrangement or connexion can be perceived amongst them. At present all that I have stated must be understood to rest on the authority of Sartorious, to whose little publications I may refer for the most detailed account of these basalts, the igneous origin of which he has the merit of having maintained, even at the time when the authority of Werner was at its height.

But the rock perhaps which exhibits the greatest combination of phænomena calculated to shake any preconceived opinion with respect to the aqueous origin of these basalts, is the Blaue Kuppe, near Eschwege, a town also in Hessia, but about twenty miles north-east of the above localities.†

---

* He however admits the uncertainty of these calculations, for the depth at which the volcanic force resided, estimated by the obliquity of one of these masses, is 125,248 feet, or upwards of four Geog. miles, whilst by another it would be only 79,104 feet. It is evident indeed that the inclination of the basalt is influenced by too many causes to afford any correct data.

† It is curious that Daubuisson in his " Account of the Basalts of Saxony," never alludes to this mountain, although he seems to have particularly examined the Meisner, which lies no more than half a dozen miles off from it, and to which he appeals as affording evidence of the aqueous origin of trap.

BLAUE KUPPE, NEAR ESCHWEGE.

In this instance compact basalt is seen associated with a substance of so light and porous a description, in its nature so analogous to the productions of modern volcanos, that it would indeed argue an excess of scepticism to refuse to attribute it to the same cause.

Unlike the other volcanic eminences, the Blaue Kuppe consists on one side of sandstone, and on the other of volcanic matter, as if the force which caused the ejection of the latter had at the same time elevated the former. As in the Pflasterkaute, the sandstone here is hardened and cracked in all directions near the line of junction, and portions of it are everywhere imbedded in the substance of the basalt.

Besides the principal mass of volcanic matter occupying one entire side of the hill, are several dykes which penetrate the sandstone, enclosing portions of it, and altering its stratification in a very remarkable manner.

One of these appears to be a prolongation of the principal mass, but two others that occur a little on one side have no connexion with it on the surface. The upper portions of this rock consist of a sort of tuff composed of fragments of cellular and compact lava, intermixed with sandstone, and cemented by wacke, whilst the nucleus is composed of basalt, which is sometimes cellular, but the cavities for the most part are filled with crystalline matter.

The quarry that has been made in this rock exposes a cavern in the midst of the volcanic matter, which serves still more fully to identify it with modern lava.

The basalt is here disposed in irregular strata, possessing a curvature corresponding with the arch of the cavern, and in the interior I have found specimens of a more cellular variety of the same rock, which seems to have depended from

It would be curious to learn whether this was the effect of accident or design, for the difficulty of explaining the phænomena of the Blaue Kuppe on the Wernerian principles furnishes at least an adequate cause for the omission.

BLAUE KUPPE, NEAR ESCHWEGE.

the roof, like the stalactites, as they are improperly termed, of the caves in the island of St. Michael.*

A few miles north of the Blaue Kuppe† stands the Meisner, which not many years back was appealed to in proof of the aqueous origin of basalt, but which will probably be viewed as affording an additional evidence of the contrary hypothesis.

The basalt here forms an extended *plateau* overlying the new red sandstone formation; though in many places it does not do so immediately, there being here and there interposed a deposit of brown coal similar to that before noticed.

The latter not only is rendered columnar, as Daubuisson admits, near the line of contact with the basalt,‡ but I am assured that it is also converted into anthracite. Daubuisson however contends that this alteration is not universal, and therefore that the incumbent mass can never have been in a melted state, an objection which will be best met in a future part of this work, when I shall have occasion to shew that even modern lavas, in flowing over the surface of a rock, do not always produce any change.

The basalt passes gradually into a granular substance, which may be called augite rock, consisting of felspar, augite, a little hornblende, and grains of titaniferous iron. All the upper part of the platform is composed of this substance, which differs from the basalt underneath only in the more distinct crystallization of the component parts.‖ This was also one of the circumstances appealed to by the Wernerians in proof of the aqueous origin of the rock, as it

---

* See Dr. Webster's description of that Island, and my Third Lecture.

† There is a drawing of this mountain by Von Hoff, together with a complete description, in the Magazin der Berliner Gesellschaft Naturforschender Freunde. 5th year. An extract from the account may be seen in Delabeche's Geological Memoirs, page 100.

‡ See Daubuisson on Basalt, translated by Neill, page 204, note.

‖ The same association of greenstone and basalt occurs at Beaulieu in Provence. See page 43.

was conceived that this crystalline structure would have
been obliterated by heat ;* and even Dolomieu was led by
this consideration to admit the Wernerian doctrine with
respect to greenstone. At that time the experiments of Sir
J. Hall, Watt, and others, had not induced Geologists to
admit that these crystals might have been the very result,
under certain circumstances,† of the process, which was at
first imagined to be incompatible with their existence.

Near Cassel is a lofty ridge of mountains called the
Habichwald, which, including the Dornberg and other hills
connected with it in character and position, forms a large
square of about four German or twenty English miles, to
the west of that city

It consists principally of beds of tuff, associated with scoriform lava and compact basalt, the whole resting on a bed
of brown coal, or, where that is wanting, upon limestone.

The tuff consists of a congeries of highly cellular and
vitreous varieties of lava, which seem to denote a recent
origin compared with that of the rocks in the neighbourhood. Its structure is well displayed near the pleasure-
grounds of the Elector, above the palace of Wilhelmshöhe.‡

The hills above particularized may serve to illustrate the
general structure of those numerous basaltic cones which
lie scattered for a considerable distance on all sides of the
Seven mountains, the Westerwald, and the Vogelsgebirge,
outliers, at it were, of that great volcanic formation, which,
in the latter places, covers the whole face of the country.

Other examples of the same kind are furnished by the
neighbourhood of Frankfort and Hanau, where small wedge-
shaped prominences of compact basalt, gradually becoming

* See Daubuisson on Basalt.
† See my Fourth Lecture.
‡ See Raspe's account of some German Volcanos, London, 1776.

NEAR FRANKFORT.

cellular near the surface, appear to have been thrust through
the midst of the sandstone formation.

At Steinheim near Hanau the cells of the basalt are occupied by a variety of sparry iron ore, called by Haussman
sphœro-siderite, which forms a number of spheres varying
from a line to an inch and a half in diameter, more generally
quite compact, but sometimes hollow, and containing within
a nucleus of a yellowish or ochreous matter, effervescing
feebly with acids, and yielding with difficulty to the knife.

I observed some circumstances relative to this mineral and
the rock containing it, which seem to deserve a brief notice.

One was the occurrence of a sort of vein of cellular
basalt passing through the substance of the compact, like
the coarse-grained granite which we sometimes see penetrating a fine-grained variety. In the midst of the cellular
portion was a cavity filled with the spathose iron above
noticed, as in the sketch underneath, where

a a are the portions of compact basalt; B B those which
are cellular; C the cavity with sphœro-siderite coating its
walls.

Another circumstance which I remarked, was, that the
external surface of the spheres had in some cases a covering
of white calcareous matter, in a powdery form, sprinkled
over it.

The third observation I made indicated still more strongly
that the whole had been in fusion, namely, the spheres being
sometimes found flattened at the points where they appeared
to have been in contact with the lava, an effect arising in all

GENERAL REMARKS.

probability from the contraction that took place in the surrounding parts whilst undergoing cooling.

Semi-opal also occurs between the interstices of this same rock, and I have even seen specks of noble opal disseminated through its substance.

The basalt both near Frankfort and near Hanau shews a remarkable tendency to concentric arrangement, and as it contains much iron, which becomes readily oxidized, the external layers decompose and peel off, leaving only a small nucleus of compact basalt retaining its original characters. In some cases the cellular and compact basalt occur intermixed, but in others the former is confined to the superficial portions. The cells however in these instances do not form that sort of net-work which is usual to them, but constitute long cylindrical tubes distinct from each other, just such as would be occasioned by bubbles of gas forcing their way upwards through a soft pulpy substance.

Enough has already been said respecting the general structure of the Westerwald and the Vogelsgebirge, which appear to be composed of extensive plateaus and cones of basalt, covering the rock of the country, but never alternating with it. Nor is there much to be learnt with regard to the manner of their formation by an examination of the districts themselves; and a Wernerian, if he were to overlook or explain away the fact of cellular products occurring amongst them, might easily persuade himself that the whole was the result of aqueous deposition, the evidence of that return of the waters, which is supposed to have given birth to the newest flœtz trap formation of his master.

When however we turn to the dykes of basaltic matter (as they may be called) which are scattered all around, we can hardly help imagining that these more extensive formations of the same rock have in reality been produced in the same manner, that the more elevated masses, which are ge-

nerally most compact, were first thrown up by the agency of
a volcano, and that the cellular matters being subsequently
ejected, arranged themselves around them in successive strata.
The volcanic operations, taking place with the greatest intensity round the area now occupied by the Vogelsgebirge
and the other basaltic groupes, would cover completely with
their products the surface of the subjacent rock; whilst at
a greater distance from the sphere of their activity, isolated
cones of basaltic matter would be occasionally thrown up,
as at Eisenach, at Cassel, in the neighbourhood of Frankfort, and on the west bank of the Rhine.

Besides the volcanos above enumerated, there occur near
the borders of the Rhine, but higher up the stream, other
rocks which are said to have a similar origin.

In the Odenwald, a group of hills in the neighbourhood
of Heidelburg, from the midst of the new red sandstone,
rise some eminences in which basalt is associated with augite rock (dolerite of Brongniart), and contains nepheline
(Katzenbuckel), mica, mesotype, olivine, and titaniferous
iron ore. It is probably analogous to the Hessian basalts.
The augite rock is seen *in situ* at Gaffstein, the basalt containing olivine at Pechsteinkopf and Durkheim.*

Near Freyburg in the Brisgaw is the group of the Kaiserstuhl, of which Dr. Boué has given an account in his memoir
on the South-west of France.† It appears from his report
to be a mass of augite rock with excess of felspar (dolerite
felspathique) thrown up from the midst of the plastic clay.

The highest mountain in the groupe is the Kaiserstuhl,
which rises to the height of about 1120 feet above the river,
and this with the other eminences composed of the same
materials are ranged in an elliptical form round a valley.

These rocks offer no trace of craters or streams of lava,

* See Leonhard Taschenbuch for 1822.
† Annales des Sciences Naturelles, August, 1824.

but are associated, especially near the surface, with cellular lava, containing calcareous spar and mesotype. Hyalite occurs in the cavities, and incrusting the surfaces of the rock. Tufaceous matters are not common, but occur along the Rhine at Breisach, where they seem, like those near Eisenach, to be contemporaneous with the augite rock.

I am also assured that the rock of Kaiserstuhl is partially covered with a calcareous deposit, the only instance I believe among the Rhine volcanos in which this occurs.

Saussure,† who visited this groupe in 1794, and appears to have been somewhat swayed by the authority of Werner, is nevertheless compelled to acknowledge the volcanic origin of the rocks about Limburg, which are in part penetrated with oval cells, in great measure void, and of the tuff about Echardberg, which contain fragments of scoriform lava. The origin of the basalt itself he considers doubtful, but there are few at the present day who will concur with him in that opinion, considering how intimate appears to be the connection between the porous and compact rocks in this locality. Upon the whole the groupe of the Kaiserstuhl may be set down as belonging to the same æra as the basalt of the Westerwald, and the trachyte of the Seven Mountains.

A few miles to the north of the Lake of Constance is the commencement of another series of basaltic and porphyritic cones, first seen at Hohentwiel and in several detached hills contiguous, which rise from the midst of the (Jura?) limestone formation.

They consist in part of clinkstone, and in part of basalt, accompanied with tuff containing fragments of trap rock (always compact) as well as of gneiss, limestone, quartz, &c. all cemented by a wacke-like paste of a ochrey colour.

In the hill of Magdeburg, a passage is said to exist be-

† Journal de Physique, vol. 44.

tween the clinkstone and the basalt;* at Hohentwiel, the former contains veins and nodules of natrolite, together with opal, pitchstone and hyalite.

In Wirtemburg basalt occurs both in overlying masses, and dykes, in various places along the chain of the Rauhe Alp south of Tubingen, and is perhaps connected with the rocks of Swabia above described. Professor Scubler † remarks, that the horizontal stratification of the limestone composing that country is destroyed, wherever the basalt approaches it.

The chief localities are Linsenhofen, Faisel, Dettingen, Jusiberg near Urach, Grabenstetten, Donstetten, the valley of Guttenburg, Rauberstege near Brachen, near Dottingen, Offenhausen, Ehningen, and Donaueschingen.

Basaltic rocks, many of which are probably referable to the same class as those already noticed, appear to be scattered over many parts of Germany, especially the skirts of the Thuringerwald, the Fichtelgebirge, the mountains of Saxony, and the Reisengebirge of Silesia. But of these the information I have been able to procure has been scanty and imperfect; topographical works on continental geology being of all others the most difficultly procurable in this country.

In order however to complete this enumeration of the principal trap formations of Germany, I shall bring together such particulars as I have been able to glean respecting their characters, position, and relations.

The Rhongebirge, a chain of mountains east of Fulda, appears to be a continuation of the same overlying formation, which constitutes the Vogelsgebirge, and is seen in so many other parts of Hessia.

Of this district I have seen no account of more modern

---

* See for this and other particulars a Memoir in Leonhard's Taschenbuch for 1823, by the Oberbergrath Selb.

† The Wirtemburgischer Jahrbucher for 1824, contains an account of these basalts, and an extract is given from it in the Bulletin des Sciences for November, 1825.

date than that of Voigt in his work entitled " Beschreibung des Hochstifts Fuld," published in 1783.

This writer speaks of the whole country as indicating volcanic action, containing in most parts rocks of basalt and clinkstone porphyry (hornschiefer).

South-east of Fulda is a circular cavity which he considers to have been a crater, and which is full of water, like· the *Maars* of the Eyfel district. It is closed in by two hills, which meet towards the east, but on their western extremity leave an aperture in which the hollow called the crater is found. The first of these hills, called the Euben, is composed of what Voigt denominates lava, probably a scoriform or amygdaloidal basalt; the second, the Pferdekopf, of porphyry slate, which seems to have been forced through the lava. The sandstone rock adjoining is hardened and otherwise altered, and where the lava is in contact with it, there is an intermixture between the two.

From this account of Voigt's one should be led to conjecture, that some of the volcanic rocks of the Rhöngebirge are more recent than those of the neighbouring country, and posterior even to the formation of the vallies, but I am not aware that the statements of this Geologist have been confirmed by any more modern observer.

The example indeed of Faujas St. Fond, who saw traces of craters in the basaltic rocks of the Hebrides, ought to render us cautious in receiving accounts given by geologists of this school, at a period when every volcano of whatever age was imagined to have been formed after the model of Etna and Vesuvius.

Dr. Boué, in his paper on Germany, notices the passage of this clinkstone rock into a kind of pearlstone at Helsburg near Coburg.

On the north-eastern limit of Bohemia at the foot of the Fichtelgebirge occurs a series of basaltic cones, extending from Egra to Parkstein. The localities are : Parkstein, where indurated marl is imbedded in the basalt, as at Eisenach ; Neustadt am Culm, Kemnat, Culmain, Friedenfels, and five different spots between Waldsassen, Redwitz, and Witterstein.

At Egra, the hill of Kammerberg * is a conical heap of scoriæ, probably of more recent date than the basalts with which it is associated.

The hot springs of Carlsbad (which, in Berzelius's opinion, † owe their temperature to the rise of carbonic acid and steam, strongly heated at a great depth, through the mineral waters of the place) may perhaps be regarded as a proof that the volcanic action goes on to a certain extent even at the present time.

From Egra to Tœplitz, and from thence to the Reisengebirge in Silesia, a chain of basaltic and clinkstone hills appears to extend in a direction nearly parallel to that of the primitive range of the Saxon Erzgebirge.

Near Tœplitz in Bohemia basalt and clinkstone occur united, as near the Lake of Constance, the latter forming the lofty conical hill of Bilin, in which fragments of gneiss occur surrounded by the volcanic matter. Beds of tuff alternating with tertiary limestone appear in the neighbourhood. The hot springs of Tœplitz are well known.

I visited this spot in 1820, but in too superficial a manner to speak with confidence with respect to the age and character of these rocks,—I believe however that they will be found to belong to the class of tertiary volcanos.

Dr. Boué states in his memoir on Germany,‡ that scoriæ occur at Friedland in the Mittelgebirge, and likewise in the circle of Pilsen, at Wolfsberg, and at Salesel.

A similar series of basaltic cones occurs likewise on the Saxon side of the Erzgebirge, from the neighbourhood of Schwartzenburg on the south-west, to the hill of Stolpen beyond Pirna on the north-east. ||

* Der Kammerberg bey Eger beschrieben, Von T. V. Goethe. Leonhard's Taschenbuch.

† Berzelius in Gilbert's Annalen d. Phys. vol. 74, p. 113. The English abridgements of the memoir do not contain the remarks to which I allude.

‡ Journal de Phys. for 1822.

|| See Daubuisson on Basalt.

Under various shapes, as in platforms, cones, and domes, basalt forms the summits of about twenty mountains, some of which are isolated, others connected below with the neighbouring mountains, the basaltic cap alone remaining separate. It forms the highest points in the country, and principally occurs near the ridge of the primitive chain. As I am not aware of the occurrence of scoriform matter in any of them, nor of their being incumbent on any very recent deposits, I should have omitted all mention of them; as unconnected with the immediate subject of these lectures, had I not been desirous of completing, so far as it was possible, this enumeration of the basaltic formations of Germany.

The localities mentioned by Daubuisson are, 1. the mountain called Schiebenburg, 1000 to 1300 feet in height, celebrated for the apparent passage of wacke, into clay on the one side, and into basalt on the other, which was long appealed to in proof of the aqueous origin of the latter rock ; 2. the Pœhlburg, near Annaberg, consisting of columnar basalt resting on gneiss; 3. the Bœrenstein, six miles to the south of Annaberg; 4. the Spitzberg, a peak consisting of mica-slate capped with basalt, 4000 feet above the level of the sea; 5. Heidelburg, near Seiffen; 6. Lichtewalde, on the frontier of Saxony and Bohemia, 3000 feet high ; 7. the Steinkopf, consisting of gneiss, capped by syenite-porphyry, covered by a small platform of basalt; 8. Geissengensberg, near Annaberg; 9. Luchauerberg, south-west of Dippols-walde; 9. Cottauer-spitze, a cone of basalt resting on sandstone; 10. Heulenberg; 11. Winterburg, near Schandau, and on the eastern side of the Elbe, the remarkable columnar basalt of the Stolpen, which is incumbent on granite.

Besides these, there occurs between Dresden and Freyburg a basaltic eminence, called the Landberg, where the fundamental rock is gneiss covered with clay-slate.

In all these cases it may be remarked that the basalt occurs as an overlying mass, and never alternating with the other rocks of the country, so that Werner was fully borne

out in his position, that it constitutes a distinct class of rock
formations; he seems also to have been correct in assuming,
that there is nothing in the characters of the substance itfelf,
or in the minerals associated with it, as found in Saxony,
which stamps it as the product of fire; and he might be
warranted, considering the vague descriptions of geological
phænomena given by travellers at the time he commenced
his career, in framing a system without reference to their
statements; but he cannot be so well excused for his obsti-
nacy in adhering to the same erroneous conclusions in spite
of the evidence afterwards brought together in contradiction
to them, and that even by some of the most eminent of his
own disciples, such as Humboldt and Von Buch, who seem
to have deviated more and more widely from the creed of
their master, in proportion to the extent of their acquaintance with volcanic phænomena.

If we follow into Lusatia the chain of the mountains of
Saxony, we find that many of the eminences are capped
with basalt, presenting the same characters and relations as
it does in Saxony.* The only difference to be observed is,
that in the latter country the basaltic hills are situated near
the summit of the mountain ridge, whereas in Lusatia they
are nearer its foot. One of the most considerable indeed,
the Lanscrone, near Gœrlitz, a conical hill near 1000 feet
in height, stands entirely isolated in the plain, and detached
from the mountain chain, from which it is about six miles
distant.

In the Riesengeberge of Silesia, which may be considered
a sort of prolongation of the Saxon chain, the hill, called
the Kleine Schneegrebe, is basaltic on its north-west side,
and the basalt appears to have been protruded through the
midst of the granite which forms the remainder of the hill.†

---

* See Daubuisson on Basalt, English translation, p. 73.

† Singer in Leonhard's Taschenbuch, 1823; and for the Wernerian view of
its formation, Daubuisson on Basalt, p. 235, Eng. Trans. Professor Jameson is
there said to regard it as an " upfilling," that is, he considers its position as
dependant upon the irregularity of the granitic surface in that part.

In one part huge fragments of this rock are entangled in, and cemented into a compact mass by a basaltic paste, the whole forming a singular species of conglomerate. The basalt is amygdaloidal, and contains zeolite and mesotype. It attains to the height of 4661 feet above the sea.

Von Buch, in his description of the environs of Landeck in the county of Glatz,* has noticed four basaltic hills superimposed on primitive rocks, the most considerable of which is the mountain of Deberschaar.

It also appears from the late work of Œynhausen on the Geology of Upper Silesia, that a numerous series of isolated basaltic cones extends from the Oder at Fachenburg to Troppau, and from thence to Freidenthal in Moravia.†

A few miles south of the latter place, but north of Olmutz, is the little town of Hof, near which, in the chain of hills called Gesenke, is a volcanic mass which exhibits indications of a more recent origin.

* Min. Description of the environs of Landeck by Von Buch, translated by Dr. Anderson. Edinburgh 1810.

† Between Michelau and Falkenburg, basalt crops out of alluvial soil.

Between Troppelwitz and Jagerndorf, are the basaltic cones of Schonweise; near Troppau those of Stremplowitz and Ottendorf; both amongst transition slate and greywacke.

Basalt of Kohlenburg amongst primitive slate.

Basalt also occurs between Tillowitz and Schiedlow, south of Falkenburg, and near Raklo.

Basalt blocks near Lipten.

At Annaberg a basaltic cone 1300 or 1400 feet high.

A basaltic hill called the Mulwitzberg extends between the towns of Michelau and Falkenburg; the basalt is prismatic, contains little olivine, but much steatite. On the sides are blocks of a quartzy sandstone, perhaps tertiary.

Near Schonweise, not far from Jagerndorf, two cones of basalt surrounded by slate clay, passing into transition conglomerate.

South of Troppau, at Ottendorff, Basalt in a transition country.

North-west of ditto, at Stremplowitz, are cones of basalt partly porous.

Between Bennesch and Raudenburg, basaltic tufa containing augite: it is a building stone.

At Kohlerberg, south of Freudenthal, occurs a basaltic amygdaloid, containing quartz, calcareous spar, chalcedony, chlorite. It resembles the basalt of the Buchberg near Landshut, and is the nearest point to the primitive range at which basalt is found.

The following account of it is extracted from a German periodical work called the Hesperus, for January 1821.

The hill called the Raudenburg, which is situated to the south-east of the village of that name, is 2250 feet above the sea, and is composed, of reddish, greyish, or blackish scoriæ, which look as fresh as those of the Puy de la Nugere or of Vesuvius, and of basalt.

These products inclose fragments of granitic rocks, and of mica or clay-slate, which are much altered, and seem to pass into the scoriform mass which envelops them. To the south-east of this hill, near Heydenpiltsch, are two hillocks of compact basalt, and more to the south near Brochersdorf is a basaltic hill called the Saunikal, and on the right of the Mora river, near Friedland, are two others. To the north-east are two funnel-shaped cavities, of which the largest is 75 feet in breadth and 18 in depth.

These basalts rise from the midst of mica-slate, are compact and sometimes columnar, and contain olivine.

Lastly, on the western border of Moravia, near the frontier of Hungary, is a small basaltic deposit near Banow, described by Dr. Boué in his geological memoir on Germany above noticed.

It is a cone of grey clinkstone, containing crystals of hornblende, and the few pores which are distributed through its substance are elongated in a vertical direction. On its western side it encloses portions of hardened clay, and sandstone, of various colours, and on its eastern side it has thrown up and cracked in various directions, a very large mass of the same kinds of rock, which are also hardened where in contact with it.

Thus it would appear that on either side of the great primitive chain which passes through the centre of Germany, the several parts of which are known under the names of the Thuringerwald, the Fichtelgebirge, the Erzgebirge, the Riesengebirge, &c. occurs a line of basaltic cones, which though detached one from the other, yet are so placed as to indicate a certain mutual connection. This notion is confirmed by observing, that similar formations occur chiefly at a certain distance from these primitive ranges, for Von Hoff has remarked,* that if a line be drawn from Upper Lusatia to Culmbach in the country of Bayreuth, and another from the same point in a north-westerly direction, so as to pass through the towns of Eisenach and Munden, no basaltic rock is to be met with north of that line, notwithstanding the abundant occurrence of it to the south.

The same author has further shewn in another of his publications,† that the shocks of earthquakes are most common in the same direction as that of the basaltic masses themselves, and round a certain distance on either side of the line in which they occur.

The importance of these observations will be more clearly perceived, when it is shewn that rocks indisputably volcanic are found to be placed in the same linear direction; as it will add one to the series of proofs by which the origin of Trap Rocks is connected with that of modern Lavas.

* See de la Beche's Geological Memoirs, p. 100.

† Geschichte der Veranderungen der Erdoberflache, vol. ii. an excellent work, which has led me to refer to several writings, in the German language especially, that might otherwise have escaped my notice.

LECTURE II.

ON THE VOLCANIC COUNTRIES VISITED BY THE AUTHOR IN 1823—24.

——

HAVING in my preceding Lecture given a detailed account
of the volcanos of France and Germany, I propose in my
present to lay before you a briefer description of those which
I visited in the years 1823 and 1824, supplying from time
to time the deficiencies of personal examination by reference
to the best authorities.

As my tour itself comprehended a portion of Hungary
and Styria, together with the greater part of Italy, Sicily,
and the adjoining islands, we may expect in the course of
the details that I shall give, if not to light upon facts relative to this branch of Natural History that have hitherto
escaped notice, at least to elucidate certain particulars less
fully developed in the structure of the preceding countries,
than in those about to be described.

Thus in order to study the Natural History of Trachyte,
Hungary is the country that should be chiefly consulted,
since from the extended scale on which the rocks belonging
to this formation are developed, we are there enabled to
follow them through all their modifications with a minute-
ness not practicable in Auvergne or in Germany.

With the assistance therefore of the elaborate work of
M. Beudant, I shall endeavour to lay before you as complete a synopsis as possible of the principal varieties, which
the trachytic formation of that country is found to present.

Beudant enumerates five distinct groups of mountains
consisting wholly of trachyte, the characters of which are
in all nearly the same, although particular parts of the formation may be more developed in one than in the rest.

The first of these groups, situated in the north-western
part of Hungary, namely, in the district of Schemnitz and
Kremnitz, occupies an elliptical space of about 20 leagues
in its greater diameter, and 15 in its smallest.

The 2nd, a smaller group, south of the preceding one,
forms the mountains of Dregeley near Gran on the
Danube.

The 3rd, is the mountain group known by the name of
Matra, situated in the heart of Hungary, east of the former.

The 4th, a chain which commences at Tokai, and extends
north to the heights of Eperies, in length 25 or 30 leagues,
and in breadth about 5 or 6.

The 5th, that of Vihorlet, east of the foregoing group,
which is connected with the trachytic mountains of Marmorosch on the borders of Transylvania.

Not only do these several groups appear unconnected
with each other, but it is Beudant's opinion, that almost
each particular mountain has been separately formed, for
their escarpments rarely correspond, as is the case with
plateaus comprised of basalt, so that it is impossible to view
them as the detached portions of one general bed cut away
by the operation of subsequent causes.

Now in the formation distinguished by Beudant under
the generic term of trachyte, that geologist has noticed five
kinds of rock, which, although possessing a common origin,
present many important differences.

These five varieties he has designated under the names
of Trachyte properly so called, Trachytic Porphyry, Pearlstone, Millstone Porphyry, and Trachytic Conglomerate.

Trachyte, properly so called, is characterized by its porphyritic structure, by the scorified and cellular aspect
which it has such a tendency to assume, by its harsh feel,
and by the presence of crystals of glassy felspar, generally
cracked, and sometimes passing into pumice. Besides
these, which may be regarded as essential to its composition, crystals of mica and hornblende are often present, and
all these minerals are either confusedly united without any
apparent *cement,* or by the intervention of a paste of a felspathic nature, sometimes compact, and sometimes cellular.
This paste is generally light coloured, though different
shades of red and brown are sometimes communicated to it
by the presence of iron, and there is one variety in which
the paste is perfectly black and semivitreous, intermediate
in its characters between pitchstone and basalt, but distinguished from either rock by melting into a white enamel.
Augite is sometimes present, and grains of titaniferous iron
are often discoverable, but olivine rarely, if ever, occurs,
and therefore appears to be the only mineral which has any
claim to be considered as peculiar to basalt.

The 2nd species, called by Beudant Trachytic Porphyry,
is distinguished from the preceding by the general absence
of scorified substances. Neither hornblende, augite, nor
titaniferous iron enter into its composition, but quartz and
chalcedony, which are wanting in the former, are commonly
present in this species. In its general aspect it bears a much
nearer resemblance to the older formations than trachyte
properly so called.

This description however applies only to the characters
of the larger portion of the mass, for Mons. Beudant is com-

pelled, in order to include all the varieties, to establish two
subspecies, the one *with*, the other *without* quartz, and in
both of these he notices a variety possessing a vesicular
structure. The subspecies indeed, which is without quartz,
even passes into pumice. Many varieties of Trachytic Porphyry contain a number of very small globules, which seem
to consist of melted felspar, having often in their centre,
a little crystal either of quartz or of mica. The assemblage of these globules, leaving minute cells between them,
sometimes gives to the rock a scoriform aspect. The chalcedony often occurs in small geodes, and sometimes intimately
mixed with the paste in which the crystals are imbedded.

Trachytic porphyry also appears to pass by imperceptible
gradations into the next species, pearlstone, which is characterized by the vitreous aspect generally belonging to its
component parts. It is evident, that this definition includes
pitchstone and obsidian, but these are of rare occurrence in
Hungary, the great mass of this formation being composed
of the mineral called pearlstone, some varieties of which
pass into pumice.

In its simplest form, this rock presents an assemblage of
globules, varying from the size of a nut to that of a grain of
sand, which have usually a pearly lustre, and scaly aspect,
and are set, as it were, one upon the other, without any substance intervening.

From this, the most characteristic variety, the rock passes
through a number of gradations, in which its peculiarities
are more or less distinctly marked. In some varieties the
globules are destitute of lustre, and exhibit at the same time
sundry alterations in their size, structure, and mode of aggregation, till at length they entirely disappear, and the
whole mass puts on a stony appearance, which retains none
of the characters of pearlstone. On the other hand the globules, becoming less distinct, either resolve themselves into
a paste resembling enamel, very fragile, in which separate
portions approaching to a spherical form are indistinctly
visible, or into a more vitreous and more homoganeous mass,

which is generally black, and presents all the characters of pitchstone or obsidian. Among these latter varieties is one which resembles the marekanite of Kamschatka.

Sometimes globules consisting of felspar occur in the rock, which are either compact or striated from the centre to the circumference, and these are sometimes so numerous that the whole mass is composed of them. Various alternations occur between the glassy and stony varieties of the pearlstone, sometimes so frequent as to give a veined or ribboned appearance to the rock, at others curiously contorted as though they had been disturbed in the act of cooling.

Lastly, all these varieties occasionally present a cellular, porous, spongy, and fibrous aspect, and pass into pumice. With respect to their chemical characters, it may be sufficient to remark that the vitreous varieties of pearlstone usually effervesce under the blow-pipe, but the stony do not. These rocks often contain geodes of chalcedony and opal, the former existing in the more vitreous, the latter in the more stony or felspathic portions. The opal is commonly opaque, but is occasionally met with more or less translucid.

The fourth species is distinguished for its hardness and cellularity, qualities which have caused it to be employed all over Hungary for the purpose of millstones, from whence the name of Millstone Trachyte has been applied to it by Beudant.

Unlike the other rocks comprised under the same generic term, it abounds in quartz, or in silex under some one of its modifications, and in proportion as the latter earth is more or less abundant, the substance puts on the characters either of hornstone or of clay porphyry. The paste is always dull and coarse looking, its colours vary from brick-red to greenish-yellow, its fracture is generally earthy, its hardness very variable, but usually considerable. It contains crystals of quartz, of felspar, lamellar, and sometimes glassy, and of black mica, imbedded. Jasper and hornstone also occur in nests, or in small contemporaneous veins, very abundantly disseminated, and siliceous infiltrations, posterior to the formation of the rock, seem likewise to occur among the cells which are every where distributed.

In examining these rocks with a glass, we discover a multitude of little globules analogous to those in the pearlstone, which seem to be of a felspathic nature, and when broken, are found to contain in their centre a little crystal of quartz, or a speck of some siliceous substance.

These globules in some cases compose the whole substance of the paste, in others they are held together by a sort of hardened clay, which here and there resembles porcelain-jasper. Notwithstanding these distinctions, there is a greater degree of uniformity in the characters of this, than in those of the other species of trachyte, and the most obvious differences between the several parts of this formation relate to the size and direction of the cells, which are sometimes so small and narrow, as to give to the rock a fibrous character, sometimes of considerable size, in which case they are in general coated internally with crystals of quartz.

The 5th and last species comprehended by Beudant under the generic term of Trachyte, consists of those heaps of pumice, and other loose materials, that occur agglutinated together on the slopes and at the base of the rocks belonging to the four preceding classes. Although the prevailing constituent is pumice, every variety of rock found in the neighbouring hills is met with amongst the fragments. The latter vary extremely in size, as well as in the mode of their aggregation; the cement which unites them is often of a porphyritic character, hardly distinguishable from the fragments themselves. Like them it often contains crystals of felspar, mica and hornblende, and sometimes grains of titaniferous iron are diffused through it, or it is coloured red by the peroxide of that metal.

The fragments of pumice are united together either immediately, or by the intervention of a paste of a vitreous character resembling obsidian, into which the pumice passes insensibly. Here and there the rock itself has become decomposed, and its destruction has given rise to beds either of a cellular nature arising from minute portions of pumice, which still preserve their fibrous texture, or (where all traces of this have been obliterated) to masses of an earthy-character similar to the trass of the Rhine volcanos or the " *tripoli* " of those in Auvergne. It is important to observe, as fixing the date of these conglomerates, that the latter variety contains, between Paloita and Prebeli near Schemnitz, marine shells of the same kind as those found in the Calcaire Grossier near Paris, and that it is covered here and elsewhere by beds, which Beudant refers to the plastic clay formation.

The changes that have taken place in the constitution of this conglomerate seem in some cases to have proceeded a step farther, the earthy beds just noticed as resulting from the reunion of the finely divided portions of the pumice, being rendered compact by the subsequent infiltration of siliceous matter ; in this state stems of vegetables of a cylindrical form, often hollow, are found in it in a silicified state, and crystals of felspar, mica, quartz, and garnet are distributed through the substance of the mass. These latter varieties often bear a considerable resemblance to the millstone.

The last stage of alteration is seen in the production of a salt composed of sulphuric acid, alumina, and potass, with excess of base, diffused through the substance of the earthy beds before mentioned. According to Beudant, this substance differs from alum in its crystallization, and is therefore distinguished by the name of aluminite or alumstone.

It appears likewise from his account, that this salt exists ready formed in the rock from whence the alum is extracted, and from thence he infers, though as I conceive somewhat precipitately, that the sulphuric acid which enters into its composition, has not been derived, as is commonly imagined, from the decomposition of the sulphuret of iron

G

that was originally present. He therefore imagines, that at some former period the rock itself formed a sort of submarine solfatara, and that owing to a continuance of the volcanic action subsequently to the formation of this deposit, the mass became penetrated with sulphurous acid, which, combining with the alumina, was in process of time converted into the particular mineral called alumstone.

Thus the process in this case will be analogous to that which is taking place at the Solfatara near Naples, and in the craters of other half-extinguished volcanos, and the same remark will apply to the formation of alum at Tolfa in the Roman States, and in other well known localities.

I have several objections to make to this mode of explanation. In the first place it is by no means universally true, that the subsulphate of alumina exists ready formed in the alum-rock of Hungary, for in some cases it is only obtained after the mass has become thoroughly decomposed. Though I did not visit the breccia near Matra or Tokay, to which Beudant principally refers in his description, I examined with some attention the works near Vissegrad, between Buda and Schemnitz, carried on by a physician named Marton, who had the kindness to explain to me the details of his process.

In this instance the stone which furnishes the alum is not a tuff or conglomerate, but a trachytic rock containing much pyrites. The object therefore of the manufacturer is to accelerate the decomposition of the latter, and thereby to furnish the acid which enters into the constitution of the alum. This is effected by exposing the stone to air and moisture for a given time, first in the open air, and afterwards in a sort of barn, the roof of which can be raised or lowered at pleasure, so as to exclude the rain, and to admit air and light. The latter agent Dr. Marton considers essential to the success of his process.

The earth is suffered to remain a sufficient time, to effect the decomposition of the sulphuret of iron, the union of the acid which results with the alumina present, and the com-

plete peroxydation of the iron, which is thus rendered in-
soluble, and no longer affeots the purity of the product.
Five years generally elapse before the whole is in a state of
readiness for lixiviation; it is then reduced to a red ochrey
powder, having on its surface an efflorescence of silky crystals, which probably consist of alum. However this may
be, it is at least certain, that the alkali originally present in
the stone, is far from sufficient to convert the whole into
crystallizable alum. Having therefore separated the saline
matter by lixiviation, Dr. Marton finds it necessary to add
about 5 per cent of subcarbonate of potass, after which the
solution being boiled to a state of sufficient concentration,
is set aside to crystallize.

This statement may be sufficient to shew, that Beudant's position, as to the alum existing ready formed in
the rock, does not hold good universally; but even where
this is the case, it by no means follows, that the salt may not
have originally proceeded from the decomposition of the
sulphuret of iron. Let us recollect, that the alum-rock of
Matra is a bed, resulting in part at least from the detritus
of the very trachyte which we have seen to be so fully
charged with iron pyrites, and so capable of yielding alum
in consequence, and that the very cause, which has brought
its materials into their present position, would be the one
most efficient in occasioning the decomposition of a metallic
sulphuret. Even if it were impossible to account for the formation of the alumstone in this manner, I should still hesitate as to adopting Mons. Beudant's explanation, because
the rock is not stated to possess any of the characters belonging to the substances found near a modern Solfatara,
and does not appear to be impregnated either with sulphur, or with the other minerals produced, wherever sulphurous acid has pervaded a mineral mass for the period,
which must be necessarily supposed.

The Trachytic formation is in general devoid of those
metallic veins which so commonly penetrate the older porphyry at Hungary, but at Konigsburg near Schemnitz, the

conglomerate belonging to this formation is richly impregnated with auriferous sulphuret of silver, which pervades
the mass, and is separated by simple washing. The gold
mines of Telkebanya, near Tokai, appear to be situated in
the same description of rock, so that it is by no means correct to say that the absence or presence of depots of these
metals serves to distinguish the newer from the older porphyry, although it may perhaps be true, that it is only in
the latter that true veins are to be met with.

Amongst the siliceous minerals so common in the trachyte
of Hungary, the different varieties of opal have principally
attracted attention. They are met with for the most part in
the trachytic conglomerate, but they occur also in the
pearlstone.

Hyalite I have myself seen, in more than one instance,
coating the fissures of the trachyte, derived perhaps from a
sublimation of the silica by the volcanic action, or from
a chemical solution of that earth by the steam proceeding
from the same source.

According to Beudant, the five species of rock included
under the generic name of Trachyte, always preserve with
relation to each other the same determinate order; that properly called trachyte occupying the central portion of the
group; the trachytic conglomerate surrounding the flanks of
the mountains; whilst the trachytic porphyry, the pearl-
stones, and the millstone porphyry lie intermediate. There
is however no appearance of stratification, or of any pause
having taken place in the volcanic operations, so that we
cannot suppose the most ancient part of the formation to
have been produced at a different epoch from the most
modern, and are therefore under the necessity of regarding
the observations that have been made with respect to the
rocks which cover the trachytic conglomerate, as determining
equally the date of the whole.

With regard to the origin of the trachytic conglomerate
in general, it is the opinion of Beudant, that as the fragments contained in it sometimes appear to be rolled, a part

of its materials is derived from the debris of the surrounding country; but as the latter supposition does not apply either to the case of the pumice which constitutes the larger proportion of this deposit, or to the angular fragments of other substances which also occur in the midst of it, it is probable that the greater part of the constituents have been ejected immediately by the volcanic action,

The following is a synopsis of the genus Trachyte, as given by Beudant.

1st Species, Trachyte, properly so called.

1st variety *granitoid,* no apparent cement, numerous crystals of glassy felspar confusedly united; crystals of black mica; hornblende rare.

2nd, *with mica and hornblende*—these crystals abundant, and generally black; paste of compact felspar, pretty pure, and fusible into a white enamel; crystals of glassy felspar.

3d, *porphyritic*—paste of compact felspar, fusible into a white enamel; crystals of felspar, glassy, lamellar, and compact; augite more or less abundant; no mica or hornblende.

4th, *black,*—the paste black, dull, fusible into a white enamel, with black spots, more or less numerous, disseminated; crystals of glassy felspar, sometimes of augite.

5th, *ferruginous,* paste ferruginous, dull, of a red or brownish colour, blackening when heated; fusible into a black or scoriform enamel; crystals of glassy felspar; numerous crystals of black mica.

6th, *earthy or domite,* paste earthy, porous, light-coloured; crystals of glassy felspar rare; crystals of black mica abundant.

7th, *semi-vitreous* (*Pseudo-basalte of Humboldt*), paste semi-vitreous, black or brown; fracture large-conchoidal, losing its colour in the fire, and melting into a white enamel.

8th, *cellular,* paste of various descriptions; contains numerous cells more or less imperfect, either round or elongated.

2d Species, TRACHYTIC PORPHYRY.

1st Subspecies, *with crystals of quartz*; base of compact felspar, with or without lustre, more or less abundant, containing most commonly a great number of small semi-vitreous globules; crystals of quartz more or less numerous; crystals of glassy felspar, generally well-defined; black mica, in small hexagonal plates, more or less numerous.

1st variety, *glistening*, base composed of compact felspar with an enamelled surface, easily fusible.

2nd, *semi-vitreous* (vitro-lithoide), almost entirely composed of semi-vitreous globules, amongst which are disseminated crystals of glassy felspar, and some of quartz.

3d, *scoriform*; paste semi-vitreous and dull, porous, or with irregular and imperfect cells.

4th, *cavernous*; paste scarcely discernible; small and very numerous cells; irregular cavities of various sizes; mass infusible.

2d Subspecies, *without quartz*: base of compact felspar with or without lustre, more or less fusible before the blowpipe; small crystals more or less numerous, often with imperfect terminations, of glassy or earthy felspar; black mica in small hexagonal plates; no crystals of quartz or semi-vitreous globules.

1st variety *glistening*; base of compact felspar, easily fusible into a white enamel; small crystals of felspar, commonly of the glassy kind.

2d *dull*; base of compact felspar, dull, difficultly fusible before the blowpipe; small crystals of felspar, commonly earthy, sometimes very rare.

3d, *cellular or pumiceous*; base almost infusible, full of cells; crystals of felspar rare and indistinct.

3d Species, *Pearlstone.*

1st variety, *testaceous*, made up of an assemblage of vitreous globules more or less distinct, generally scaly (testacès) and with a pearly lustre; mica and felspar very rare.

2nd, *spherolitic*, paste of pearlstone not testaceous, with an enamelled lustre, and a grey colour; numerous crystals of very
brilliant black mica; glassy felspar in small crystals, ordinarily
with their terminations imperfect.

3d, *pitchstone*, vitreous paste approaching to obsidian, often with
a fatty lustre; crystals of glassy felspar with imperfect terminations; little geodes of chalcedony more or less numerous.

4th, *globular stony*, stony mass, composed of globules with a compact or radiated structure, semi-vitreous or altogether stony.

5th, *stony in mass*. The whole mass semi-vitreous, or altogether
stony; the structure passing sometimes into porphyritic.

The Millstone Porphyry is so uniform in its composition, as not to
admit of being distinguished in the manner of the preceding species; but

The 5th Species, Trachytic Conglomerate, is divided into
1st, *the conglomerates made up of the debris of Trachyte*, cemented
by an earthy or more or less crystalline paste.

2d, *the conglomerates consisting chiefly of the trachyte and millstone
porphyry*, rounded or angular.

3d, *the pumiceous conglomerates*, composed of fragments of pumice
and obsidian, agglutinated either immediately, or by the intervention of some cement more or less earthy.

4th, *the porphyritic conglomerates*, resulting from the decomposition of the pumice.

5th, *the aluminous beds*, consisting of tufaceous or conglomerated
rocks impregnated with alum.

I have now, by the assistance of Beudant's laborious and
apparently accurate treatise, presented you with an account
of the trachytic formation as it exists in the country where
it is most fully developed, and have been induced in the
present instance to enter more into particulars than I shall
in general do, as I am not aware of any detailed account
of this class of rocks existing in the English language. This

deficiency must, I suppose, be ascribed to the total absence of the formation itself from all parts of the British dominions, for though we possess for the most part excellent descriptions of whatever relates to the physical structure of our own country, we are not so well supplied with respect to that of the Continent, in this respect falling considerably short of our neighbours the Germans, who with a laudable zeal and impartiality instantly avail themselves of every addition to the existing mass of knowledge, without regard to the source from whence it proceeds.

Not but there are some rocks even among ourselves which remind us of the trachytes of Hungary, although not absolutely referable to the same class. The clay porphyry associated with red sandstone in the Isle of Arran, and that of Sandy Brae in the county of Antrim, present at least numerous analogies, and the latter rock not only passes into pitchstone, sometimes resembling the pearlstone of Hungary, but also contains nests and veins of wax opal.

The two formations are indeed distinguished by the constant absence of scorified matter from the porphyries of Ireland and of Arran, and its occasional presence in those of Hungary, but this difference will I hope be satisfactorily explained by considering the several ages of the two rocks, and the manner in which the agency of heat has been modified by this circumstance. The porphyry of Arran is indeed interstratified with red sandstone, and that of Sandy Brae, though an overlying mass, is probably referable to the same æra as the basalt of the Giant's Causeway, which seems to be about the date of the chalk; whereas the trachyte of Hungary, though it is covered by plastic clay, seems from its containing shells belonging to the Paris basin, not to be older than the tertiary formation.

Though therefore all these rocks may have been produced under water, yet we may readily suppose the pressure exerted by the ocean to have been greater in the one case than in the other, and may thus explain the absence of scorified matter without supposing the Arran and Irish porphyries to have been produced by a cause essentially different.

Now the igneous theory, as applied to the trachytes of Hungary, has appeared to me, since the publication of M. Beudant's work, to be placed on so solid a basis, that I should have considered it unnecessary to undertake a journey into that remote country, if I had had no further object than that of satisfying myself with respect to its nature.

There is however yet another question relative to this subject which still divides the scientific world, namely, whether the resemblance between the trachytic formation and any of the rocks on which it reposes, be so intimate, as to oblige us to adopt the same opinion with respect to the origin of the latter, which we entertain as to that of the former.

The trachyte of Hungary, like that of the Rhine and of Mexico, reposes on a class of rocks which appear to belong to the transition series, and which indeed cannot without violence be attributed to a more modern epoch, as they sometimes support a rock which seems to correspond with the old red standstone, and in other places certain limestones at least as old as the earlier members of the secondary class.

Among the rocks thus circumstanced, the most prevailing is a greenstone porphyry consisting essentially of compact felspar and hornblende, which alternates with syenite, mica slate, and even with granite. It is therefore evidently a matter of considerable interest to determine whether a substance so related exhibits any close analogies to trachyte, since whatever conclusions we arrive at with respect to its origin, must be extended to that of all the other rocks associated with it.

My attention therefore whilst in Hungary was particularly directed to this circumstance, and I must admit that there is an occasional resemblance between the older and newer porphyry; but when viewed upon the great scale, the characters of the two formations seem to me sufficiently distinct. It is indeed true that where the trachyte rests upon the older porphyry, the two rocks appear to pass one into the other, and the latter to a considerable distance is found to contain crystals of glassy felspar; but even those who contend for the igneous origin of the greenstone porphyry, will probably consent to attribute this to the change produced by the contact of the melted trachyte, for it is evident that there can be no real transition between rocks of such different ages.

I am therefore ready to give the Plutonists all the advantage which their argument can derive from the analogies that subsist beween the two formations; but I do not consider that argument to be stronger in the present case from the circumstance of the rocks being thus associated, than it would be if they occurred in two different localities, since there is in fact just as great a separation between them, from the long interval of time that divided the formation of the transition from that of the tertiary class of rocks, as there would be if they existed in two different quarters of the world.

I shall therefore content myself with observing, that the resemblance which the greenstone porphyry bears to the trachyte may be accounted for, by supposing the latter rock to have been produced by the action of heat from the materials of the former, and that, although the older porphyry may possibly have been affected by fire, the evidence in favour of such an hypothesis does not seem stronger as applied to this rock, than it is with reference to the other primitive formations.

The trachytic formation is sometimes accompanied by detached cones of basalt, as at the Calvarienburg, near Schemnitz; but the principal masses of that rock lie at a considerable distance near the lake of Balaton, where they occur in the midst of a sandy plain, either in detached cones or elevated tabular masses.

Not having visited that part of Hungary, I am unable to state whether they are posterior to the formation of the vallies, but the quantity of highly porous and scoriform substances, with which they are said to be accompanied, seems to indicate their recent origin. It is certain at least from Beudant's account, that they were not anterior to the tertiary class of rocks, for they are seen in several places resting on the molasse.

Volcanic Rocks of Transylvania.

In Transylvania,* volcanic rocks of undoubted tertiary origin occur in the eastern part of the country alone,

The above formation constitutes a range of hills covered with thick wood, which separates from Transylvania the Szeckler land, or the valley in which the Hungarian tribe of that name reside.

The chain itself extends from the high hills of Kelemany, north of Remebyel, to the hill of Budoshegy, about ten or twelve miles north of Vascharhely. The wild tract included within this mountainous range is so broad, that it requires a day's journey to cross it in a carriage ; at its northern extremity, however, it gets gradually narrower. Its limits are, to the east the river Marosch from Toplitza upwards, the Aluta from Varosch to Tuschnad and Kasson ; to the

---

* I am indebted for the whole of this account of Transylvania to my friend Dr. Boué, the author of a Geognostical Essay on Scotland, and of several interesting papers on the Geology of France and Germany. Its value is enhanced from the circumstance that no individual, so far as I am aware, has communicated to the public any description of this remote country, since the branch of natural history which relates to the physical structure of the earth has began to assume its present form. It is to be regretted that Dr. Boué did not complete his undertaking, which comprehended the whole of the Bannat, and the provinces of the Austrian empire, as far as Trieste, but a severe illness, occasioned by the villainy of a servant, who attempted to poison him, in order the more readily to make off with his money and property, brought his researches to an abrupt termination. Before however this event occurred, he had examined a great part of the southern and eastern portion of Hungary, including the trachytic formation of Transylvania, of which he has sent me the subjoined account.

west, a line passing through Kasson, Tulle, Udyarhely, Parayd, Libonfalva, and Pata. It is for the most part composed of various kinds of trachytic conglomerate; of which the best sections are presented along the course of the Marosch, for elsewhere a most impracticable forest of pine and oaks covers it nearly throughout. From the midst of these vast tufaceous deposits, the tops of the hills composed of trachyte, a rock which forms all the loftiest eminences, here and there emerge. Of these the most elevated is called Kelemany; the other principal ones are Fatatschion, Pritzilasso, Hargala, Barot, the hills south of Tuschnad, &c. &c. The trachyte is ordinarily reddish, greyish, or blackish; it mostly contains mica. In the southern parts, as near Tschik Sereda, the trachyte incloses large masses, sometimes forming even small hillocks, of that variety of which millstones are made, having quartz crystals disseminated through it, and in general indurated by siliceous matter in so fine a state of division that the parts are nearly invisible. The latter substance seems to be the result of a kind of sublimation, which took place at the moment of the formation of the trachyte.

Basalts were no where observed, although black trachyte abounds. Distinct craters are only seen at the southern extremity of the chain. One of the finest observed by Dr. Bouè was to the south of Tuschnad; it was of great size, and well characterized, surrounded by pretty steep and lofty hills composed of trachyte. The bottom of the hollow was full of water. The ground near has a very strong sulphureous odour. A mile in a S. S. E. direction from this point there are on the table land two large and distinct " maars," like those of the Eyfel, that is to say, old craters, which have been lakes, and are now covered with a thick coat of marsh plants; the cattle dare not graze upon them for fear of sinking in.

Some miles farther in the same direction is the well-known hill of Budoshegy (or hill of bad smell), a trachytic mountain, near the summit of which is a distinct rent, from which exhale very hot sulphureous vapours. The heat of the ground is such as to burn the shoes. A deposition of sulphur has taken place there, and the rock is converted into alum-stone by the action of the vapors upon the constituents of the trachyte. In this manner hollows are formed in the rock. At the base of the hill are some very fine ferruginous sulphur springs. much resorted to for various diseases by the inhabitants, who encamp near them in the open air during summer. Chalybeate sulphur springs generally abound at the base of this volcanic range, and chalybeates with carbonic acid still more. Some of these appeared as good as those of Pyrmont, and the most famous, that of Borsah, a bathing place much resorted to by the Transylvanian nobles, contains more carbonic acid than Pyrmont water itself.

The craters last described have thrown out a vast quantity of pumice, which now forms a deposit of greater or less thickness along the Aluta and the Marosch from Tuschnad to Toplitza. Impressions of plants and some siliceous wood are likewise to be found in it, as is the case in Hungary. These fragments of pumice have been deposited under water. Some, says Dr. Bouè, might be disposed to set down a more considerable portion of Transylvania as trachytic, than I have done, but I have satisfied myself that many rocks which may appear to be trachyte are nothing but some of the newer transition or coal-sandstone porphyries, which are here and there more scorified than elsewhere, or of which the scorified portions have stood the action of the weather better than the rest. This may be the case with the most recent porphyries of the two great deposits of that formation, the one of Marmorosch, the other in the Gespannschaft (comitat) of Hunyad, and the Stuhl of Muhlenbach. On these I shall dwell at full length in my general account of Transylvania.

To this account of the volcanic rocks of Transylvania, I have only to add that a basaltic cone is mentioned by Beudant as occurring in Schlavonia near Peterwaradin, and that I have myself seen specimens from that province, and probably from the same locality, in the possession of Professor Schuster at Buda, which from their scoriform aspect I should judge to be of modern formation.

Dr. Boué also informs me, that between Ober-Pullendorf and Stoop, near Güns in Hungary, south of the lake of Neusiedel, is a flat conical hill about 100 feet in height, half a league in its greatest diameter, and a quarter of a league in its smallest, which rises from the midst of the upper tertiary deposits, or amongst the marly beds lying above the blue shelly marl common to Austria and the Apennines. The rock itself is composed of a blackish or greyish felspathic basalt, which is sometimes compact, and contains oval nodules, partly of mamillary or botryoidal iron ore, and partly of arragonite; sometimes very porous, and with the cavities either entirely empty, are coated with globules of sphœro-siderite.

The direction of the cells is from east-north-east to west-south-west, and the same is the direction of the range itself. It is decidedly a tertiary basaltic cone, having its base only covered by recent marls.

On the Volcanic Rocks of Styria.

On my way from Vienna to Italy I deviated a little from the direct road, in order to look at some rocks of a volcanic nature that occur near Friedau in Styria, a little to the south-east of Gräbz, of which the only account which has been published, is one by Von Buch, in the Transactions of the Academy of Berlin.*

The formation in question may be briefly stated, as consisting of a central nucleus of trachyte, which consitutes the lofty conical hill called the Gleichenburg, round which on all sides are mantle-shaped strata of volcanic tuff, alternating with beds belonging apparently to the tertiary class.

* Vide Leop. von Buch in der Abh. der physical Classe der Kön. Akademie zu Berlin, 1818.

This tuff consists in general of a congeries of very minute fragments of volcanic matter, which seem to have been immediately ejected from the volcano, mixed up and loosely agglutinated with small quartzy pebbles. In the midst of it are fragments of cellular and compact basaltic lava, sometimes containing *nests* of olivine. Masses of the same substance of a globular form, not imbedded in any matrix, are found also distributed amongst the tuff. Specimens of augite, and of a substance looking like altered granite likewise occur. The tuff becoming more and more mixed with particles of clay and sand, passes at length into a loamy earth, at first dark, and afterwards, where it is unmixed with volcanic matter, of an ash-grey colour. The constituents are in a state of very fine divison, and a number of minute specks of silvery mica impart a. sparkling lustre to the general mass.

Besides this, which looks like a bed of silt deposited tranquilly at the bottom of a lake, we find, at a somewhat greater distance from the central trachyte, strata of limestone, full of shells, belonging to the recent order of deposits, and especially abounding in that minute fossil, the miliolite, which imparts to the stone an oolitic appearance. At a village called Khelig, a little to the south of the former locality, I observed that the tuff, which here contained decided scoriæ, was superimposed on a rock which no wise differed from ordinary basalt, but in the existence of minute internal pores. It formed a number of concentric lamellar concretions, of which the external have become decomposed, whilst the internal retain their solidity. The exterior surface of the balls is coated with asphaltum. The whole rests upon a bed of marl without any traces of volcanic matter.

Two hypotheses present themselves with respect to the age of the trachyte of the (leichenburg; for it may either be said, that having been first thrown up by volcanic action, the beds of tuff and of marl collected by degrees around its base ; or that after the latter had been formed in a position approaching to the horizontal, the rock of the Gleichenburg, being forced up through the midst of them, imparted the inclination which they are now seen to possess.

For my own part I am most disposed to adopt the latter opinion, on the same ground on which I assented to M. Bertrand Roux's ideas with respect to the rock of the Mount Mezen; for it seems probable that if the trachyte had been formed in the first instance, fragments of it ought to appear intermixed with the other materials of the tuff, which I did not discover to be the case. The inclination likewise possessed by the strata of tuff seems to me too considerable to be consistent with the former hypothesis, but accords very well with the latter.

The following sketch may give an idea of the disposition of the central trachyte.

Where a. & b. are alternating beds of tuff and loam or sand, C is the trachyte, and D a valley of denudation separating the two rocks.

EUGANEAN HILLS.

On entering Italy by the side of Venice, we have not far to go before we meet with a very extensive and interesting volcanic district.

To the south of Padua lie the Euganean hills, an isolated tract of high ground in the midst of a level country, consisting of a trachytic formation, not unlike that of Hungary, which, from its cellular structure in some cases, and its semi-vitreous aspect in others, would at once be taken for a volcanic product. Like the formation too of the latter country, it consists of several kinds of rock, which however are so allied, and so connected by mutual passages, as to shew that they have been all derived from a modification of the same process.

The most characteristic variety is a rock of an ash-grey colour and uneven fracture, very like the porphyry of Mont Dor, or the first species of Beudant's trachyte formation, (Monselice). It contains numerous crystals of glassy felspar, sometimes decomposed, sometimes fresh, and occasional specks of black mica, which is also accumulated in *nests*, the several parts of which have a slaty structure, like that of mica slate. Crystals of augite are also found under the same circumstances. Associated with this is a rock possessing a splintery fracture, waxy lustre, and vitreous appearance, which may be called an hornstone porphyry. Some varieties are cellular, and contain infiltrations of quartz and chalcedony, like the millstone trachyte of Hungary. Others approach very nearly to the characters of pearlstone, presenting, together with the vitreous aspect of that substance, an approach to a similar concentric arrangement. (Monte Siave). In these cases the crystals of glassy felspar, which distinguish true trachyte, are either absent, or very rarely occur.

This formation is associated at Monte Venda with basalt, the relation of which to the trachyte is as obscure as in the parallel case of the Siebengebirge. It is also surrounded,

H

at Castelletto, by strata of tuff, and of pumiceous conglomerate, in a manner analogous to what I have described
as taking place near the Gleichenburg in Styria, but disposed more vertically.

In some parts a conglomerate or breccia occurs, (Monte
Nuovo) which seems to be principally made up of the
hornstone above described, intermixed with a white powdery siliceous substance, which fills up the interstices. The
whole of this mass might be imagined, as well from its
vitreous appearance, as from the intimate union of its
parts, to have been consolidated by fusion, or at least by
the action of heat.

The trachyte of the Euganean hills rests upon a calcareous rock, which appears to correspond with the chalk of
Great Britain. It is called Scaglia, from its slaty structure,
being disposed in thin horizontal layers. Its colour is commonly white, now and then with a shade of red, and its
compactness usually is quite equal to that of our hardest
chalk, though softer varieties are sometimes met with.

The points however chiefly to be insisted on, as establishing the identity of the two formations, are, the kidney-
shaped masses of flint disposed in beds throughout the
Scaglia, as in the chalk of England, and the nature of the
petrifactions that occur in it, which, from the list given in
the Abbé Maraschini's late work,* appear to consist of am-
monites, terebratulites, and various species of the echinus
family; viz. the echinoneus, galerites, ananchytes, spatangus,
cidaris, nucleolites, and echinus *proper*, of Lamarck.

By comparing this list with the one given in Messrs.
Conybeare and Phillips' Geology of England and Wales,
p. 73, it will be seen, that the analogy between the two formations is in this respect considerable.

I know not whether the redness and brittleness of the
flints in a part of the rock which lies near the trachyte, not

* Sulle Formazioni delle Rocce del Vicentino. Padova, 1824. p. 122.

EUGANEAN HILLS.

far from the village of Battaglia, is to be explained in the same manner in which Messrs. Buckland and Conybeare have accounted for a similar change in some of those near the Giant's Causeway, namely, by the influence of the melted matter upon them; but it is at least certain, that this vol-canic rock has sometimes produced, upon the surface of the subjacent bed, alterations, which afford additional evidence of its igneous origin.

Thus at the village of Schevanoya on the southern slope of the Euganean hills, the trachyte is incumbent on an argil-laceous variety (as I presume) of the scaglia, which is natu-rally so incoherent as to be softened by every shower of rain. As we trace it upwards however, we find it gradually be-coming more and more compact, until at last, where it touches the incumbent trachyte, it becomes perfectly hard and splintery in its fracture.

Other indications of volcanic action may perhaps be ga-thered from the springs of hot water impregnated with sul-phuretted hydrogen, which gush out from the rock near the village of Battaglia,* and are still in repute, as they were in the time of the Romans, for their medicinal qualities.

Perhaps the fable of Phæton,† who was said to have fal-len from heaven, or to have been struck by lightning on the

---

* It was the Fons Aponi mentioned by Lucan, Lib. 7.

> Euganeo, si vera fides memorantibus, Augur
> Colle sedens, *Aponus terris ubi fumifer exit*
> Venit summa dies, geritur res maxima, dixit;
> Impia concurrunt Pompeii et Cæsaris arma.

Claudian also celebrates it.

† Tzetzes, in his Schol. on Lycoph, says, that some supposed the Lake Avernus to exist among the Euganean Hills, and the circumstances that gave rise to the fable of Phæton, to have happened there. Martial too has these lines:

> Æmula Baianis Altini littora villis,
> Et Phaethontei conscia sylva rogi;
> Quæque Antenorio, Dryadum pulcherrima, Fauno
> Nupsit, ad Euganeos sola puella lacus!

borders of the Po, may refer to some tradition that existed
of volcanic phenomena, which may have continued here as
they now do in Transylvania, long after the formation of
the trachyte.

The neighbouring country to the north of Vicenza is interesting to the volcanist, as enabling him to trace the differences that exist between the ignigenous rocks of the very
same country, according to their relative degrees of antiquity.

It appears that all the formations of that country from the
talc slate, which is the fundamental rock, up to the scaglia,*
which is an equivalent of our chalk, are accompanied by
trap rocks, both in beds and in dykes, having an uniformly
compact structure, or cells completely filled with crystalline
matter; whereas the tertiary beds that lie above them all
alternate with a tuff, consisting of materials, the volcanic nature of which is more plainly attested by the scoriform and
vitreous aspect which so often belongs to them. It is impossible to imagine any combination of phenomena more in
accordance with the idea, that the compactness of lavas is
regulated cæt. par. by the pressure which they have undergone, and that the absence of vacuities in the case of all
those formed during the deposition of the older rocks, arose
from the mass of water superimposed; since it is seen that

---

* The Abbe Maraschini was good enough to shew me a hill near Recuaro,
north of Schio, where according to him the greater part of the formations met
with in that neighbourhood are seen united. In his Memoir entitled " Obser-
vationi sopra alcuni Località del Vicentino," which was published in the
Biblioteca Italiana, and in his late work referred to above, a full enumeration
of the series is given; I shall therefore content myself with stating, that on
this hill are seen, resting on the talc slate, which appears to be the fundamental rock of the country, 1st, a red sandstone, 2d, an augite rock (dolerite),
3d, another red sandstone with seams of slate coal, and above, three alternations of sandstone and limestone, which the Abbe is inclined to refer, I know
not how correctly, to distinct formations analogous to those in England and
Germany.

the products of volcanos in action subsequently to the date of the chalk, at a time when we have reason to believe the ocean to have sunk to a lower level than it had then stood at, approximate in their characters more nearly to the ignigenous formations of the present day.* It has even been remarked, that a difference appears to subsist between the volcanic rocks in the Vicentin of anterior date to the scaglia, corresponding to their respective ages : † the dykes which penetrate the older formations producing a greater hardening, and in general a more marked alteration in the part contiguous to it, than those which traverse the more modern. I do not however believe that this remark can be generalized, for the effects of dykes on the adjoining surfaces of the rocks they traverse, are no where more marked than at the Giant's Causeway, where they are at least as modern as the chalk.

I shall therefore omit all mention of the older formations of the same description found in this country, and content myself with comparing together the trap rocks associated with the scaglia, and the volcanic tuff which accompanies the tertiary beds that rest upon it.

Near Schio, north of Vicenza, the scaglia occupies the lowest part of the valley of Cengiette, and rises to a considerable distance on the hills of either side. At the hill of Belmonte, a rivulet exposes five stratiform masses of basalt, often changed by decomposition into wacke, which I am disposed to consider, with my friend Dr. Boué, as dykes parallel to the stratification of the chalk. Dykes of basalt are also frequently seen traversing this formation, at Chiampo, Valdagno, and Magre, but without altering the adjacent rock ; the external position of the dyke is frequently so much decomposed as to be converted into a sort of clay.

Above this is a thick and extensive formation of greenstone porphyry, or porphyritic augite rock, which the Abbé

* See Maraschini in the work above quoted, p. 130.
† Maraschini.

VICENTIN.

Maraschini sets down as corresponding with the trachyte of the Euganean Hills. It lies in a sort of basin, filling up all the pre-existing hollows between the older rocks. Thus in some places it rests immediately on chalk, and in others on rocks of an older date.

This porphyry has generally a claystone base with crystals of augite disseminated. It is of various shades of brown, with reddish or greyish spots, sometimes more or less vitreous in its fracture, passing even into pitchstone or obsidian porphyry. It is generally very tough, but where it has undergone decomposition, has passed into the state of kaolin.

The most remarkable circumstance attending it is its containing veins of metallic matter, for it is not usual to find the latter accompanying rocks of a volcanic nature, or of a date so recent as that which must be assigned to a formation covering the scaglia. Near Schio, where its superposition is distinctly seen, the porphyry is penetrated by veins of blende, galena, arsenical pyrites, sulphate, carbonate, molybdate, and (according to Professor Catullo) chromate of lead, accompanied with quartz crystals, calcareous spar, sulphate of baryte, and manganesian epidote.* The upper part of this formation becomes amygdaloidal, containing cells which are for the most part filled with calcareous spar, chalcedony, and various species of zeolites. The preceding rocks are covered with numerous alternations of calcareous with brecciated or tufaceous deposits.

The former are marked as tertiary by the occurrence of nummulites and other shells enumerated by Brongniart in his Memoir on the Vicentin;† the latter are made up of fragments not only of basalt, but also of volcanic sand and scoriform lava, thus indicating the commencement of a new order of volcanic products. The tuff is often as full charged

---

* Maraschini, p. 133, et seq.

† Brongniart sur les Terrains calcareo-trappiens du Vicentin. Paris. 1823.

VICENTIN.

with shells as the limestone rock itself, abounding in nummulites, &c. &c., it also contains large masses of brown coal, and even of silicified wood. In the midst of the calcareous beds above mentioned are some of a bituminous slaty marl containing impressions of fish; they occur at Monte Bolca; at Monte Novale near Valdagno; and at Monte de Salzedo.

At Monte Bolca, the only locality which I visited, the Ichthyolite limestone, as it may be called, rests upon a calcareous rock with nummulites, and is covered by the same; whilst a deposit consisting of volcanic tuff lies both under and above. The alternations indeed between these two classes of deposits are often extremely numerous; at a place called Ronca alone we have in a very short compass no less than six, but the lowest volcanic bed is not tufaceous, but consists of cellular basalt. The occurrence of this substance, sometimes cellular, sometimes amygdaloidal, and sometimes even compact, interstratified with the other rocks, renders the structure of Vicentin less simple than it would otherwise be considered, and inclines one to think that streams of lava were thrown out during the formation of the tufaceous and calcareous beds. That the whole indeed of the basaltic, as well as the materials of the tufaceous rocks are referable to igneous action, I cannot bring myself for a moment to doubt, although aware that Brocchi, the first of Italian Geologists, has in his Memoir on the Val de Fassa expressed himself with some degree of hesitation on the subject.*

Admitting even that there are whole beds of tuff which exhibit no traces of igneous action, yet these are so associated with others containing volcanic products of the most unequivocal kind, that I know not how we are to separate the one from the other.

---

* This Memoir was published several years ago, so that it may not represent his present opinions.

Even the compact basalt passes, at Monte Glosso near Bassano, into a vitreous rock, which approximates to obsidian; nay we discover in the tuff itself at Chiampo, large detached masses of cellular lava, the cavities of which have often that glazed surface which so strongly indicates fusion.

At the same time the presence of shells in the tuff itself, and its alternation with regular beds of unaltered shelly limestone, prove that the sandy matter and loose fragments of which this aggregate is composed, were originally deposited under the surface of water, at the period during which the calcareous beds were in the act of forming. That the accumulation of the materials of which the tuff consists was a slow and gradual process, I infer, among other reasons, from a specimen in my possession, in which a rounded fragment taken from one of these beds is seen covered by serpulæ, a plain proof that the stone remained for some time under water, uncovered by any of the matter which afterwards formed above it.

The occurrence therefore of beds of volcanic tuff alternating with strata of shelly limestone seems in this instance capable of explanation, by supposing showers of ashes and lapilli to have proceeded from some adjacent volcano, which, as they sunk to the bottom of the water then covering the face of the country, would become intermixed with the fragments washed down from the adjoining rocks, and be consolidated like mud in a stagnant pool, acquiring additional consistency in proportion to the mass of matter superimposed.

That the volcanic action indeed was going on in this very spot, is proved by the hills of cellular lava, or of basalt, that occur in the midst of this formation, and the effects of these operations upon the tuff itself may be traced in the inclined position of its beds, so different from what would occur in a mass of matter deposited tranquilly under the surface of water.

VICENTIN.

The structure of a hill called Montecchio Maggiore near Vicenza illustrates this, as well as some other points in which I have been insisting. The rock is here composed of tuff having fragments of amygdaloid disseminated through a paste composed of wacke. It is remarkable that this paste contains no crystalline matter, though the imbedded portions have their cells filled with calcareous spar, sulphate of strontian, mesotype, and other minerals.* On the other hand it often encloses shells, which have never yet been detected in the fragments. We may therefore fairly conclude that these two sorts of rock were originally distinct; the amygdaloid having been produced by some volcano antecedent to that which gave rise to the wacke. The amygdaloid after being thus formed into a coherent stratum by the volcano at one period, may have been broken into fragments and ejected by it at another, like the loose materials found so frequently round the craters of all active volcanos; and if a shower of ashes occurred at the same time, and had become intermixed with the fragments, whilst they lay under the surface of water, the whole might have been consolidated into a conglomerate, possessing the appearances of the rock at Montecchio.

* Among the Monte Berici near Montecchio Maggiore, as well as in the neighbourhood of Bassano, are found in the cavities of a cellular volcanic rock those curious geodes containing water, which have received the name of *enhydrous agates.* They appear to have been known and prized by the ancients. Pliny, Lib. 37. l. 73, defines it " enhydros semper rotunditatis absolutæ, in candore est lævis, sed ad motum fluctuat intus in eâ, veluti in ovis, liquor." Propertius seems to refer to it under the name of " crystallus aquosa," " *crystallusque* suas ornet *aquosa manus;*" and Claudian has celebrated the gem in several epigrams, for it seems probable that he refers to this stone, and not to rock crystals containing water, as in Ep. 12, 13, he represents it as globular, and in Ep. 10, as convex.

Clauditur immunis convexo tegmine rivus,
Duratisque vagus fons operitur aquis.

VICENTIN.

The materials however which compose this tuff, after being thus made to coalesce, would seem to have been brought into a more intimate union by some subsequent process, and the entire mass must be supposed to have been heaved up, and thrown upon its edges, inasmuch as the next bed of limestone rather abuts against than rest upon it.

The above imperfect description may perhaps serve to convey a general notion of the geological features displayed in this volcanic district; for a more detailed account I must refer you to the works of Fortis, Marzari, and especially the later production of the Abbe Maraschini, which I have already so frequently noticed. All these writers concur in attributing to the volcanic formation of the Vicentin, a tertiary origin, and with this the general absence of craters, as well as the mixed character of the products, completely accord.

There appears indeed to be only one spot in which any vestiges of a crater are to be met with, and that on the summit of a hill called Montebello, on the road between Vicenza and Verona.

But in this case the character of the rocks that compose the mountain, so far from being basaltic, is in all respects analogous to that of recent lavas, so that the distinction between these, and the other volcanic products in the neighbourhood is calculated to confirm our belief in the non-existence of craters elsewhere. Proofs of volcanic action are not by any means confined to the immediate neighbourhood of Vicenza, they extend to Verona, Brescia, and perhaps to the Lake Lugano,[*] between which and the Lago Maggiore, near Grantola, is a spot, the igneous origin of

---

[*] See in Biblioteca Italiana, Vol. 5, a Memoir by Pallini, on the Lake de Garda, from which it appears that this piece of water gives out sulphuretted hydrogen gas.

MONTE CIMINI, &c.

which has been much contested.* Lastly, on the west of
the Lago Maggiore, near the town of Intra, Mount Sim-
molo is composed of a trap rock, which may perhaps be connected with the preceding volcanic formation.

The state of the weather, at the time I was in this
country, prevented me from visiting these latter spots, as
well as from examining a still more interesting chain of
volcanic hills which I passed over on my road from Sienna
to Rome. The most elevated spots of this district, such as
the Monte Cimini near Viterbo, and the Monte Amiata
near Radicofani, are said by Brocchi † to consist of the rock
called by him necrolite, which seems to correspond with the
trachyte of other geologists. It is associated with basalt, as
at Viterbo, ‡ where it is columnar, and rests on a bed of
pumice and tuff, containing the bones of quadrupeds, the
whole more modern than the trachyte. In the valley of
Trepenzio, not far from the above locality, the prisms are
quadrangular, and the rock is covered by tuff. Hence there
would seem to be alternations between the two.

The slopes of the central trachytic mass are likewise
covered by strata of marl containing shells referable to the
tertiary class, and by beds of cellular and scoriform lava
containing augite and leucite. This latter mineral, which
has been before noticed as occurring at the Lac of Laach
in Germany, is however nowhere so common as amongst the
lavas in this part of Italy, although it is found likewise, as
will afterwards appear, at Vesuvius. What the circumstances are which contribute to its formation, has not been

* See Beudant's Hungary, Vol. 2. p. 588, and Vol. 3. p. 541, where the
aqueous origin of the pitchstone and claystone porphyry of Grantola is
asserted. It appears, that Mons. Fleurian de Bellevue in France, and Pini in
Italy, (Memor. sur alcuni fossili della Lombard, 1790,) have supported the
contrary theory.

† See his Catalogo ragionato de Rocce, 1817.

‡ See a Memoir by Brocchi in the 3rd Volume of the Biblioteca Italiana.

altogether ascertained, though it is Von Buch's opinion,* that a certain degree of *repose* in the mass during the period of its cooling is required. Thus the streams of lava that flowed from Vesuvius in 1767 and 1777, which contain leucite, having been received in a nearly level plain between the cone from which they issued, and the Monte Somma, moved onwards but slowly; whilst those of 1760 and 1794, which are destitute of this mineral, flowed with great impetuosity down the slope of the mountain, until they reached the sea. It might be shewn that the volcanic products around Viterbo, where leucite so abounds, were probably inthe former predicament; this however would be anticipating the subject of a future lecture.

But whatever may be the conditions upon which the production of leucite depends, it has been shewn I think by Von Buch, that in its origin it is contemporaneous with the lava itself, and that it neither has *pre-existed* in the materials, nor been formed like some other crystals by subsequent infiltration.

The rocks at Civita Castelletto and Borghetto contain a number of oval cavities, all of them elongated in the same direction, as must always happen in a substance which disengages elastic fluids, whilst its parts are in progressive motion.

Now the leucite crystals, when they intervene between the cavities, retain their usual octohedral figure, but where they make a part of the walls of the cells, are found to be elongated in the same direction as the latter. This latter fact it is difficult to explain, except on the supposition that they were formed at the very time the lava was undergoing consolidation, for granting that they had existed previously, it must at least be admitted that they were elongated by the

* See Von Buch's Memoir in the Journ. de Phys. Vol. 49. It must be confessed however that his remark seems equally true with respect to the formation of every other kind of crystal.

heat applied, and in that case their crystalline form would have been obliterated.

Near Viterbo is a small lake, which emits a sulphureous odour, and seems from the rise of bubbles through it to be in a state of continual agitation, and a little farther on the road to Rome, is the Lake of Vico, formerly the Lacus Cimini, which has all the appearance of a crater, and according to some of the antient writers was caused by a sudden sinking of the earth,* in further proof of which, they say that the ruins of a town that formerly existed on this site might be seen at the bottom of the lake, when the water was clear. At all events the whole physical structure of the country favours the idea of a volcanic origin.

The Lake of Bolseno, between Viterbo and Sienna, possesses the shape of an ancient crater, and its being bounded by volcanic rocks is consistent with such an hypothesis; but we must hesitate in admitting it to be so formed without more conclusive evidence than we possess, considering that it is more than twenty miles in circumference.

On its borders was situated the antient Volsinium, one of the principal towns of Etruria; and the analogy of the modern name with the word Vulcan, especially according to the old spelling,† may lead us to imagine that it derived its name from the homage paid to that God, originating in the *volcanic* phænomena, which excited the fears of the earlier inhabitants. It is curious that the Volsci, as well as the

---

* See Amm. Marcell. L. 17, c. 7. Servius in his note on the line in the 7th book of the Æneid, in which this Lake is mentioned, (" Et Cimini cum monte lacum, lucosque Capenos,") alludes to a fable grounded upon this tradition.

† The old spelling of the word Vulcanus seems to have been Bolcanus. At least there is in the Vatican, a Roman altar dug up at Ostia, with an inscription, BOLCANO. SAC. ARA. and at Tivoli, in a wall of an antient building, BOLCANO AEDES. REF. COERAV. C. CAEPJO L. F. and there are many more that run in the same way. (See Sickler. Ideen zu einem vulc. Erd-globus. Weimar. 1812.)

VITERBO.

Volsinii, inhabited a volcanic country;* and it is known that particular homage was paid to *Vulcan* all over Latium.†

Volsinium therefore may have been originally written volcanium,‡ ór rather *bolcanium*, as the V and B are easily convertible, and we may recognize the original spelling in some measure restored in the modern name Bolseno. Such a view however would imply that volcanic appearances have existed in the country at no very remote æra, and as this is a question of some importance, it would be well if some geologist, who visits that part of Italy, would ascertain whether such be the case. Pliny mentions that Volsinium was burnt down by lightning, a statement which might possibly have arisen from its having been overwhelmed by some neighbouring volcano; but of course until it is fully proved that there was one in action since the foundation of Rome, we can have no excuse for adopting such an opinion.∥

Some appearances of the same description seem likewise to be displayed even at the present time among the Lagunes of Tuscany, from which sulphureous vapours, chiefly consisting of sulphuretted hydrogen, often arise, and produce those effects upon the surrounding rocks, which I shall more

---

* The Volsci inhabited a country between Albano and Terracina, which will be shewn afterwards to be volcanic.

† See my account of the neighbourhood of Rome.

‡ Sickler suggests that volcanus or bolcanus may have been derived from the Greek words ϭωλος gleba, and καιω uro—perhaps χαινω hisco, might be the more probable derivation. My learned friend Dr. Pritchard, the author of a work on the Egyptian Mythology, has however suggested to me a Celtic Etymology, which he thinks more probable, as many Oscan words are derived from that language.

" The Welsh word Bwlchau or Vulchai, he says, signifies a break in a mountain, and probably a crater, from the adjective Bwlch, broken. B is mutable into V, and from Bwlchau would be formed a Latin word Vulca, whence Vulcanus."

∥ Livy mentions that the Lake was tinged with blood—meaning probably, bitumen. Lib. 27.

particularly notice when speaking of the solfatara of Naples. But besides the sulphur and the sulphuric salts that effloresce round these lagunes, a substance is collected that has been found only in one other spot of Europe; I mean the boracic acid, which is sublimed alike from the lagunes of Tuscany, and from the crater of the Island of Volcano, attended too by similar sulphureous exhalations, and therefore probably connected as to its origin in both these localities.

The lagunes in question are represented as being little cratershaped cavities formed on the surface of the ground, by the continual escape of sulphuretted hydrogen gas from fissures in the rock. These cavities, according to Prystanowski,* (a German, who has published the most modern account of this phænomenon) are at the bottom of a valley, and are therefore often filled with water, either by the rain, or by the overflowings of an adjoining brook.

This water is raised to a boiling temperature by the passage of the heated gas through it, and hence it is that the lagunes generally emit a lofty column of steam, which first arrests the traveller's attention, and has consequently led to the adoption of the name Fumacchie, by which the lagunes are often designated. The sulphuretted hydrogen carries up with it in a gaseous state some boracic acid, but this is condensed by the water, and is found amongst the mud, when the pool has dried up in consequence of the evaporation from it exceeding the supply from without.

The lagunes are situated a few miles to the S. W. of Volterra, near Monte Rotundo, and near Monte Cerboli, the rock from whence the vapour issues is calcareous, and is considered by Brongniart† as subordinate to the diallage or gabbro rock of the Appennines. This is worth remarking, as it serves to distinguish the above phænomenon from a similar one which occurs among the tertiary rocks of

* Prystanowski über den ursprung der Vulkane in Italien. Berlin, 1822.

† Annales des Mines, 1821.

Sicily, and of the Modenese, known under the name of
salses or air volcanos.* As I did not visit the latter locality,
I shall defer my account of this phænomenon until I speak
of the rocks of Sicily, referring you at present, to Mons.
Menard de Groye's " Memoire sur les feux de Barigazzo,"†
for an account of the phænomenon, and to Brocchi's Conch.
Subapp.‡ for the best description of the rock, in which as
I conceive, it will be found to occur.

The same cause, which I have assigned to account for my
overlooking the volcanic rocks of Tuscany, circumscribed
considerably my excursions amongst those near Rome,
where indeed the traveller, surrounded as he is by antiquities
of such extreme classical interest, can hardly help being
frequently called away from subjects of scientific inquiry.
It has been said, that what Vesuvius is to Naples, the
Coliseum and St. Peter's are to Rome, and as the scholar
almost necessarily imbibes somewhat of the spirit of a
naturalist during his stay in the former city, from his attention being so frequently directed to the movements of the
volcano, so it is equally to be supposed that the study of
nature will give place to that of art, whilst we are in the
midst of the monuments of Roman taste and magnificence.

I saw enough however of the physical structure of the
neighbourhood, to be persuaded, that the interpretation
which Breislac ‖ has put upon some well-known fables or
traditions handed down to us by antient writers, in proof of
his idea that antient Rome occupied the site of a volcano,

* Brocchi however in his Conch. Subapp. V. 1. p. 71. thinks that it origi-
nates among tertiary rocks.

† Journ. di Physique. Vol. 85.

‡ I allude to his description of the marl of Italy, Vol. 1, p. 66, to which I
shall have occasion to refer in speaking of a similar formation that occurs in
Sicily.

‖ Breislac, in proof of his opinion as to the existence of volcanos on the site
of Rome, appeals to the worship especially paid to Vulcan, whose temple,
according to Plutarch and Dionysius of Halicarnassus, overlooked the Forum.

is altogether untenable, and that his assertion as to the Capitol of the eternal city " Capitoli immobile saxum"

This spot, he supposes to have been affected by the same agent even subsequently to the foundation of the city, for the chasm in the midst of the Forum, into which Marcus Curtius precipitated himself, was probably caused, if we believe our author, by volcanic action.

The principal crater he supposes to have existed in the circular space between the two summits of the Aventine, where the church of St. Balbina now stands, and the fable of Cacus, whose cave stood on this hill, furnishes him with a confirmation of this opinion.

The description given by Virgil of this mountain, applies very well to the picture of the phænomena of a volcano.

> Huic monstro Vulcanus erat pater. Illius atros
> Ore vomens ignes, magná se mole ferebat.
> . . . .
> Faucibus ingentem fumum, mirabile dictu,
> Evomit, involvitque domum caligine cæcâ,
> Prospectum eripiens oculis ; glomeratque sub antro
> Fumiferam noctem, commixtis igne tenebris.

And his account of the Cave corresponds equally with the idea of a Crater.

> At specus et Caci detecta apparuit ingens
> Regia, et umbrosæ penitus patuêre cavernæ.
> Non secus ac si quâ penitus vi terra dehiscens
> Infernas reseret sedes, et regna recludat
> Pallida, diis invisa, superque immane barathrum
> Cernatur, trepidentque immisso lumine Manes.

Virgil also mentions a tradition respecting the Capitoline Hill, which Breislac converts into an allegorical representation of volcanic phænomena.

> Hinc ad Tarpeiam sedem et Capitolia ducit,
> Aurea nunc, olim silvestribus horrida dumis.
> Jam tum religio pavidos terrebat agrestes
> Dira loci ; jam tum silvam saxumque tremebant.
> Hoc nemus, hunc, inquit frondoso vertice collem,
> Quis deus, incertum est, habitat deus ; Arcades ipsum
> Credunt se vidisse Jovem, quum sæpe nigrantem
> Ægida concuteret dextrâ, nimbosque cieret.

It is a pity that so many ingenious analogies should be thrown away, but the existence of a volcano on the spots alluded to is quite irreconcilable with their known physical structure.

I

having been erected on the tottering edge of a crater, however well suited it may be to point an antithesis, or to illustrate the vanity of human pretensions, rests on too slender
grounds to deserve a place in a scientific treatise.

The soil of Rome, as an eminent Italian geologist* has
since fully proved, is in reality composed of an alternation of
sandy or calcareous beds, with a tuff containing fragments of
scoriform as well as compact lava, often rolled, and accompanied likewise with pebbles of the Appennine limestone,
that display evident marks of attrition. There is however
no proof that these fragments of lava were ejected by any
volcano which occupied the immediate site of Rome, on the
contrary the nearest spot from which we can suppose them
to be derived, is the Lake of Albano, more than twelve miles
distant.

There are indeed some passages in antient writers, which
might lead us to suppose a volcano to have existed among
these mountains, even at a period within the limits of authentic history; for Livy notices a shower of stones which
continued for two entire days from Mount Albano, during the
second Punic war,† and Julius Obsequens in his work de
Prodigiis remarks, that in the year 640 ab U. C. the hill
appeared to be on fire during the night.‡

There are likewise several other traditions preserved by
classical writers, which speak to the same point, one of them
relating to a king of Alba, Aremulus Sylvius, who, according to some accounts, was precipitated, together with his
palace, into the Alban Lake, by an earthquake, and according to others carried into it by a whirlwind, after being
struck dead by a thunderbolt.‖

---

* Brocchi Suolo di Roma, 1820.

† Albano monte biduum continenter ep:dibns pluit. Lib. 25, cap. 7.

‡ Mons Albanus nocte ardere visus   Julius Obs. c. 110.

‖ Aurelius Victor de origine Gentis Romanæ.

NEIGHBOURHOOD OF ROME.

These accounts indeed, if not confirmed by other testimony, might be rejected as fabulous, but they may perhaps suffice to establish the comparatively modern date at which the volcanic action continued, when viewed in connexion with the physical structure of the lake itself, which, from its circular form, the absence of any *natural** outlet for its waters, and the volcanic materials surrounding it, might at once be taken for the crater of a volcano.

It appears indeed to have given off a current of lava, known as that of the Capo di Bove, which may be traced to within two miles of Rome, terminating on the Appian road, near the mausoleum of Cecilia Metella.

This Coulée is usually compact, though small cells and cavities are here and there scattered over it, especially near its surface. It is very hard and sonorous, and appears to consist of an intimate mixture of augite and leucite, either in crystals or granular masses, the former of a bottle-green colour passing into brown, the latter white or azure.

Besides the above, the rock of the Capo di Bove contains several other minerals, which may be regarded in the light of accidental ingredients.

Amongst these are, according to Brocchi,†—1st, Black dodecaedral crystals with rhomboidal planes, which consist of melilite or pseudo-nepheline.

2. Fine thread-like crystals, of a white colour, which melt before the blowpipe into a vitreous or opaque globule, emitting a phosphoric light. They are called by Hauy pseudo-nepheline.

---

* I say *natural*, because it is well known that an artificial canal does exist penetrating the side of the internal cavity, and thereby preventing the waters from rising beyond a certain point. It was cut, according to Livy, (Lib. 5. c. 15.) by the Romans, during the siege of Veii, in compliance with the injunctions of an oracle, which assured them that the city would never be taken, until the Lake had its water drawn off.

† Biblioteca Italiana, vol. 7.

3. Other needle-shaped crystals of a brown coffee colour, or of a lively red. Exposed to the air they lose their natural colour, and become, first yellow, and afterwards white. Before the blowpipe they melt into a blackish globule. These crystals appear to be the same with some found by Breislac* in the crater of the Solfatara near Naples, which have been named in honour of him. Calcareous spar is likewise found, as well as the new substance, called abrazite by Gismondi, and distinguished by its octaedral crystallization, by not effervescing with acids, and fusing without ebullition in the flame of a candle.

The breadth of this current near its termination at the tomb of Cecilia Metella, is sixty feet; it rests on rapilli and tuff, and may be traced along the plain nearly up to the hills about Albano.

The whole of the country for several miles of Albano, abounds in volcanic appearances. Amongst the mountains in this group are no less than four lakes, which appear originally to have been craters, the one already mentioned, that of Nemi, Joturna, and Vall. Aricia.† With respect to the latter place, Pliny mentions a report that the ground would set fire to charcoal,‡ and Livy notices a shower of stones that fell there, as well as the bursting out of a warm spring, having its water mixed with blood, which Heyné supposes to have been bitumen.||

Yet the differences of mineralogical character between the volcanic rocks of these mountains, and those found at Rome itself, oblige us to abandon the idea that the latter can have been derived from the same quarter. The hills in

* Breislac Campanie. Tom. 2. p. 132.

† It has likewise been said that the Lake Fucino, in the Abruzzi, is a volcanic crater, but this has been shewn by Brocchi not to be the case.

‡ Plin. Hist. Nat. Lib. 2. c. 111.

|| Heyne Opusc. Acad. vol. 2. p. 263.

the immediate vicinity of Rome, consist of that aggregate of volcanic materials which all are agreed to designate as tuff, whilst the neighbourhood of Albano is constituted of a material which the Italian geologists have chosen to mark as a separate rock under the name of Peperino.* It is easy, says Von Buch,† to distinguish these two substances; in Peperino nearly the whole mass is fresh, undecomposed, and bright to the eye, whereas in tuff the greater part is dull, and appears weathered. The former resembles a porphyry, the latter a sandstone and other similar aggregates. The substance, of which Peperino consists, preserves almost uniformly an ash-grey colour, but the tuff of Rome is generally darker. With respect to its fracture too, Peperino is less friable than tuff, and the mica, which is distributed over it either in detached plates, or collected into masses, sometimes as large as a cannon-ball, mixed with crystals of augite and magnetic ironstone, preserves its original black colour and lustre, which in the tuff is not the case.

A still more marked difference perhaps between the two formations, is the entire absence of pumice from the neighbourhood of Albano, whereas it is very frequent in the tuff about Rome.

The latter therefore seems with more probability referable to the chain of volcanic mountains north of Rome, known by the name of the Mount Cimini, since we find on the slope of these hills similarly constituted beds, to which indeed, according to Brocchi, the tuff of Rome may be traced without interruption.

The characters of this rock in both localities correspond with those of the tuff of Auvergne and of the Vicentin, rolled masses and freshwater shells are occasionally found in it, so that though the materials of which the aggregate is composed are chiefly volcanic, the rock itself betrays marks

---

* See Brocchi Catalogo Ragionato. passim.

† Reise durch Deutschland und Italien. Berlin, 1809. vol. 2. p. 70.

of a mechanical origin. Like the tuffs before noticed, it contains mica, augite, and felspar, but is distinguished from them by the presence of leucite, which rarely however preserves its lustre and crystallization, but is generally more or less decomposed, and of a mealy consistence, like the pumice in the trass of the Rhine.

Brocchi* distinguishes two principal varieties of this tuff, lithoide found at the Capitoline and the Esquiline hills, and granular met with on the Pincian and Quirinal.

Lithoide tuff is generally of a reddish brown colour, with specks of a darker or orange tint derived from fragments of scoriform lava, approaching in texture to pumice. Its fracture is earthy in the small, conchoidal in the great, and its hardness sufficient for a building stone, for which purpose it was employed by the antient Romans. Granular tuff, on the contrary, is generally of a blackish-brown, dark violet, or russetty yellow colour. It is light, friable, and composed of largish grains, weakly cohering, having white specks of mealy leucite disseminated over the mass, together with fragments of augite, scales of mica, and rolled pebbles of grey or black lava. It is in short nothing but an aggregate of lapilli, and by its decomposition it gives rise to the only other variety, which is distinguished by Brocchi under the name of earthy tuff.

The origin of all this formation must have been antecedent to the commencement of the present order of things, as is evinced, not only from the vallies by which its once continuous beds are at present intersected, but likewise from the Neptunian deposits with which it is seen repeatedly to alternate. The principal of these is either a sandy marl containing fragments of older rocks, or that calcareous deposit known under the name of travertine stone, which has furnished the material for most of the edifices of antient as well as modern Rome.

* Suolo di Roma.

This rock, it may be observed, is not produced by the waters of the Tiber at the present moment, but we have ocular proof of the manner of its formation in the Lago di Solfatara, near Tivoli, where the same process is constantly going on. The water is there impregnated with sulphuretted hydrogen, and carbonic acid gases, which rise up in bubbles through it, and to the presence of the latter substance it owes the property of dissolving calcareous earth, which is again deposited round the lake in the form of travertino, in proportion as the carbonic acid escapes.

The same process appears to have taken place formerly on a much more extensive scale in the neighbourhood of Rome, and, if we are disposed to theorize, we may attribute it to a general disengagement of these two gases at a period more nearly approaching to that at which the volcanos of the neighbouring country were in activity. For we shall see as we proceed, that the extrication of elastic fluids, of which sulphuretted hydrogen and carbonic acid are the most common, takes place round the site of a volcano for a long period after the cessation of its more violent action, so that there is no absurdity in supposing that the same operations, which, when in their greatest intensity, produced the materials of the volcanic tuff itself, may, during the periods of their partial intermittence, have given rise to these gaseous exhalations, which imparted, to water impregnated with them, the property of dissolving calcareous matter; nor shall we want analogies to support us, if we assume, that a feeble remnant of the same action may even at this distance of time continue, and manifest itself in the sulphureous exhalations near Tivoli.

It would appear, that these indications (if they may be so considered) of languid volcanic action, were more extensively distributed about the neighbourhood in earlier periods than at the present. Thus Varro makes mention of warm baths near the temple of Janus, whence the spot obtained

the name of *Lautolæ* "*à Lavando ;*" a spot on the Esquiline Hill was called *Puticulæ,* from the sulphureous smell
which it emitted ; and the wood consecrated to the Goddess
Mephitis renders it probable that a noxious gas arose
from that place. All these have now ceased, and nothing
remains but the Lago de Solfatara to remind us of their
existence.

It is remarkable that no kind of animal is seen near this
water, a circumstance which can only be attributed to the
noxious qualities of the sulphuretted hydrogen, for the
Lago de Tartaro near, so well known for its calcareous
incrustations, contains abundance of molluscæ. Shells are
also rare in the ancient travertine near Rome.

The existence of masses of this latter substance, on the
very summits of the Seven Hills proves, that at the period of
its formation, the site of Rome must have been covered
with water to the depth of at least 140 feet.

From the character of the shells sometimes contained in
the Travertine, which Brocchi has ascertained to belong to
existing species, we may conclude that the water, which
deposited it, was not impregnated with salt, and are consequently enabled to fix the date of the volcanic tuff which
accompanies these Neptunian deposits, as corresponding
with that of the latest freshwater formation.

Brocchi has further shewn, that the beds above noticed all
rest upon a formation containing oysters and other marine
shells, which is seen underlying the rest at the Monte
Mario, and in the excavations made at the foot of the Capitoline Hill.

On the road between Rome and Naples, the first indications of volcanic action, after passing the Pontine Marshes,
occur a little to the south-west of Mola de Gaeta, near the
River Garigliano, the antient Liris.* We there find our-

* See Breislac Campanie, vol. 1. p. 86.

selves between two chains of hills, that to our right the
Mount Massico, composed of the Appennine limestone, the
other on our left, consisting entirely of volcanic materials.
The town of Sessa itself stands on an eminence composed
of tuff, which seems to be of comparatively recent date,
since, like that of Herculaneum, it covers the remains of an
antient city, for on digging the foundation of an house some
time ago, there were discovered, many feet beneath the surface, a chamber with antique frescos, and the remains of an
amphitheatre.   Yet so far from any account existing of such
a town having been destroyed, we have no tradition even of
an eruption having taken place from any volcano near it
since the memory of man.   Not far from the same place, the
bed of a rivulet, which has washed away the tuff, exposes
two subjacent streams of lava, one of which closely resembles
in composition that with which the Appian way was paved.
Several other masses of lava, having the appearance of
Coulées, occur either near Sessa or the neighbouring village
of Casale, all of which seem referable to the volcanic mountain of Rocca Monfina, and one indeed from its freshness
would seem to have proceeded from it at no very distant
epoch.

Rocca Monfina seems to retain the vestiges of the great
original crater from which these volcanic masses proceeded.
In many parts indeed its sides have fallen in, but enough
yet remains to enable the eye of the traveller to fill up the
outline.   The now detached hills, which appear to have
resulted from the destruction of the walls of the crater, must
have enclosed a circumference of no less than nine miles,
but it is probable that the actual section is much below its
former elevation, and that its height was at first considerably greater.

Within the space occupied by the original crater, two
other volcanic cones have since been thrown up, each provided with its crater; the magnitude of one of them may be

judged of by the fact, that on the summit of the cone is a plain near a mile in circumference, bounded by two lofty eminences, which are the remains of it.

It appears therefore that the latest eruptions of this volcano have taken place since the country was inhabited by man, but it will hardly be considered probable that the whole of the tuff, which extends from the River Garigliano, beyond Capua, and connects itself with that of the neighbourhood of Naples, is in the same predicament with that of Sessa, which is of such recent date. It is not likely indeed that the Rocca Monfina was the source from whence the greater part of the tuff was derived, since its position on the summits of detached hills, as well as at their base, seems to shew that it has been produced antecedently to the general revolution to which we attribute the formation of the vallies.

It will not however be necessary to inquire further into its origin at present, as it is clearly the same with that near Naples, which will be afterwards more fully considered.

A few miles west of Mola de Gaeta lie the Ponza Islands, four of which appear to be entirely volcanic, the fifth (Jannone) in part Neptunian. No tradition exists of their having been in activity, nor was Dolomieu able to discover any traces of a crater. The most recent description of them is that communicated by Mr. Poulett Scrope, to the Geological Society, of which I extract the following notice from the Annals of Philosophy for July, 1824.

The whole of these islands, with the exception of Jannone, are composed of rocks, of the trachytic series, fine sections of which are presented along the coast. The Isle of Ponza is long and very narrow, and is eroded by the sea into deep concavities. Harder masses left along its shores show that it once was broader, and protruding ledges mark its former connexion with Quannone and La Gabbia. Prismatic trachyte, variously coloured and disposed, forms

MOUNT VULTUR.

the ossature of the island. It is constantly accompanied by, and alternates with, a semi-vitreous trachytic conglomerate, formed of minute pulverulent matter enclosing fragments of trachyte. The prismatic trachyte seems to have been forcibly injected through the conglomerate, and wherever it touches the latter, its earthy base is converted into a pitchstone porphyry; sometimes it becomes a pearlstone, at others it incloses a true obsidian. These rocks are connected with a siliceous trachyte, resembling in appearance the siliceous buhrstone of Paris.

Resting on the semi-vitreous trachyte, and forming the base of the Montagna della Guardia, is a rock 300 feet thick, which the author distinguishes mineralogically from common trachyte, and proposes to call greystone. In Jannone the trachyte overlies a limestone, which Brocchi describes as transition limestone; at the point of contact the latter becomes a dolomite.

On looking over Mr. Scrope's specimens, I was struck with the resemblance between many of them and the constituents of the Euganean Hills already described.

In the province of Basilicata, near the town of Melfi, rises in the midst of the chain of the Appennines a large isolated hill, the Mount Vultur, which Horace has celebrated as the scene of some of his early poetical adventures.*

This mountain, both from its conical figure, and the rocks composing it, is easily recognised as volcanic. On its slope, and scattered over various parts of its surface, are no less than twelve protuberances,† thrown up by the action of

---

* Me fabuloso Vulture in Appulo
　Altricis extrâ limen Apuliæ
　Ludo fatigatumque somno
　Fronde novâ puerum palumbes
　Texere.　　Carm. Lib, iii. Ode 4.

† This account is extracted partly from a memoir of Brocchi in the Biblioteca Italiana, partly from Romanelli Topografia del regno di Napoli, vol. 2. p. 232, and partly from the Calendario di Napoli, for 1824.

subterraneous fire. Each has its respective cone, and in the midst of them are two craters, one of which is no less than 2000 paces in depth, and has two lakes at its bottom. These lakes still are said to be impregnated with carbonic acid, and to emit an inflammable gas—probably sulphuretted or carburetted hydrogen.

The lava of Mount Vultur is on its surface of a greyish brown colour, half scorified, hard to the feel, and easily frangible, but in its interior has a blackish tint, and is much more compact, so as indeed to approach in character to basalt. The most remarkable substance contained among its productions is *Haüyne,* which indeed, according to Brocchi, is the principal ingredient of the lavas, and especially of that stream which constitutes the greater part of the eminence, on which the city of Melfi is situated. Nature here is so lavish of this mineral, that there is no need of searching for pieces containing it, which present themselves every where. It occurs in masses of various sizes, some larger than a filbert, their colour generally of an azure green, passing by various gradations into a dark blue, and their texture granular.

The lava has also disseminated over it another substance, in a still larger proportion. It is blackish, opaque, or translucent only in very thin pieces, with a bright lustre, a vitreous aspect, and a fracture generally conchoidal, but sometimes lamellar. It generally appears on the lava in the form of irregular spots, or else has the porous and scoriform aspect of a substance that has undergone fusion. At first sight it might be mistaken either for augite or blackened leucite, but Brocchi is persuaded that it is in fact a variety of Haüyne. With this the form of the substance, when it can be procured crystallized, completely accords, it being a dodecahedron, with rhomboidal faces.

The blue and black varieties in some of the lavas are intermixed, but in others only the black is seen.

MOUNT VULTUR.

The Haüyne also occurs massive in rectangular pieces of the size of the thumb, and is accompanied with leucite, melilite, hornblende, pseudo-nepheline, mica, and other minerals.*

Near the town of Melfi occur several alternations of strata, consisting of puzzolana, pumice, and ferruginous sand, covered with vegetable mould. In another place of the same neighbourhood the volcanic tuff is seen alternating with rolled pebbles.

About a mile to the east of Mount Vultur, in a place called Rendina, is a *Moffette,* or an exhalation of some noxious vapour, which produces a sharp, smarting sensation on the organs of sight, smell, and taste, and causes fainting in those who breathe it too freely. Near Atella, on the western side of Mount Vultur, are waters impregnated with sulphuretted hydrogen, and carbonic acid gases. I know not whether the neighbouring town of Acherontia, now Achera, derived its name from any appearances of the same kind, like the Lake Fusaro, near Naples.†

The magnitude of Mount Vultur, which is stated differently at 22 and at 30 miles in diameter at its base, indicates the extent of the volcanic operations that formerly must have taken place, yet all records of its eruptions are lost in the darkness of antiquity, and we must even attribute them, if we believe one writer,‡ to a time when the physical structure of the country was different from what it is at present, and the low land between Melfi and the Adriatic constituted a sort of gulf, extending from Taranto upwards, the waters of which washed the foot of this volcano.

---

* Calendario.

† Horace calls it " Nidus Acherontiæ," which seems to imply that the town was placed in an elevated and circular cavity on a mountain, such as the crater of a volcano would exhibit. This analogy of appearance suggested, no doubt, the name of " Nid de la Poule," applied to a crater near the Puy de Dôme in Auvergne. I think however it is somewhere stated, that Achera has for its foundation a calcareous stratum.

‡ Cagnazzo Congettura di un antico sbocco dell' Adriatico. Napoli, 1807, as quoted in his Romanelli's Topografia.

Not having seen the work referred to, I am unable to state in what degree his hypothesis is borne out by fact, and shall only remark that it seems favoured by the direction of the Appennines, as laid down in common maps,* where they are represented as dividing about Melfi into two branches, one of which takes the direction of Bari to the east, the other that of Calabria to the south; thus inclosing the greater part of the province of Basilicata in a kind of basin. What this intermediate tract of country may consist of, I have not been able to ascertain, but should it be such as bears out this conjecture as to extension of the Gulf at one period in the direction contended for, we may derive, from the present extinct condition of Mount Vultur, an additional proof of the theory which I shall propose in another part of this work with respect to the necessity of the access of the sea, or other large bodies of water to feed the fires of every volcano. At present the distance of Mount Vultur from the Adriatic cannot be less than five and thirty miles, whilst from Naples it is nearly twice as remote.

Between the two volcanos of Mount Vultur and Rocca Monfina is the Lago di Ansanto, in which name the classical reader will directly recognize the Lake and Valley of Amsanctus, alluded to by Virgil, into which the fury Alecto plunges, after having excited discord between Turnus and Æneas :

> Est locus Italiæ medio sub montibus altis
> Nobilis, et famâ multis memoratus in oris,
> Amsancti valles: densis hunc frondibus atrum
> Urget utrimque latus nemoris, medioque fragosus
> Dat sonitum saxis, et torto vertice torrens:
> Hic specus horrendum, sævi spiracula Ditis,
> Monstratur, ruptoque ingens Acheronte vorago
> Pestiferas aperit fauces; quis condita Erynnys
> Invisum numen, terras cælumque levabat
>
> Æneas, lib. vii. l. 563.

* See the excellent map attached to Mr. Cramer's useful Description of antient Italy, Oxford, 1826.

This description, and the existence of a temple on the spot dedicated to the goddess Mephitis,* naturally suggest to us the occurrence of volcanic phænomena, and the account given of the spot by modern observers is such as fully confirms this conjecture.

The valley of Ansanto † is bounded by Rocca S. Felice on the south-east, Villa Maina on the west, and Frigento on the north; it contains several small lakes, through which escape bubbles of very fetid air. The largest of the lakes has a circumference of about 160 feet, and is no more than five or six in depth; its waters are from 7 to 21 degrees of Reaumur above the temperature of the external air, the excess being least in winter and greatest in autumn; it is in continual and violent ebullition from the rise of much sulphuretted hydrogen gas, the odour of which is very perceptible at a distance. Besides this there are given out from clefts in the rock near the lake much sulphurous acid, carburetted hydrogen, and carbonic acid gases. These being wafted to different places, according to the direction of the wind, become fatal to the animals in the lower parts of the valley, the specific gravity of the sulphurous and carbonic acid gases causing them to accumulate near the surface of the ground. As no injurious effects are caused to the *windward* of the spots from whence the vapour issues, we may readily explain the seemingly capricious action of the *mofette* upon animals in different parts of the valley, by the direction towards which the wind blows. The waters of the lake being impregnated with hepatic air are celebrated in many diseases of cattle, and provided there be the slightest movement in the atmosphere, the gases do not accumulate around its borders in sufficient quantity to be pernicious.

There is one spot however in the midst of a torrent which

* See Pliny, lib. 2. l. 93.

† See Brocchi in Biblioteca Italiana for 1820, and Calendario de Napoli for 1822.

flows along the valley, called the " *Vado Mortale*," from
the nature of the *mofette* existing there. This, which consists entirely of carbonic acid, attains usually to the height
of four or five feet, so that it is constantly fatal to animals
that pass the stream at that point.

A vast accumulation of sulphur takes place in this valley,
owing doubtless to the decomposition of the sulphuretted
hydrogen, which is emitted in such quantities, that it has
been proposed to collect it for commerce; and petroleum
has likewise been met with intermixed with the former combustible. Volcanic products occur in the neighbourhood.

Let us now proceed to consider the volcanic appearances
that present themselves in the immediate vicinity of Naples,
where we have the advantage of comparing their effects in
various stages of activity, as well of contrasting those produced in earlier periods of the world, with such as are taking
place before our eyes.

To the east of the Bay rises the most recent of the volcanos near Naples, and the only one at present in complete
activity. The date of that part of the mountain properly
called Vesuvius, or rather of its cone, perhaps does not go
farther back than the period of the famous eruption of 79
after the Christian æra, in which Herculaneum and Pompeii were destroyed; for the ancient writers never speak of
the mountain as consisting of two peaks, which they probably would have done, if the Monte Somma had stood, as
at present, distinct from the cone of Vesuvius. It is also
remarked that the distance mentioned in ancient writers as
intervening between the foot of Vesuvius and the towns of
Pompeii and Stabiæ, appears to have been greater than
exists at present, unless we measure it from the foot of
Monte Somma, so that this affords an additional probability, that the latter mountain was then viewed as a part of
the former, and that no separation between them at that

time occurred. We may also be sure from the semicircular figure which the southern escarpment of the Monte Somma presents towards Vesuvius, that it constituted a portion of the walls of the original crater, and Visconti, it is said, has proved by actual admeasurements, that the centre of the circle, of which it is a segment, coincides as nearly as possible with that of the present crater.

There seems therefore little room to doubt, that the old mouth of the volcano occupied the spot now known by the name of the Atrio del Cavallo, but that it was greatly more extensive than this hollow, as it comprehended likewise the space now covered by the cone, which was thrown up afterwards in consequence of the renewal of the volcanic action that had been suspended during so many ages.

This view * likewise tends, as I think, to reconcile the accounts which ancient writers have given of the structure of the mountain, antecedently to the period before mentioned. Florus, for instance, in his narrative of the insurrection of Spartacus, describes the manœuvre by which that General contrived to escape from the Roman army, which besieged him in this mountain, in the following manner :

" Prima velut arena viris Mons Vesuvius placuit. Ibi cum obsiderentur a Clodio Glabro, per fauces cavi montis vitineis delapsi vinculis, ad imas ejus descendére radices; et exitu invio, nihil tale opinantis ducis subito impetu castra rapuere."

FLORUS, l. 3. c. 20.

Vesuvius was the spot pitched upon for their first enterprize. Being besieged there by Clodius Glaber, they descended through the defiles of this mountain by means of vine-twigs, and reached its very bottom, where they sur-

* Whilst this sheet was passing through the press, a statement has appeared in the public prints respecting an antient picture of Vesuvius recently dug up at Herculaneum, which represents the Monte Somma as forming a part of Vesuvius. I am anxious to obtain further particulars of a discovery so corroborative of the view I have taken.

VESUVIUS.

prised by a sudden assault the camp of the general, who anticipated nothing of the kind.

Plutarch, who evidently refers to the same event, notices it in a manner, which perhaps will enable us to ascertain what the real structure of the mountain at that time must have been. After describing the first successes of Spartacus and his army, he says; " Clodius the Prætor was sent against them with a party of three thousand men, who besieged them in a mountain (meaning evidently Vesuvius) having but one narrow and difficult passage, which Clodius kept guarded; all the rest was encompassed with broken and slippery precipices, but upon the top grew a great many wild vines; they cut down as many of their boughs as they had need of, and twisted them into ladders, long enough to reach from thence to the bottom, by which, without any danger, all got down except one, who stayed behind to throw them their arms, after which he saved himself with the rest." *

But how are these representations consistent with the account given by the accurate Strabo respecting the structure of Vesuvius, which, he says, is surrounded on all sides by fine fields, except on the summit, which is in great measure flat,† but barren and desolate.

I know no other method of reconciling these accounts, than that of supposing Spartacus to have encamped within the crater, which occupied what is now the Atrio del Cavallo; that the walls of this crater were at that time en-

---

* Life of Crassus. Plutarch's op. Reishe Vol. 3. p. 240.

† The following is the entire passage referred to.

" Ὑπερκειται δε των τοπων τουτων ορος το Ουεσσουιον, αγροις περιοικου-μενον παγκαλοις, πλην της κορυφης· αυτη δ'επιπεδος μεν πολυ μερος εστιν, ακαρπος δ'ολη· εκ δε της οψεως τεφρωδης, και κοιλαδας φαινει συραγγωδεις πετρων αιθαλωδων κατα την χροαν, ως αν εκβεβρωμενων υπο πυρος· ως τεκμαροιτ'αν τις το χωριον τουτο καιεσθαι προτερον, και εχειν κρατηρας πυρος, σβεσθηναι δ'επιλιπουσης της υλης.—Ταχα δε της ευκαρπιας της κυκλω τουτ 'αιτιον. Strabo. Ed. Falc. Vol. 1. p. 355.

tire, except on one part where the army of Clodius established themselves; and that the insurgents found their ladders useful in descending some of the steep precipices which existed on the external slope of Monte Somma, as well as in the first instance in climbing up to the brim of the then existing crater. The western front looking towards Naples being broken away, Strabo might naturally have considered what was once the interior of the then extinct crater as the summit of the mountain, and it is by no means unlikely that the former may at that time have possessed that level surface, which he notices as the general character belonging to it. At all events no alternative seems to exist between this explanation, and our regarding the passage as inaccurate, for the structure of the Monte Somma is such as plainly demonstrates that it has not been thrown up by any subsequent eruption, but existed at at least its present elevation in the period to which Strabo alludes.

Mons. Neckar de Saussure, who has published in the Transactions of the Society of Geneva * the most recent account of this mountain, remarks that the base of Somma as well as of Vesuvius consists of tuff, which is covered over in many places by the modern lavas ejected by Mount Vesuvius. From underneath this tuff rise the older lavas, of which the whole of Somma is composed. The beds, into which the latter are divided, as observed from Vesuvius, appear horizontal, for they are seen to range at an equal elevation in every part of the semicircular wall of lava, which bounds the Atrio del Cavallo. This appearance of horizontality is however deceptive, and arises from the circumstance of the beds dipping equally in all directions from the centre of the mountain; for it is found, that they are every where inclined at an angle of 30°, and rise towards the cone, which, as I have before observed, is placed in the exact centre of the original volcano.

* T. 2. pt. 1. p. 155.   Genève. 1823.

VESUVIUS.

The older lavas of Somma consist of leucite and augite, and are separated by occasional beds of tuff. They are also intersected by vertical dykes of a rock containing much leucite, which seems to resemble in most respects, except in compactness, the rock which it penetrates. The dykes are sometimes observed to terminate above, and at other times below, nor are there wanting cases in which they seem altogether inclosed in the substance of the rock. It may be remarked, that, unlike most other dykes, they neither produce any change in the position of the beds, nor in the character of the rock with which they are in contact.

Mons. Neckar imagines, that, if we suppose the semi-circular escarpment of Monte Somma to be a part of the walls of the original crater, we may readily explain the direction which its strata are found to assume. Every mountain of this description, he maintains, has been originally produced by a series of operations succeeding each other in the following order. When once the violence of the volcanic operations has arrived to such a pitch as to create a rupture of the strata of the earth, the elastic vapours, hitherto pent up, throw out portions of the liquid lava, through which they force their way, just as takes place when a mass of melted metal happens to fall into a vessel containing water. These portions, projected into the air, descend again either in the form of scoriæ or sand, and collect into an aggregate, which when agglutinated together will form tuff. But the projection of these fragments is soon followed by the overflow of the melted lava itself, which by degrees reaches the brim, spreads over the tuff, and forms a regular bed encircling the original aperture. The repetition of these successive operations causes that alternation of beds of lava and tuff which compose the substance of most volcanic mountains, and it will be at once perceived, that the direction in which they are found to lie, rising on all sides towards the crater, is a necessary result of this mode of formation.

VESUVIUS.

Without entering into the question as to the soundness of such an explanation with reference to other volcanic mountains, I am disposed to object to it as applied to the particular case before us, on account of the difficulty it involves with respect to the mode in which the dykes that traverse the rock are supposed to have been formed. If the crater existed previously to their production, as according to this view must have been the case, I cannot imagine that the lava could have *created* the fissures which it now occupies, since there would be ample scope for its more ready escape through the vast area existing in the centre. I do not deny indeed, that the pressure of an immense column of lava on the walls containing it does in many cases cause fissures, which are immediately filled with liquid lava, thus constituting dykes in the surrounding rock; but I contend, that such dykes ought to be most frequent in the lower parts of the column where the pressure is greatest, and that it must very rarely happen that they would be formed so high up as the crater, since if the rock possessed any thing like an equal consistency throughout, the greater force of the column of lava below would effect a passage through the sides of the mountain, before it could *create* fissures in the part where its pressure was least. Accordingly most writers have supposed, that the fissures having existed previously, the lava boiling up within the crater insinuated itself into them, and Mons. Neckar remarks, that those dykes which appear inclosed in the substance of the rock can hardly have been formed in any other manner. But if that be the case, why do not we find them occurring indifferently in all directions, instead of cutting the beds, in most instances, nearly at right angles? Why do not some traces of the same leucitic lava occur in the bottom and sides of the crater, instead of being circumscribed so exactly within the limits of these fissures? or why has not a mass of melted matter, capable of filling to the very brim, the vast area of

the ancient crater forced a passage outwards, and descended
the external slope of the volcano? or, if it had done so,
where are the traces to be found of an eruption, the magnitude of which must have been such, that all those of more
recent date would appear quite insignificant by its side?

These circumstances lead me to think that the dykes of
Monte Somma, like those found in other situations, were
ejected through the substance of a solid rock, and consequently that the crater was not coeval with the formation of
the strata with which it is surrounded. This perhaps may
have happened in the following manner :

Let us suppose a succession of nearly horizontal strata
composed of any kind of rock, to cover a portion of the
earth, in the interior of which volcanic operations are
going on. It is evident, if the elastic vapours obtain sufficient force to elevate a portion of these beds, the latter will
acquire in consequence a position corresponding with that
belonging to the rocks of Monte Somma, for the elastic vapours, acting with the greatest violence round the point at
which they have created for themselves an aperture, will
upheave the strata equally for a given distance round, and
thus cause them to dip on all sides away from this central
point. Von Buch, in his description of some of the Canary
Islands, published in the Transactions of the Academy of
Berlin, has shewn that many volcanic mountains have been
raised in this manner from the sea,* and has distinguished
craters to which he ascribes this origin, by the name of
Craters of Elevation, " Erhebungs Cratere." It is evident
that such craters as that of the Mosenburg in the Eyfel,
which shew no vestiges of volcanic matter, can only have
been formed in this latter way.

Adopting then this theory as applicable to the origin of
Monte Somma, we shall have less difficulty perhaps in re-

* See my third Lecture.

conciling the existence of the dykes in this case to the received opinions with respect to their origin in general.

We may suppose for instance, that the strata which constitute the substance of the mountain were originally deposited in an horizontal direction, and under the surface of water; * that whilst in this position, a renewal of volcanic action took place, the first effect of which was the production of these dykes formed by the lava struggling to make its escape. A continuance and increase of the same action may afterwards have caused an upheaving of the strata in all directions round the area since occupied by the crater, and those dykes, which now appear inclosed in the substance of the rock, may have once been continued in that portion of it, which has been broken away or destroyed by the subsequent operations of the volcano.

The preceding remarks may be considered as bringing down the history of the volcano to the commencement of the Christian era, when so long an interval had elapsed since the period at which it was last in activity, that no certain record of preceding eruptions appears to have existed.

It is true that Diodorus Siculus, who, from having visited Etna, must have been familiar with volcanic appearances, was struck with the indications of a similar origin which he perceived in Vesuvius,† and we also find that Vitruvius, when speaking of the Puzzolana near Naples, which he supposes to have been formed by heat, notices a tradition that Vesuvius also in former times emitted flame. Lucretius likewise has been supposed to refer to this moun-

---

* This is the opinion of Breislac, Gioeni, and other naturalists. Breislac's Campanie, Vol. i. p. 125.

† Ωνομασθαι δε και το πεδιον τυτο Φλεγραιον, απο τυ λοφου του το παλαιον απλετον πυρ εκφυσωντος, παραπλησιως τη κατα την Σικελιαν Αιτνη. καλειλαι δε νυν ο τοπος Ουεσουιος, εχων πολλα σημεια του κεκαυσθαι κατα τους αρχαιους χρονους.

VESUVIUS.

tain in his 6th Book, where he speaks of a spot near Cumæ
which sends forth sulphureous fumes; but little ought to be
built on a passage which has such different readings,* and
it seems just as possible that the poet may have meant to refer to the exhalations of the Solfatara, which lies much
nearer Cumæ : nor is Silius Italicus, who has been quoted
in proof of an eruption having taken place during the 2d
Punic war, an authority altogether unquestionable, since
having witnessed the famous eruption of 79, he may, with a
very natural poetical licence, have introduced a similar
event as happening at an earlier period without any direct
sanction from history.

On the other hand, it may be remarked, that Strabo seems
rather to infer the igneous origin of Vesuvius from the nature of the rocks found upon it, than from any received
tradition, and that it seems natural to attribute to the
mountain a long period of tranquillity, both from the
cultivated state in which it existed,† and from the circumstance of the crater having fallen in on one side, as must
be understood to have been the case, if we suppose that
this was the spot having only one outlet in which the army
of Spartacus was besieged.‡

* This passage is very differently read—
      Is locus est Cumas apud ; acri sulfure montes
      Oppleti calidis ubi fumant fontibus aucti:
   or,
      Qualis apud Cumas locus est montemque Vesuvum
      Oppleti calidis ubi fumant fontibus auctus.
It is evident that even the latter reading affords no proof that Vesuvius had
experienced any recent eruption, or was in a more active state than the Solfatara of Puzzuoli, which will be afterwards mentioned.

† See Virgil's Georgics, Martial, Varro, and others.

‡ If Mr. Hayter be correct in deriving Herculaneum from Her and Koli,
the burning mountain, we might obtain an additional argument in favour of
the ancient eruptions of Vesuvius; but it is an evident affectation to substitute
a far-fetched etymology for one so obvious as that which derives it from Hercules: and even admitting the former to be the correct derivation, it affords

VESUVIUS.

This period of apparent security was however at length to cease ; in the year 63 after Christ, the volcano gave the first symptom of internal agitation in an earthquake, which occasioned considerable damage to many of the cities in its vicinity. A curious proof of this is exhibited by the excavations made at Pompeii, which shew that the inhabitants were in the very act of rebuilding the houses overturned by the preceding catastrophe, when their city was finally overwhelmed in the manner I am about to describe.

On the 24th of August of the year 79, the tremendous eruption took place, which has been so well described in the letters of the younger Pliny.* It was preceded by an earthquake which had continued for several days, but being slight was disregarded by the inhabitants, who were not unaccustomed to such phænomena. However on the night preceding the eruption the agitation of the earth was so tremendous, as to threaten every thing with destruction.

At length about one in the afternoon, a dense cloud was seen in the direction of Vesuvius, which after rising from the mountain to a certain distance in one narrow vertical trunk, spread itself out laterally in a conical form, in such a manner, that its upper part might be compared to the branches, and the lower to the trunk of a pine. It was descried from Misenum, where the elder Pliny, as commander of the Roman fleet, was stationed, with his family, among whom was his nephew the younger Pliny. The latter, who seems already to have imbibed somewhat of the spirit of the Stoical philosophy, which inculcated rather an indifference to the course of external events, than an inquiry into their nature, pursued his usual train of studies as before ; but the former, with the zeal and enterprize of a mo-

perhaps no stronger proof, than the names of some of the mountains in Auvergne and Hungary do of the modern date of their eruptions. In both cases the aspects of the spots themselves might be sufficient to suggest the application of their names.

* Plin. Epist. Lib. vi. Ep. 16. 20.

dern naturalist, prepared in defiance of danger, to obtain a
nearer view of the phænomena.

Accordingly he first repaired to Resina, a village immediately at the foot of Vesuvius, but was soon driven back
by the increasing shower of ashes, and compelled to put in
at Stabiæ, where he proposed to pass the night. Even here
the accumulation of volcanic matter round the house he
occupied rendered it necessary for him to remain in the open
air, where it would appear that he was suddenly over-
powered by some noxious effluvia, for it is said that whilst
sitting on the sea-shore under the protection of an awning,
flames, preceded by a sulphureous smell, scattered his atten-
dants, and forced him to rise supported by two slaves, but
that he quickly fell down, choaked, as his nephew conjectured, by the vapour, which proved the more fatal from
his previous weak state of health. The absence of any external injury proves, that his death was caused by some
subtle effluvia, rather than by the stones that were falling at
the time, and it is well known that gaseous exhalations,
alike destructive to animal and vegetable life, are frequent
concomitants of a volcanic eruption.

The other circumstances of this memorable catastrophe
are sketched by the younger Pliny with a rapid but masterly
hand. The dense cloud, which hovered round the mountain,
pierced occasionally by flashes of fire more considerable
than those of lightning, and overspreading the whole neighbourhood of Naples with darkness more profound than that
of the deepest night; the volumes of ashes which encum-
bered the earth, even at a distance so great as that of Misenum; the constant heaving of the ground, and the recession
of the sea, form together a picture, which might prepare us
for some tremendous catastrophe in the immediate neighbourhood of the volcano.

Yet the covering of three entire cities under an heap of
ashes from 60 to 112 feet in depth, would seem an effort

VESUVIUS.

almost too gigantic for the powers of this single mountain,
if we were not aware of the vast depth at which the volcanic
operations are going on, and the immense extent to which
their influence may therefore be supposed to reach. It has
been calculated indeed that the masses ejected at different
times from Vesuvius vastly exceed the whole bulk of the
mountain ;* and yet the latter seems upon the whole to undergo no diminution, for the falling in of its cone at one
period appears to be balanced by the accumulation of ashes
at another.

The cities of Stabiæ, Pompeii, and Herculaneum, which
were destroyed in the course of this eruption, appear to
have been overwhelmed, not by a stream of melted matter,
but by a shower of cinders and loose fragments ; † for the
various utensils and works of art that have been dug from
thence nowhere exhibit any signs of fire, and even the deli-
cate texture of the Papyri appears to have been affected
only in proportion as it has subsequently been exposed to
air and moisture. Thus in those at Pompeii, which was covered by a mere uncemented congeries of sand and stones,
decomposition has proceeded so far that their contents are
illegible, whereas at Herculaneum, where they have been
preserved under a species of tuff, their characters often admit of their being decyphered. Now the formation of this
latter substance is explained on the supposition of a torrent
of mud having accompanied in this quarter the ejections of
the volcano, which, favouring the agglutination of the loose
materials, reduced them to a state, which though less consistent than tuff generally is, was capable of preventing in
some degree the access of air and humidity to the substances
underneath. Sir W. Hamilton notices a fact, which, shews
very conclusively both that the tuff of Herculaneum was

* This was remarked even by the antients, and Seneca, Letter 79, after
starting the difficulty, solves it by remarking, that the fire of the Volcano
" in ipso monte non alimentum habet, sed viam."

† The stones that fell at Pompeii are said many of them to weigh 8lb. the
largest of Stabiæ only an ounce.

VESUVIUS.

once in a pasty state, and that it owed its softness not to heat but to moisture, the head of a statue that was dug up, having left a cast in the tuff which had formed upon it, without appearing to be itself in the least scorched.

It is not my purpose to go through a detail of all the subsequent eruptions which have taken place from the volcano, since the accounts that have been handed down respecting them present for the most part little more than vague pictures of the terror and desolation occasioned, without serving to increase our knowledge of the phænomena themselves.

The second eruption appears to have happened in the year 203, under the emperor Severus, and is described by Dion Cassius, and Galen; the third in the year 472, which is said by Procopius to have covered all Europe with ashes, and to have spread alarm even at Constantinople.* Other euruptions are recorded in the years 512, 685, and 993. That of 1036 is said to have been the first which was attended with an ejection of lava; in preceding accounts we hear only of ashes and lapilli being thrown out. Between that period and the commencement of the seventeenth century the mountain appears to have been only five times in a state of activity, and in 1611, the interior of the crater, according to the report of Braccini, was covered with shrubs, and every thing indicated the greatest tranquillity. Yet only twenty years afterwards, in 1631, one of the most terrible of its eruptions took place, which covered with lava the greater part of the villages lying at its foot, on the side of the Bay of Naples. Torrents of water also issued from the mountain, and completed the work of devastation. The volcano is likewise said to have been in activity in the years 1660,

---

* Silius Italicus aptly compares the flight of Hannibal's soldiers after the great battle of Zama into regions of Africa where the Carthaginian name had never been heard of, to the appearance of the ashes of a volcano, in places far removed from the site of its eruptions. Lib. 12.

1682, 1694, and 1698, from which time till the present its intervals of repose have been less lasting, though its throes perhaps have diminished in violence; for the longest pause since that time was from 1737 to 1751, and no less than eighteen distinct eruptions are noticed in the course of little more than a century, several of which continued with intermission for the space of four or five years. That of 1737 gave rise to a stream of lava, which passed through the village of Torre del Greco, continued its course until arrested by the sea, at which time its solid contents were estimated at 33,587,058 cubic feet.

Of the latter eruptions one of the most formidable was that of 1794, recorded by Breislac, himself an eye-witness of it, in his travels in Campania. The torrent of lava that proceeded from it again destroyed the town of Torre del Greco, and advanced into the sea to the distance of no less than 362, with a front of 1127 feet. The cubic contents of of this single current are estimated by that naturalist at 46,098,766 cub. feet.

The eruption of 1813 has been described by Menard de Groye, and the last which happened in 1822, by Monticelli, to whose observations I shall have occasion to refer in a future part of these Lectures, the more especially as notwithstanding the opportunities for scientific inquiry thus afforded by the frequent recurrence of these eruptions, it is only of late that any facts appear to have been obtained that promise us the least insight into the real nature of the phænomena.

Our forefathers indeed appear to have followed quite a mistaken course, in taking account only of the grander operations of the volcano; whereas these in reality are of all others the least instructive, as being accompanied with circumstances of terror and risk, which preclude the possibility of near and accurate research. As the chemist in the investigation of a new substance, will often light upon pro-

perties when operating on a single grain, which had escaped
his research when embarrassed with a larger quantity; so
it may often happen, that the minor eruptions of a volcano,
which vulgar observers overlook, reveal to us phænomena
that had been unperceived amidst the terror and confusion
of its more violent commotions; and if the Naturalist be
disappointed in witnessing the effects on a less magnificent
scale than he could have desired, he will be amply repaid
by being enabled to investigate them at a time when no considerations of a personal nature control or interrupt his
proceedings.   Nevertheless, as every *degree* of energy in the
volcanic processes may be expected to give rise to certain new
phænomena, it is highly proper, that an account should be
taken of the circumstances that present themselves on the
mountain from a state of the most languid to that of the most
vigorous action, and it is only by combining observations
made during a considerable period, such as may include all
these varieties of condition, that we can ever hope to obtain
correct data on which to build a theory of volcanic operations.

In considering then the natural history of a mountain of
this kind with reference to the above object, three distinct
classes of substances present themselves to our notice, namely, the lavas, the ejected masses, and the gaseous exhalations
given off.

In the case of Vesuvius I should suggest a division of
the first and second classes into two distinct heads; namely,
into such as were produced by the mountain before the
renewal of its activity in the year 79, and such as are of
posterior formation; and I should also propose that the
elastic vapours should be likewise divided into those
emitted during a period of comparative quiescence and of
activity.

It may be difficult to pronounce decidedly what lavas
have been of anterior formation to the eruption of 97, but we

VESUVIUS.

observe on the flank of Mount Somma, in the hollow way termed the Fossa Grande, several beds superimposed one on the other, which appear to be derived from that mountain, and to have no connexion with the present cone of Vesuvius. The differences however that exist between the lavas, have not hitherto at least been shewn to be dependent on their age, but wherever the mass has cooled under circumstances favourable to the developement of crystalline arrangement, the result appears to have been a granular mixture of felspar, leucite, augite, and titaniferous iron.

Von Buch remarks,* that the streams of 1760 and 1794, resemble basalt in colour, fracture, hardness, and weight, but he admits that the majority both here and elsewhere, bear more analogy to trachyte.

The ejected masses derived from the antient volcano are of great variety and interest. They are accumulated in the Fossa Grande, and other hollow ways on the slope of the mountain, where torrents have exposed a section of the several beds. I have collected in these situations a series of granular limestones, nowise inferior in the fineness of their grain to those found in primitive districts; yet the only calcareous rock met with in the neighbourhood is the coarse limestone of the Appennines, and it seems therefore not altogether improbable that the conversion of this stone into marble has been effected by volcanic agency.† Breislac mentions, that Dr. Thomson observed the same change to take place sometimes in calcareous stones exposed to the heat of a lime kiln, and that a specimen of that kind, one half of which is common limestone, the other marble, is at

---

* Reise, vol. 2, p. 180.

† Cornelius Severus, in his spirited poem on Etna, has happily expressed the distinction between the substances to which the volcanic fire has given new properties, and those which it has only partially affected—

" Pars igni domitæ,—pars ignem ferre coactæ."

VESUVIUS.

present among the specimens bequeathed by that naturalist to the University of Edinburgh.*

A list of the various minerals found among these ejected masses would comprehend more than one third of all the species that are known, so that a description of them would be altogether incompatible with the limits of this under-taking; I am happy, however, to refer you to an account of them which is soon expected to appear from the hands of two intelligent naturalists at Naples, who devote themselves to the study of the phænomena and productions of this volcano.†

I must remark however that these minerals are, for the most part, confined to the ejected masses of Monte Somma, those at present vomited forth by the volcano containing only crystals of felspar, augite, titaniferous iron, leucite, mica, and olivine. Besides the above we may observe, among the loose fragments of the Fossa Grande, portions of volcanic matter that seem to have been detached from some torrent of lava that lay within the sphere of the volcanic operations; these disrupted masses are called Erratic Lavas, to express at once their supposed origin, and to separate them from such as constitute a part of some existing current.

The gaseous exhalations most commonly given off from Vesuvius appear to be sulphurous and muriatic acid gases, the former chiefly produced during a period of calm, the latter at, and immediately subsequent to, an eruption. Nitrogen gas has likewise been detected, and much aqueous vapour is generally present. The latter indeed often constitutes the sole product of those Fumaroles that surround the external slope of the crater, when the mountain is in a tranquil state.

---

* Breislac, Institut. Geol. § 255.

† Sign. Monticelli and Covelli have announced their intention of publishing an account of these minerals, to appear in parts, under the title of Prodromo della Miner. Vesuviana.

Besides these, fatal *Moffettes*, or exhalations of noxious gas, are given out from crevices in all parts of the mountain; they are frequently found in the cellars of Portici and Resina after an eruption, and, when they occur in the fields, prove speedily destructive to vegetation. They are supposed to consist chiefly of carbonic acid gas.

Reserving however for a future occasion a more full account of the gases exhaled from this as well as other volcanos, I shall proceed to examine the remaining phænomena connected with our present inquiry, which Naples offers to our regard.

And first, with regard to those related to the actual condition of the volcano, and derived from its operations.

I shall say nothing of the numerous earthquakes which are of such common occurrence in Calabria, as this is a subject which will be noticed in another place, but shall mention two circumstances which have been urged in proof of the fact that the volcanic action extends even at present to the opposite side of the Bay.

The first relates to the temple of Serapis, near the town of Puzzuoli, the phænomena of which appear at first sight to indicate, that the sea has twice changed its level with reference to this coast since the Christian æra.

The facts that have led to this conjecture are :

1st. That the floor of the temple is at present somewhat below high-water mark, and consequently exposed to the rushing in of the sea, which it is presumed was not the case when it was founded :

2dly. That the pillars, which yet remain standing, are perforated by pholades at a height of about sixteen feet from their base, which implies, that at some intermediate period the sea water must have stood at this level.

Now if we admit that these facts afford sufficient evidence of the points contended for, it would certainly involve a less

TEMPLE AT PUZZUOLI.

difficulty to suppose that the change of level was owing to the movements of the land, than to those of the sea, since a general rise of the latter fluid (and no *permanent* rise could be partial) to the extent supposed, would have left on other coasts of the Mediterranean, evidences which could not have been overlooked.

But even this latter hypothesis seems to involve considerable difficulties. It implies that the ground has undergone at least two changes; the first a depression of nearly 30 feet to have brought it down to the level of the sea at low water; the second an elevation of nearly as many, by which it was restored to about its original elevation.—Had such been the case, it is probable that not a single pillar of the temple would now retain its erect posture to attest the reality of these convulsions, and the hot springs which gush out on one side now, as they did sixteen hundred years ago, would either have been dried up, or turned into some other channel.

It seems therefore most probable, that the effect depended on some purely local cause, and that the existence of Pholades at that height arose from the formation of a small lake on the site of the temple, which was ponded up considerably above the level of the sea by accidental circumstances.

This accordingly is the hypothesis of Goethe,* and many other naturalists, who are nevertheless embarrassed to account for the continuance of salt-water at so great a height. Nor do I conceive that Goethe has much lightened the difficulty by supposing that the salt was obtained, like that of the Dead Sea, from volcanic exhalations, for these are commonly so charged with noxious ingredients, that their presence would render the water even more unfit for the abode of Pholades, than it would be without any salt at all.

It appears to me that the observations of the Canonico

---

* See Goethe über Naturwissenschaft, Band 2, translated in Edin. Phil. Journ. vol. ii. p. 93.

TEMPLE AT PUZZUOLI.

di Jorio,* if correct, are calculated to throw some light upon this question, as he has shewn that the proofs of an overflowing of the sea may be gathered not only from the temple itself, but from the land on either side of it, which is covered with pebbles and other alluvial matter, to the same height as the parts of the pillars which have been perforated. Now granting this to be the case, we have only further to explain, in what manner it happened, that after the sea had resumed its former level, the water still continued round the temple of Puzzuoli for a sufficient length of time to allow of the perforation of the pillars.

For my own part I here concur with Goethe, in believing that the circumstance best admits of explanation by calling to our assistance the volcanic action, of which the neighbourhood bears sufficient proofs.

The collection of ashes and scoriæ, under which the whole temple was buried up to the point at which the Pholades seem to have begun their operations, proves that an eruption did take place from some neighbouring mountain, which covered with its ejected materials this particular spot, and if we suppose the latter to have been so disposed as to form a kind of bason, which really appears to have been the case,† it is evident that the sudden rise of the sea, which the Canonico di Jorio speaks of as certain, would have the effect of leaving a salt-water lake surrounding the upper portion of the pillars, the lower being buried under the volcanic products.

Even the inundation itself may perhaps have arisen from the very same cause, for we know that a sudden and considerable rise of the sea is in fact one of the most common as well as the most fatal consequences of similar convulsions of nature.

* Ricerche sul tempio di Serapide in Pozzuoli. di Can. And. di Jorio. Napoli, 1820.

† See Goethe's Memoir, Ed. Journ. p. 94.

TEMPLE AT PUZZUOLI.

Thus it is particularly mentioned in Pliny's account of the famous eruption of 79, that the sea and land appeared to have changed places, and in the earthquake of Messina we hear of an immense wave rising far on the opposite coast of Calabria, and sweeping away 2000 people.

The difficulty of supposing the lake to have continued salt long enough for the effect produced, is lessened by an observation of Brocchi's, who states that Pholades will live in water very slightly saline, so that the lake might have continued slowly to drain off, and yet, if supplied, in a nearly equal ratio from the hills above, with fresh water, the Pholades might long continue their labours.

I may add, that de Jorio seems to have removed the other difficulty which has been started respecting this locality, that, namely, of the pavement of the temple being below the present high-water mark,[*] by shewing, that this very inconvenience was plainly contemplated by the original builders, who erected a dyke, of which there are some remains, evidently intended to prevent the ingress of the sea.[†]

The site of the temple was indeed, in all probability, determined by that of the warm baths, which gushed out from the spot, and the Romans were too much in the habit of controlling the movements of the sea near Baiæ by artificial barriers, to be deterred from erecting the temple, on the spot which superstition dictated, by the circumstance of its being a few feet below high-water mark.

The second, and a much more decided case, which I shall adduce of a change produced on the physical structure of the country round Naples by volcanic agency, was the rise

[*] Breislac has mentioned other facts which seem to shew a similar rise of the sea, but as they occur within the circuit of the Bay of Naples, that is within a space exposed to frequent earthquakes, they will be hardly viewed as establishing his point. At all events, the consideration of them is foreign to the present purpose.

[†] Di Jorio, p. 55—6.

MONTE NUOVO.

of a new mountain on the northern side of the Bay in the
16th century. Vesuvius had at that time been for a long
interval tranquil, but a succession of earthquakes had taken'
place in the country for two years previously. At length on
the 28th of September, of the year 1538, flames broke from
the ground between Lake Avernus, Mount Barbaro, and the
Solfatara, followed by several rents of the earth from which
water sprung, whilst the sea receded two hundred feet from
the shore, leaving it quite dry. At last, on the 29th, about
two hours after sun-set, there opened near the sea a gulph,
from which smoke, flames, pumice and other stones, and
mud were thrown up, with the noise of thunder.

In about two days the ejected masses, formed a mountain
413 feet in perpendicular height, and 8000 feet in circumference. The eruption finally ceased on the 3d of October.
On this day the mountain was accessible, and those who
ascended it reported, that they found a funnel-shaped opening on the summit—a crater, a quarter of a mile in circumference.

Some suppose that the Lucrine lake was filled up or destroyed by this eruption, but others maintain, that the latter,
which was a basin originally separated from the sea by a
dam, had been merged in the bay again upon the decay of
its embankment. It is certain that cotemporary writers, in
their account of the formation of the Mount Nuovo, say
nothing concerning this lake.

The Monte Nuovo is composed entirely of fragments of
scoriform matter, or of a compact rock of an ash grey colour,
sometimes resembling trachyte, and at others approaching
to porphyry slate. The scoriform matters include pumice,
and most other varieties of volcanic substances, intermixed
with a white sand, but never agglutinated so as to form a
tuff. Its form is that of a compressed or oblong cone, and
it has in its centre a crater almost as deep as the mountain
is lofty.

SOLFATARA.

Near the bottom of the crater are one or two small caverns, the interior of which I found covered here and there with an efflorescence, having an alkaline taste. The sand near the foot of the mountain, even under the sea, possesses so high a temperature when brought up from a little below the surface of the water, that we are led to conclude that the volcanic action is still going on to a certain extent, and the same inference may be drawn from the extreme heat of the water which gushes from the rock in a cavern not far distant, called the Baths of Nero, which is sufficient in a very few minutes to boil an egg.

The nearest approach however to the phænomena of Vesuvius is exhibited in a hill between the Monte Nuovo and Puzzuoli, called the Solfatara, which, although considered an extinct volcano, as no ejections of solid matter have ever been known to proceed from it, is continually giving off gaseous exhalations, mixed with aqueous vapour.* I am indebted to the kindness of my friend Mr. Herschell, and to Signor Covelli of Naples, for an opportunity of

---

* It appears to have been precisely in the same state near 1600 years ago, for the description, which Cornelius Severus gives of it, applies exactly to its present condition.

> Neapolim inter,
> Et Cumas, locus est multis jam frigidus annis,
> Quamvis aeternum pinguescat ab ubere sulphur.

Strabo also notices it under the name of Ηφαιϛου αγοϱα, and Petronius Arbiter gives the following animated and correct description of the spot.

> Est locus exciso penitus demersus hiatu
> Parthenopen inter magnæque Dicharchidos arva
> Cocyta perfusus aqua.   Nam spiritus extra
> Qui furit, effusus funesto spargitur æstu.
> Non hæc auctumno tellus viret, aut alit herbas
> Cespite lætus ager: non verno persona cantu
> Mollia discordi strepitu virgulta loquuntur:
> Sed chaos, et nigro squallentia pumice saxa
> Gaudent ferali circum tumulata cupresso.
>
>   SATYRICON, cap. 120.

examining the nature of those that were emitted at the time
I visited Naples in 1825; these gentlemen having been good
enough, as I was incapacitated by illness from revisiting the
spot for that purpose, to condense a portion of the steam
given out from one of the fumaroles in the crater of that
mountain, which of course would contain all the gases
which water is capable of dissolving. These appear to be,
sulphuretted hydrogen, and a minute portion of muriatic
acid, and from the known chemical properties of these two
bodies, it is easy to explain the presence of most of the substances that occur about the Solfatara.

The rock of this mountain is a sort of trachyte, which,
besides a little potass, consists essentially of silex and alumine, with an occasional admixture of iron, lime, and
magnesia.

The muriatic acid, acting upon these ingredients, forms
severally with them a quantity of saline matter proportionate
to that in which it is emitted, but the most abundant salt of
this class is the muriate of ammonia, the formation of which
may perhaps be thus accounted for.

When muriatic acid is suffered to act upon an alkaline
hydrosulphuret, it combines with the base and separates the
sulphuretted hydrogen; very little however of the latter
exhales in a gaseous condition, but it is for the most part
precipitated in the form of an heavy oil, which is found by
analysis to consist of 1 atom hydrogen and 2 atoms sulphur.
Now, as sulphuretted hydrogen consists of 1 atom of each
ingredient, it follows that the formation of this body must
be accompanied by the disengagement of an equal volume
of hydrogen gas. But what becomes of this latter body
since it is not to be detected afterwards in a separate state?
It is probable that it has united with the oxygen of the
atmosphere, or with its nitrogen, perhaps indeed with both;
in the latter case the presence of the ammonia is explained,
in the former it is rendered more comprehensible, since we

have many examples in which nitrogen in its nascent state
is known to unite with hydrogen, held in combination by
weak affinities.

Let us now consider what becomes of that (by far the
largest) proportion of the sulphuretted hydrogen, which the
small quantity of muriatic gas present does not influence.

Uniting with the different earths and other bases, of
which the Rock of the Solfatara consists, it will form with
these in the first instance the different combinations known
by the name of Hydrosulphurets.

Now this class of compounds undergo decomposition,
when exposed to air and moisture, in two ways :

1st. The presence of carbonic acid slowly separates the
two ingredients, the acid combining with the base, and the
sulphuretted hydrogen set at liberty being resolved into its
elements, of which the hydrogen forms water with the
oxygen of the atmosphere, and the sulphur is partly precipitated, and partly converted into the hyposulphurous
acid. The hyposulphurous acid again unites with the
different earthy bases, but these combinations are not permanent, but are finally resolved into sulphur and sulphates.

Hence the hyposulphurets, so far as they are influenced
by the presence of carbonic acid, give rise to the deposition
of sulphur, and the formation of sulphuric salts.

2dly. The same ultimate result appears to be obtained by
the mere action of atmospheric air upon the same compounds, part of the hydrogen of the sulphuretted hydrogen
abandoning the sulphur, combining with the oxygen of the
atmosphere, and forming water, whilst the remainder exists
in combination with a double proportion of sulphur, and
produces an hydroguretted sulphuret. This substance is
however finally decomposed, the hydrogen slowly combining
with oxygen, and the sulphur being either oxygenized, or
deposited in a solid form.

Should any of the sulphuretted hydrogen remain in its

uncombined state, it will be speedily resolved into its elements by the action of air and water; such at least I have
found to be the case on confining a portion of this gas, either
with or without an admixture of air, over water, which becomes turbid from the sulphur separated from the gas.

Over mercury, on the contrary, it seems to remain for a
long time unchanged.

It appears then that in both these ways the compounds
formed by the union of the sulphuretted hydrogen with the
constituents of the rock, would be finally resolved into
sulphur and sulphuric salts; we have only therefore to compare this with the phænomena actually exhibited at the
Solfatara.

The rock, of which this mountain is composed, is naturally
hard and dark coloured, but in proportion as it is exposed
to the action of these vapours, its texture and colour undergo a remarkable alteration. The first stage of the process seems to be a mere whitening of the mass, in consequence
doubtless of the removal of the iron to which its colour is
attributable; it then is seen to become porous and fissile;
when the process is yet further advanced, it becomes honey-
combed like a bone that has been acted on by the weather;
and at length it crumbles into a white powder, consisting
almost entirely of silex.

The saline substances that appear efflorescing on the
surface of the rock, correspond with the above statements;
they consist (in addition to the muriatic salts before alluded
to) of the sulphates of iron, lime, soda, magnesia, and above
all, of alumine. Breislac has given some interesting details
respecting these salts, in which however I have not time to
follow him; it is sufficient for our present purpose to show,
that they appear in almost the same proportion, as that in
which they would result from the action of the same gas
upon a stone similarly constituted.

Meanwhile the sulphur deposited diffuses itself through
the rock, and lines the walls of its cavities: but Breislac

SOLFATARA.

appears to have shewn very satisfactorily, that it results entirely from the sulphuretted hydrogen, and is not sublimed in an uncombined state; for we only find it in those parts of the mountain, which are near enough to the surface to admit of the ready access of atmospheric air. It is sometimes accompanied with sulphuret of arsenic, which has probably also been disengaged, combined with hydrogen, and may be the cause of the destructiveness of those exhalations, which frequently succeed an eruption. I have been assured, that the paradoxical statement of Breislac's with respect to the pear and olive being proof against the deleterious influence of these gases, when all other kinds of vegetables are destroyed, is not unfounded. As selenium is now ascertained to exist among the products of the crater in the Island of Volcano, it appears probable, that a gas composed of that metal with hydrogen, will, ere long, be detected together with the gases before enumerated.

The influence of these exhalations is by no means confined within the compass of the Solfatara, for it extends to a considerable distance on the hills which bound that volcano towards the north, as is evinced by the whiteness and decomposed condition of the rocks, in consequence of being acted upon during so many ages by sulphureous vapours. It would appear that the antients designated them from this circumstance under the name of the Colles Leucogæi, for Pliny* mentions medicinal waters existing between Puzzuoli and Naples, called Leucogei Fontes, and there is still found a hot spring at a place called Pisciarelli, situated on the slope of these hills.

The Solfatara returns an hollow sound when any part of its surface is struck, and appears not to be made up of one entire rock, but of a number of detached blocks, which hanging as it were by each other, form a sort of vault over the abyss, within which the volcanic operations are going on. Hence there would seem, not to be one aperture larger

---

* Plin. Hist. Nat. Lib. 31. c. 8.

than the rest, such as we are wont to suppose existing in the centre of every active volcano, but a series of fissures which allow of the escape of elastic fluids, but would prevent the fall of any bulky body.

It has given off however from its south-eastern side a stream of lava, which extends in one unbroken line to the sea, forming the promontory, called the Monte Olibano, on the road between Naples and Puzzuoli. It is remarkable from its want of resemblance to the lavas of most other volcanos, as it consists of a rock which can hardly, I think, be separated by its mineralogical characters from trachyte.

It consists essentially of crystals of glassy felspar, imbedded in a basis of a felspathic nature, having an uneven fracture, and ash-grey colour. It differs in these respects from the rock of the Solfatara itself, which I have found in general to possess a darker colour, and a conchoidal fracture. Both however agree in being composed essentially of felspar, and containing augite only as an occasional ingredient, so that we are obliged to class them equally under the head of trachyte. The stone of Mount Olibano is in general compact, but cells are sometimes present, especially in the upper part of the rock; it is seen to rest immediately on a thin bed of loose fragments of lava, and is covered by a sort of tuff, which, being placed on a part of the hill difficult of access, I left unexamined, but which I conjecture to consist of scoriæ, and sand, that succeeded the ejection of the lava.

The whole rests upon the extensive formation which reaches from Puzzuoli to Cumæ, and appears to be continuous with the rock found in the immediate neighbourhood of Naples. This, which has long been known by the name of Puzzolana, is a formation of volcanic tuff, bearing many analogies to the trass of the Rhine, and the pumiceous conglomerates of Hungary.

Its basis is generally of a straw yellow colour, dull and harsh to the feel, with an earthy fracture, and commonly a loose degree of consistence. It contains imbedded fragments of pumice, obsidian, trachyte, and many other varieties of compact as well as cellular lava, the softer kinds often rounded, the harder mostly angular. I did not succeed in detecting any kind of rock except of a volcanic nature, imbedded in it, and as the construction of a new road round the promontory of Gaiola, gave me a good opportunity of examining this point, I do not believe that such specimens are common. Shells have been met with in this rock as is noticed by Sir W. Hamilton, and I have heard of the bones of ruminating animals having been likewise discovered.

This mass of tuff is sometimes separated into beds by strata of loam, pumice, or of a ferruginous sand, and Von Buch notices a bed consisting of fragments of limestone, cemented by calc-sinter, occurring in the midst of it between Naples and Puzzuoli.

The height of this tuff, in many places near Naples, is very considerable; the hill of the Camalduli, the loftiest eminence next to Vesuvius in the whole country, is composed of it, and to the west of Naples it forms a sort of wall, so lofty and abrupt, that the former inhabitants of the country apparently found it easier to avail themselves of the soft and friable nature of the stone, and to cut through, than to make a road over it.

This is the origin of the celebrated Grotto of Posilippo, a cavern 363 toises, or 2178 feet in length, 50 feet in height, and 18 in breadth, which strikes every stranger with surprise from the mass of rock cut through, until he reflects at the ease with which a stone of such a description admits of being hollowed out.

This immense mass of Puzzolana forms some considerable hills round Naples, many of which, as the Monte Barbara, Astroni, and others, have very regular craters, but do not appear to have thrown out any currents of lava.

LAGO D'AGNANO—GROTTO DEL CANE.

The Lake Agnano has every appearance of occupying the original site of a crater, as the strata, whose edges are visible in its interior, instead of having a corresponding dip on the opposite sides of the cavity, as is commonly the case with vallies formed by diluvial action, seem on the contrary to slope in all directions away from the crater, just as would happen, if the strata had been thrown up by a force acting from below.

Perhaps the deep fissures, that are found so commonly in the neighbourhood of this and other hills near Naples, may owe their origin, like the Barancos of the Canary Islands,* to the upheaving of the contiguous rocks, rather than to the action of torrents.

The volcanic action here, as in many other places round Naples, seems hardly extinct, for there are exhalations of warm vapour constantly rising near the Lake, which are much esteemed in various complaints.

The celebrated Grotto del Cane, situated on its borders, is constantly giving out volumes of carbonic acid gas, containing in combination much aqueous vapour, which is condensed by the coldness of the external air, thus proving the more exalted temperature of the spot from whence it proceeded.

The mouth of the cavern being somewhat more elevated than its interior, a stratum of carbonic acid goes on constantly accumulating at bottom, but upon rising above the level of its mouth, flows like so much water over the brim. Hence the upper part of the cavern is free from any noxious vapour, but the air of that below is so fully impregnated, that it proves speedily fatal to any animal that is immersed in it, as is shewn to all strangers by the experiment with the dog.

The sensation I experienced on stooping my head for a

* See the next Lecture.

LAKE AVERNUS.

moment to the bottom, resembled that of which we are sometimes sensible on drinking a large glass of soda water in a state of brisk effervescence. The cause in both instances is plainly the same.

The quantity of carbonic acid present in the cavern at various heights, was shewn by immersing in it various combustibles in a state of inflammation. I found that phosphorus would continue lighted at about two feet from the bottom,* whilst a sulphur match went out a few inches above, and a wax taper at a still higher level.

It was impossible to fire a pistol at the bottom of the cavern, for though gunpowder may be exploded even in carbonic acid by the application of a heat sufficient to decompose the nitre, and consequently to envelop the mass in an atmosphere of oxygen gas, yet the mere influence of a spark from steel produces too slight an augmentation of temperature for this purpose.

It is probable that the Lake Avernus may likewise have been the crater of a volcano, and afterwards a Solfatara, which will account for the noxious properties attributed to it by the ancients.†

---

* Forsyth has erroneously stated that though torches of gunpowder lose their inflammability there, yet phosphorus resists the carbonic acid.

† When we consider the considerable specific gravity of sulphuretted hydrogen gas, it is very reasonable to suppose that the thick woods which in former times surrounded the Lake Avernus would favour materially the accumulation of this noxious vapour. The surface of the lake screened from the access of the winds in every quarter, must have been covered with a thick stratum of unrespirable gas, which would be very slowly dissipated. If carbonic acid were present, the same thing would take place with this, in even a much greater degree than with the former. But when the woods were cut down, as we are told was done by Agrippa, the air of the lake would become continually intermixed with the surrounding atmosphere, so that, unless a pretty rapid disengagement of gas took place, the noxious qualities would cease. This coincides with the accounts of Strabo, Silius Italicus, and other classical writers.

PHLEGREAN FIELDS.

Quam super haud ullæ poterant impuné volantes
Tendere iter pennis, talis sesc halitus atris
Faucibus effundens supera ad convexa ferebat,
Unde locum Græci dixerunt nomine Aornum.

At present no such exhalations are given out, and birds
seem to resort to it as much as to any lake near Naples.

The Lakes of Agnano and Avernus then are examples of
craters converted into lakes, and may be compared with those
of the Eyfel district noticed in the last lecture.

The circumstances indeed under which the volcanic action was manifested in both these countries seem to have
borne to each other a considerable analogy, as what we observe is [not a single lofty conical mountain, like Etna or
Vesuvius, from which streams of lava have successively been
derived, but either a series of circular depressions in the
midst of the Puzzolana, as those of Avernus and Agnano,
or certain comparatively low hills with craters rising out of
them, which however are not found in any case to have emitted more than a single stream of lava, and in most instances
appear to have given vent solely to exhalations of gaseous
matters, or to ejections of scoriæ.

The Solfatara supplies us with an instance of the former
description, the Monte Barbara and Astroni of the other.

The Monte Barbara* is probably the most elevated vol-

* This mountain is frequently noticed in antient writers under the name of
Gaurus, and was famous for its vineyards. The following elegant lines by
Aurelius Symmachus, one of the later Latin poets, allude to this, as well as to
the heat of the water near its foot.

> Ubi corniger Lyæus
> Operit superna Gauri;
> Vulcanus æstuosis
> Media incoquit cavernis;
> Tenet ima pisce multo
> Thetis, et vagæ sorores:
> Calet unda, friget Æthra;
> Simul innatat choreis
> Amathusias renidens,
> Salis arbitra, et vaporis,
> Flos siderum, Dione.

canic hill on this side of Naples, it has a crater on its
summit, one side of which is broken away, and its extreme
antiquity is manifested by the circumstance of its surface
being covered with verdure. A single farm house indeed is
seen in the interior of the crater, in the most solitary situation probably that could be chosen within so short a distance
of a great capital.

The perfect condition of the crater of Astroni has caused
it to be selected by the King of Naples as a preserve for his
wild boar and other animals destined for the chase; it is a
circular cavity, nearly a mile in diameter, the walls of which
are formed of a congeries of scoriæ, pumice, and other
ejected materials.

If we believe Breislac, the number of craters, of which
indications exist in the neighbourhood of Naples, will
amount to no less than twenty-seven, but I am inclined to
think that many of the hollows, which he supposes to be
derived from this cause, have in reality been produced by
water; nor does this Geologist appear to have applied the
only *test* by which hollows that establish the existence of
craters, can be clearly distinguished from those derived from
other causes—I mean the direction of the strata composing
their walls.

These, it is evident, ought in the case of a valley of denudation to possess a similar inclination and dip on the opposite sides; whereas in a crater they will diverge in all directions from the original centre of the cavity.

In the following sketches *A* represents a valley originally
formed by water, *B* one resulting from a crater partially
destroyed.

PHLEGREAN FIELDS.

Now I was unable to satisfy myself, that this latter arrangement was exemplified in any of those hills, which
Breislac supposes to have been portions of craters once
existing on the immediate site of Naples.

On the other hand it may be remarked, that if we only
grant the Puzzolana to have been formed at a period antecedent to the excavation of the vallies, the existence of
hollows and other irregularities of surface, which Breislac
regards as the remains of craters, is readily explained ; for
it is evident, that a rock of so soft and yielding a texture,
ought, of all others, to exhibit the most decided evidences
of diluvial action.

Now, independently of other circumstances which will be
afterwards insisted on, the manner in which the Puzzolana
has insinuated itself into the bottoms of vallies for a considerable distance round Naples seems to prove this opinion.
We observe it so situated in the valley of Maddelona, near
the aqueduct of Caserta, and in several of the mountain
gorges near Sorrento, on the opposite side of the bay of
Naples.

The tuff of Sorrento indeed is not precisely similar to
that of the coast opposite, but its origin is probably the
same; and the regular distribution of this material in both
cases over so extensive a surface, and amongst rocks that do
not appear at any time to have been affected by volcanic
action, seems to indicate the operation of water.

But even if we limit the craters that existed in the
Phlegrean fields to those of which present appearances
leave no doubt, their number will be sufficient to give us a
frightful picture of the condition of the country at an early
period of history, and serve to account for the fables of the
Poets, who imagined the entrance to the Infernal Shades to
lie among these recesses.

It was not then, as at present, a single mountain which
sent forth flames and melted matters at certain intervals, and

M

secured a comparative immunity to the rest of the district ;
but there was a constant exhalation of noxious vapours from
a variety of orifices, attended with earthquakès, and other
phænomena, which bespeak the operation of volcanic agency
over a widely extended surface.

If then the early settlers in Sicily were so alarmed at the
eruptions of Mount Etna, as to fly to some other part of the
island, and if in modern times, among the Canaries, the
inhabitants of Lanzerote were compelled to migrate on
account of the ravages made upon their possessions during a
succession of years by subterranean fire,* it is not unnatural
that the picture which Homer had received of the Phlegrean
fields should have been so terrific, as to have led him to
describe them† as placed at the utmost limits of the habit-
able world, unenlightened either by the rising or setting sun,
with groves consecrated to Proserpine, rivers with streams
of fire, and enveloped in an eternal gloom. These ideas
would be confirmed, if we imagine that the Cimmerians,
who first peopled the country, lived in those caverns and
hollows of the rock which now exist, and were thus by the
very nature of their habitation shut out from the light of
day.‡

Such a picture indeed accords very little with the ideas

---

* See my third Lecture.

† Odyssey K. Λ.

‡ I have noticed in my memoir on Sicily (Edinb. Journal) the existence of
artificial caverns, which we hardly know whether to consider as the dwellings
or burial places of the early inhabitants. Though wretched abodes, they
would at least be preferable to those from which I have myself seen the
Cyganis of Hungary creep out, which were literally holes in the ground, in
which they burrow like the Troglodytes of old. Capt. Lyon has described
similar residences in Northern Africa. See his Travels, p. 25. But there can
be no doubt from the number of caverns noticed by travellers in the East, (see
particularly Buckingham's Travels in Palestine, p. 113.) as well as from history, sacred and profane, that the practice of hollowing out the rocks of the
country for habitations in many instances long superseded the use of them for
erecting regular houses. No rock however affords facilities for this practice
equal to the tuff of volcanos, and accordingly, whereever this occurs, we find
even at this day that dwelling places are excavated in it. Such is the case in
Ischia, in Auvergne, and in the Vivarais. That the Puzzolana of Naples
therefore should be hollowed out for this purpose, is no more than we might
reasonably expect.

suggested by the luxuriance of modern Campania; but it must be recollected, that at the time when Homer wrote, that luxuriance had not yet been developed by cultivation, that the recent occurrence of the eruptions had probably devoted many parts to a temporary sterility, and that others were overshadowed with thick and gloomy forests.

The neighbouring islands of Procida and Ischia are likewise composed of volcanic rocks bearing a considerable resemblance to those of the Campi Phlegrei.

The former seems to consist entirely of tuff separated by beds of cellular lava. In one part where the coast exposes a section of the strata, I observed that they were so contorted as to represent an arch, whilst in the places intermediate, as well as on either side, they were horizontal.

This effect, happening as it does in a volcanic country, seems attributable to the pressure of elastic vapours from below, heaving up the strata round a given area, and it will be readily perceived, that if the force applied had been considerable enough to cause a disruption of the beds, we should then have had them dipping in all directions away from the opening, as happens where a crater is produced.

In the following sketches No. 1. represents the actual section, and No. 2. that which would have resulted, had the force been somewhat greater. In either case A represents the beds of tuff, and B. those slaggy lava interposed.

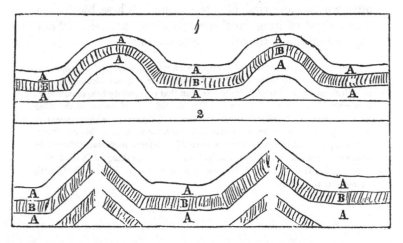

The Island of Ischia is somewhat more varied in its composition. It is composed for the most part of a rock which seems to consist of very finely comminuted pumice, re-agglutinated so as to form a tuff, sometimes resembling the Puzzolana of Naples.

From the very fine state of division however into which it was reduced at the time when it underwent consolidation, a rock has often resulted of so homogeneous a texture, as to be considered a variety of felspathic lava; but I am, upon the whole, disposed to believe it to be rather a substance resulting from the reaggregation of fragments of pumice and other analogous products.*

This formation is often separated by beds consisting of loose portions of pumice and obsidian, and with this exception nearly all the island may be said to consist of tuff, which extends even to the summit of Mount Epomeo, its loftiest point.

I could discover nothing like a crater on the summit of this hill, but conceive its superior elevation to arise merely from the rock in this part having resisted decomposition more than it has done elsewhere.

Although the pumiceous conglomerate, as I shall venture to call this rock, is seen in every part of the island, yet at Monte Vico, near the town of Foria, we observe intermixed with it huge blocks of trachyte, sometimes 30 feet in diameter, consisting of a congeries of crystals of glassy felspar, often without any kind of intermedium. These blocks are angular and of irregular shape; they are scattered without any order through the substance of the tuff.

* Although I give this as what appears to me the most probable opinion, I think it right to mention, that Brocchi and other eminent Geologists, consider it as an earthy variety of trachyte, and that it certainly bears a considerable resemblance to the Rock of the Puy de Dôme. There are on the other hand varieties of the micaceous conglomerate of Hungary, which possess in quite as great a degree the characters of a trachytic rock; and this consideration, coupled with the frequent presence of scoriform portions, different in colour from the matrix, is my principal reason for speaking of it as I have done. Other Geologists will do well to consider the rock particularly with reference to this point.

A little beyond the village of Casamiccola, is a conical hill, called the Monte Thabor, composed entirely of trachyte, one variety of which seems to approach to clinkstone porphyry.

This trachyte rests upon a bed of clay, sometimes red and ferruginous, at others blue, in which are imbedded several species of arca, murex, turbo, and trochus, enumerated by Brocchi.* Thus the date of the trachyte cannot be anterior to that of the tertiary class of rocks.

As we proceed from thence in an easterly direction round the coast of Ischia, we meet with evidences of volcanic operations of a still more recent date.

At Castaglione the ground is covered with loose fragments of pumice and obsidian, which I did not succeed in tracing to their source. Still further we cross the stream of lava, which issued from the side of the mountain in the year 1302, remarkable from the large crystals of glassy felspar which are imbedded in it. Its surface is still undecomposed, and consequently barren, moss alone growing upon it, and that only in a few parts, a proof of the number of ages required for bringing some lavas into a state fit for cultivation. This current may be readily traced up the mountain to the point from whence it issued, which is marked by the existence of a crater, originating apparently from the eruption itself.

The castle of Ischia itself stands on a projecting mass of lava, which appears to have made a part of a current that may be traced to the neighbouring heights of Campignano, where it constitutes a sort of ridge resting upon the pumiceous conglomerate. Its high antiquity is evident from the changes that must have since taken place in the figure of the island, for not only has the promontory been separated from the island by some subsequent convulsion, but we immediately perceive that a stream of lava at the present time would pursue a very different direction, and, instead of

* Conch. Subapp. p. 354.

reaching the promontory, would fill up the vallies and in-
dentations in the coast which the present current overlooks.

Thus Ischia appears to have been subjected to volcanic
action of as many different periods as the neighbourhood of
Naples itself, its pumiceous conglomerate corresponding
with the Puzzolana, its trachytes to the rock of the Solfatara, and the lava of the Capo d'Arso to those of Vesuvius.
Even the antients were fully aware of its volcanic nature,
attributing it to the Giant Typhœus being confined under
the mountain, and Strabo relates that a colony sent over by
Hiero, tyrant of Syracuse, was so alarmed by the frequent
earthquakes, that they deserted the island.*

At present the only immediate indications of volcanic
action are, the high temperature of the sand on the shore
near the Monte Vico, which, two feet below the surface, I
found to be 110° of Fahrenheit, and the hot vapour which
issues from the ground in various places of the same neighbourhood. The fissures through which the steam passes
are often coated with a white siliceous incrustation, which
Dr. Thomson, I believe, was the first to notice under the
name of fiorite.

Dr. Macculloch† has noticed a similar phænomenon as
occurring in the graphic granite of the Isle of Rona, where
the surface of the quartz, or chalcedony, as it has been
otherwise called, obtains from exposure to the weather a
glossy enamel, arising apparently from a partial solution of
the silex. A similar enamel is to be observed investing the
sandstones of Jura and of Schihallien, and in the granite of
Rockall.

The circumstances of the case are certainly more favourable to chemical action in the vapour baths of Ischia, as the
aqueous particles are presented to the silex at a high temperature, and in a minute state of division, at the moment of
their deposition from the state of vapour.

* Strabo. Lib. v. He tells us that Ischia was torn by some convulsion of
nature from the main-land, but this is not probable.

† Geolog. Trans. vol. 2. p. 392.

STROMBOLI.

I need not insist upon the analogy between the *above* phænomena, and those presented at the Geysers in Iceland, but I may remark that similar concretions are noticed as occurring among the volcanic rocks of the Solfatara and of Santa Fiora in Tuscany; by Humboldt at Teneriffe; and by Von Buch in Lanzerote. In these cases the alkali, which in the Iceland Geysers is supposed to assist in dissolving the silex, does not appear to be present.

The Lipari Islands, between Naples and Sicily, are made up of a class of volcanic products, very analogous to those we have been just considering.

Like the neighbourhood of Naples too, they allow us to compare the phænomena of active and half-extinguished volcanos, with those which have arisen from the same cause at earlier periods.

The Island of Stromboli consists of a single conical mountain, having on one side of it several small craters, one of which is in a state of activity, the rest extinct. The volcano is in this respect peculiar, that although it rarely has its periods of intense energy, yet it as rarely enjoys any intervals of repose, no cessation having ever been noticed in its operations, which are described by writers antecedent to the Christian æra, in terms which would be well adapted to its present appearances.*

---

* Callias, a cotemporary of Agathocles, who made himself absolute in Sicily from B. C. 317 to 289, is quoted by the Scholiast to Apoll. Rh. III. 41, as mentioniug the fires af Lipari, meaning probably Stromboli, and it is probable that Theocritus may refer to the same island, when he says—

Εϱωs δ'αϱα καὶ Λιπαϱοιο
Πολλακιs Αφαιϛοιο σελαs φλογεϱωτεϱον αιθει.

As likewise Aristotle, where he states that in Lipari the fire, after a cessation of 16 years, returned on the 17th. Stromboli (Στϱογγυλη) is mentioned by name by Diodorus Siculus, as giving rise to explosions of air and to ejections of sand and heated stones. The same, he says, is the case in Hiera, now Volcano. Strabo, Lib. 6, speaks of Lipari, Volcano, and Stromboli, as emitting flames, and says that those in the latter island are inferior to the two former in point of violence, but superior in brightness. Cornelius Severus, on the other hand, speaks of the volcano in Stromboli as greater than that of Etna.

Fæcundior Etna
Insula, cui nomen facies dedit ipsa *rotundæ*.

STROMBOLI.

These consist in ejections, repeated at very short intervals, of stones, scoriæ, and ashes, which either fall back within the crater, or are carried in one or the other direction, according to the drift of the wind. As however the crater is placed on the slope of the precipice, and not upon its summit, the ejected matters tend but little to increase the accumulation of substances in its immediate neighbourhood, but are for the most part carried into the sea.

I reached with considerable difficulty the summit of the mountain, which rises at an angle often of nearly 40°, and is covered completely with volcanic sand, consisting of titaniferous iron, amongst which I found numerous crystals of augite, and masses of black pumice, or of an highly scoriform and fibrous description of lava, which seems to approach nearly to that mineral.

On looking down from that elevation upon the volcano, I perceived that its minor explosions were in general almost continuous, but that the greater ones, which alone were audible below, take place at intervals of about seven minutes. The latter were sufficiently terrific to give me an idea of what takes place during an eruption of Etna or Vesuvius, but as the wind did not blow the stones in our direction, we should have incurred no considerable risk in approaching it nearer. On expressing however this wish to my guides, I was reminded, by their refusing to accompany me, of the remark which Spallanzani makes in respect to the superstitious horror entertained in his time by the Liparotes of the crater of Volcano, which obliged him to procure a Calabrian for his attendant ; and finding that no one would venture to accompany me nearer, I thought it prudent to abandon the attempt.

* Humboldt remarks (Pers. Narr. vol. i. p. 226.) that he has seen black pumice stones, in which augite and hornblende are easily recognized : they are less light, of a spongy texture, and rather cellular than fibrous ; we might be tempted to think, that these substances owe their origin to basaltic lavas. He has observed them in the volcano of Pichincha, as well as in the tufa of Pausilippo, near Naples.

I contented myself therefore with visiting those parts of
the island which were not in the immediate vicinity of the
Volcano, and found them chiefly composed of a tuff, in some
places not unlike that which I have described as occurring
in Ischia. In one place the cavities are lined with specular
iron in very minute laminæ.

This tuff is in some places penetrated by dykes of a
cellular description of rock, which approaches in its mineralogical characters to trachyte.

These dykes often pursue so regular and horizontal a
course through the tuff, that they may be readily mistaken
for beds alternating with that rock, but when we trace them
to any distance, the deviations that occur from their original
direction, and their origin from two or three roots, that rise
vertically from below, sufficiently betray their real nature.

Tuff penetrated by Dykes of Slaggy Lava—Stromboli.

No. 1.

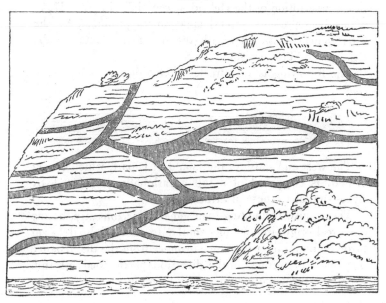

Tuff penetrated by Dykes of Slaggy Lava—Stromboli.
No. 2.

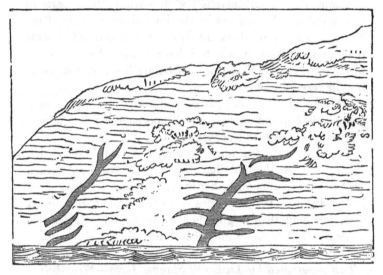

Dr. Macculloch has noticed some dykes of a similar kind in his description of the Hebrides, especially in plate 17, in which he represents several appearances produced by the interference of the trap with the secondary strata on the east coast of Trotternish.

The following is a contracted sketch of one of his drawings :

The west and south of the Island of Lipari, which I next visited, is composed of a tuff occasionally penetrated by dykes similar to those of Stromboli.

These, notwithstanding their general horizontality, betray their real origin in a manner similar to that of the dykes at Stromboli, and likewise in one instance from the disturbance they have occasioned in the direction of the beds of tuff, which are thrown up obliquely by the dyke.

Section of the Tuff and Slaggy Lava in a place called Vulcanello, Isle of Lipari.

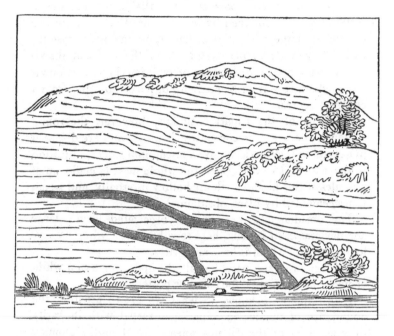

The tuff in the west and south of Lipari contains imbedded fragments of obsidian and pumice; but on the south the whole surface of the country is covered with the latter substance, which forms several considerable hills, and extends to the furthest point of the island.

Dolomieu* has remarked, that this pumice seems to be derived from the fusion of granite, for not only did he observe in the midst of this substance fragments consisting of quartz, mica, and felspar, but when such fragments were exposed to heat, they were converted into a substance, resembling in its general characters the pumice surrounding it.

---

* Dolomieu sur les Isles de Lipari, p. 67.

The pumice of Lipari is found to rest on a bed either of obsidian, or of a semivitreous substance nearly allied to it. Some of the varieties possess a remarkable resemblance to certain products obtained by Mr. Gregory Watt* during the cooling of large quantities of basalt, an incipient crystallization beginning to manifest itself in the midst of the vitreous mass by the appearance of white or lighter coloured spots, which appear to be made up of parts radiating from a common centre.

In many of the Lipari obsidians, however, the round spots are composed of concentric laminæ, and are disposed in general in lines, so as to give a resemblance of stratification to the mass. In other cases the whole mass is made up of globules of this kind, which are hollow internally, and sometimes cemented by black obsidian.

The obsidian also occurs in a brecciated form; large angular masses of it being held together by a white earthy-looking paste, which is hard and gritty.

In other cases the paste has all the appearance of a white enamel, such as is used for china or pottery.

It is difficult to know, to what we are to attribute the abundance of pumice in Lipari contrasted with its rarity in other parts of the Mediterranean, for though Dolomieu is disposed to attribute it to the kind of material on which the fire had operated, I am more inclined to the opinion of Humboldt, who considers pumice rather a particular state into which many minerals may be brought, than a separate species; and am therefore obliged to look to the different modes in which the heat was applied for an explanation of the fact.

For the formation of pumice it seems requisite, that a considerable disengagement of vapour should have taken place, during the time at which the body acted upon was in a plastic, though not in an altogether fluid, condition.

But why this should have occurred at Lipari and among

* See Phil. Trans. for 1804.

the older volcanic rocks near Naples, rather than at Etna and Vesuvius, still remains unaccounted for.

It is also difficult to explain why vitreous lavas, such as obsidians, should be of such common occurrence in Lipari,* for the latter are not, as we should be disposed to consider them, loose ejected masses, which from the great relative extent of their surface, were soon coolled in the atmosphere, and therefore put on a vitreous form, but they constitute extensive beds, which ought, it should seem, to have been subjected to the same laws of congelation as the lavas of other volcanos.

The only indications of volcanic action at present existing in Lipari are the hot springs, situated about four miles west of the town. I observed however a rock, which, from the changes it had undergone, seems evidently to have been acted upon by sulphureous exhalations, so that it is probable, that at no very remote period, some of the less equivocal effects of subterranean fire may here have manifested themselves. The antients indeed speak of Lipari as emitting a fiercer fire than Stromboli,† and Strabo particularly mentions an eruption of mud, attended with smoke and flame, which took place in the sea between Hiera (the Island of Volcano, near Lipari), and Euonymos, now (according to Cluverius) Lisca Bianca.‡

---

* Is the formation of obsidian at all connected with the presence of boracic acid? This I believe has been conjectured by some Swiss naturalist, but the experiments of Dr. Turner, detailed in the Edinburgh Journal of Science for January, 1826, seem to discountenance such a notion, as he finds that neither the pumice nor obsidian of Lipari appear to contain that substance.

† Τὴν δὲ Λιπαραν καὶ τὴν Θερμισσαν εἰρηκαμεν. ἡ δὲ Στρογγυλη, καλειται μεν απο του σχηματος, εστι δε και αυτη διαπυρος, βια μεν φλογος λειπομενη, τω δε φεγγει πλεονεκτουσα. Strabo, lib. vi.

‡ May not the comparatively recent origin of the Isle of Lipari be inferred from the sterility ascribed by Cicero to the country; (see Orat. 3. in Verrem) for as it is at present very fertile, its barrenness may have arisen from the circumstance, that sufficient time had not elapsed to cause a suitable decomposition of the masses ejected.

Before this occurred, the sea between these two islands rose to an unusual height, and became so warm, that the fish died in numbers sufficient to taint the air.* Strabo indeed adds, that flames have been observed rising from the sea in the neighbourhood of those islands, and Pliny notices the same as happening for several days during the Social War.†

Homer seems to allude to something of the same kind in the 12th book of the Odyssey, where Circe relates to Ulysses the dangers he is to undergo near the coast of Sicily;‡ and even the epithet of " floating," (πλωτη) which he has applied to the island in which king Eolus reigned, may be supposed to refer to the earthquakes,‖ with which the country was agitated at a time when the volcanic operations were in greater activity.

This may seem far-fetched in a country like our own, happily but little subject to these convulsions of nature; but to an American it might appear an obvious allusion, as Humboldt remarks, that on the coast of Peru earthquakes are so frequent, that we become as much accustomed to the undulations of the ground, as the sailor is to the tossings of

* Julius Obsequens, who evidently alludes to the same event, places it in the Consulship of Æmilius Lepidus and Aurelius Orestes, or 125 years before Christ.

† Near a century before Christ.

‡ Especially in the following lines:

Αλλα θ'ομя πιναχας τε νεων και σωματα φωτων
Κυμαθ' αλος φορεнσι, πυρος τ'ολοοιο θνελλαι.

μ. 68.

‖ Αιολιην δ'ες νησον αφικομεθ'; ενθα δ'εναιεν
Αιολος Ιπποταδης, φιλος αθανατοισι θεοισι
Πλωτη ενι νησω. Odyssey K.

VOLCANO.

the ship, caused by the motion of the waves.* As however
there are no vestiges of any thing like a crater in this island,
it is probable that these and similar phænomena were derived for the most part from the Island of Volcano, which
is separated from it only by a narrow channel.†

This, which appears at a period antecedent to the Christian era to have been in a state of activity at least equal to
that of Stromboli, still emits gaseous exhalations from the
interior, as well as from several parts of the external surface,
of a crater situated on the highest part of the island.

These vapours, acting upon the rock they penetrate,
decompose it, and form with its constituents large quantities
of alum and other sulphuric salts.

---

* From our infancy, the idea of certain contrasts fixes itself in our minds ;
water appears to us an element that moves ; earth, a motionless and inert mass.
These ideas are the effect of daily experience; they are connected with every
thing that is transmitted [to us by the senses. When a shock is felt, when the
earth is shaken on its old foundations, which we had deemed so stable, one
instant is sufficient to destroy long illusions. It is like awakening from a
dream; but a painful awakening. We feel, that we have been deceived by
the apparent calm of nature ; we become attentive to the least noise, we mis-
trust for the first time a soil, on which we had so long placed our feet with
confidence. If the shocks be repeated, if they become frequent during successive days, the uncertainty quickly disappears. In 1784 the inhabitants of
Mexico were accustomed to hear the thunder roll beneath their feet, as we are
to witness it in the region of the clouds. Confidence easily springs up in the
human breast, and we end [by accustoming ourselves on the coast of Peru to
the undulations of the ground, like the sailors to the tossings of the ship, caused
by the motion of the waves. Humboldt's Pers. Narrative, vol. 3. p. 321.

† We may collect from the old chronicles, that the last indication of volcanic
agency in Lipari took place about the sixth century, for we are told that St.
Calogero, the patron of the island, put to flight the Devils, which, like the
Typhon of old, inhabited the recesses of the island, and that the latter first took
refuge under the mountain from whence the warm springs issue, but being
driven from thence repaired to Vulcanello, and finally were chased into the
crater of Vulcano. Later writers always speak of the flames of Lipari as
extinct. See Dolomieu sur les Iles de Lipari, p. 71.

VOLCANO.

The process is somewhat different from that which takes place at the Solfatara, since the vapour *here* consists of sulphurous acid, *there* of sulphuretted hydrogen, as I ascertained by procuring portions of the condensed vapour from either source.

The ultimate result however in both instances is the same, the compounds arising from the union of this acid with the earths contained in the rock, with which it comes into contact, being all ultimately resolved into sulphuric salts.

There is one product however that seems peculiar to this volcano, or at least has not been found belonging to any other in the south of Italy. I mean the boracic acid, which lines the sides of the cavities in beautiful white silky crystals, and combined, it is said, with ammonia.

Ammonia is likewise found united with muriatic acid, or in the form of sal ammoniac; its production I have attempted to explain when speaking of its occurrence under similar circumstances at the Solfatara.

In a mixture of this salt with sulphur, Stromeyer has lately detected the presence of selenium.* It is probably sublimed in combination with hydrogen, just as the sulphur and arsenic which accompanies it are supposed to be.

The operations of this volcano appear to be going on with much greater vigour than those of the Solfatara, and exhibit perhaps the nearest approximation to a state of activity, during which a descent into the crater would have been practicable.

Nor can I imagine a spectacle of more solemn grandeur than that presented in its interior, or conceive a spot better calculated to excite in a superstitious age that religious awe

---

* See, for a full account of this discovery, the Göttingen Gelehrte Anzeigen for February, 1825. The English Journals contain an abridgment.

VULCANO.

which caused the island to be considered sacred to Vulcan, and the various caverns below as the peculiar residence of the God.

> Quam subter, specus, et Cyclopum exesa caminis
> Antra Etnea tonant, validique incudibus ictus
> Auditi referunt gemitum, striduntque cavernis
> Stricturæ Chalybum, et fornacibus ignis anhelat,
> Vulcani domus et Vulcania nomine tellus.

To me, I confess, the united effect of the silence and solitude of the spot, the depth of the internal cavity, its precipitous and overhanging sides, and the dense sulphureous smoke, which, issuing from all the crevices, throws a gloom over every object, proved more impressive than the view of the reiterated explosions of Stromboli, contemplated from a distance, and in open day.

Close to Vulcano, is an isolated rock called Vulcanello, which, though without a crater, emits from its crevices vapours of a sulphureous nature, a feeble remnant of the volcanic action, by which it was formerly itself thrown up from the bosom of the sea. It is probable, at least, that it is to this event that Aristotle* refers, when he states that in the Island of Volcano part of the ground swelled up, and rose with a noise into the form of an hillock, which burst and gave vent to a great quantity of air, carrying along with it flame and ashes, the latter in sufficient quantity to cover all the town of Lipari.

The time at which this event happened, seems to be fixed by other writers, for Pliny† mentions an island which

---

* Vide Aristot. περι Μετεωρων. Lib. 2. c. viii.

† Ante nos et juxta Italiam inter Æolias insulas, item juxta Cretam emersit è mari M. M. D. passuum cum calidis fontibus, altera Olympiadis CLXIII anno tertio in Tusco sinu; flagrans hæc violento cum flatu. Pliny, lib. 2. c. 87.

VULCANO.

emerged from the sea among the Lipari Islands, in the 144th
Olympiad, or about 200 years before Christ, whilst Orosius
fixes the event as happening in the Consulship of Marcellus
and Labeo, which answers to the year 182 B. C.*

Whichever of these dates be preferred, it is equally clear
that the island now called Volcano cannot have been
referred to, for Thucydides, who flourished at least four
hundred years before Christ, mentions Hiera (Vulcano) as
being *cultivated,* which implies that a certain time had
elapsed since its production.† It is probable therefore, that
the rock which Aristotle states as having been thrown up
from the sea, was that now called Vulcanello, which lies at
a short distance from Vulcano.

The Lipari Islands are so placed with reference to Naples
and Sicily, that they seem to form a link between the two
countries, whence some have inferred that a subterranean
communication passes through them, extending from Etna
to Vesuvius. It would be necessary however to shew in a
more satisfactory manner than has hitherto been done, that
the condition of any one of these volcanos is influenced by
that of the rest, before we venture to adopt any such opinion;‡ at present I shall content myself with pointing out
by a detail of the structure of Sicily, so far as the latter is
connected with the present subject, what degree of resem-

* Orosius, lib. 4, c. 20.

† Και οι μεν εν Σικελια Αθηναιοι και Ρηγινοι, του αυτου χειμωνος,
τριακοντα ναυσι στρατευουσιν επι τας Αιολου νησους καλουμενους. νεμονται
δε Λιπαραιοι αυτας, Κνιδιων αποικοι οντες, οικουσι δεν μια των νησων ου
μεγαλη, καλειται δε Λιπαρα. τας δε αλλας εκ ταυτης ορμωμενοι γεωργουσι,
Διδυμην, και Στρογγυλην, και Ιεραν. νομιζουσι δε οι εκεινη ανθρωποι, εν τη
Ιερα ως Ηφαιστος καλκευει, οτι την νυκτα φαινεται πυρ αναδιδουσα πολυ, και
την ημεραν καπνον. Thuc, lib. 3. c 88·

‡ See the Chronological Table of the eruptions of Etna and Vesuvius, ap-
pended to this Lecture.

blance may subsist between the volcanic phænomena of that island and those of Naples.*

Nearly all the central portion of Sicily is occupied by a vast deposit of blue clay or marl, in which are contained numerous and thick beds of selenite and gypsum, of common salt, of sulphur, of combinations of that mineral with iron and copper, and of the sulphuric acid with most of the earthy bases. The crystals of sulphate of strontian found in the sulphur mines are unrivalled for their beauty; and are intermixed with those of sulphur, which occur lining the fissures, often in large and regular octaedra. The latter mineral is always of a bright yellow, and never of that liver colour which occurs in some other localities, a distinction worth attending to, as Brocchi[†] conceives that all sulphurs of the former description have undergone sublimation. The sulphate of lime occurs lining the fissures in beautiful and regular crystals, but, when it is met with in beds, it is frequently found in large transparent plates, nearly a foot in length, and six or eight inches in breadth, which seem to be the result of an irregular crystallization.

This blue clay deposit is associated with beds of white calcareous marl, and of a calcareous conglomerate which becomes more abundant as we proceed south, and is there accompanied likewise with beds of limestone without any brecciated structure.

The formation, it may also be remarked, contains all the substances that are at present sublimed from volcanos in activity. The sulphur, the various sulphuric salts, the muriate of soda, are products found equally at Etna, at Vesuvius, and at the Solfatara, so that if it were once established that the rock itself rested upon a volcanic formation, we might suppose its imbedded minerals to have been produced, like those in the craters of existing volcanos, by exhalations from beneath.

* See, for a more detailed account of the Geology of Sicily, my Memoir inserted in Jameson's Journal, vol. xiii.

† Conchiol. sul app. p. 67.

SICILY.

This too would account for the absence of shells, so common among tertiary rocks in general; for we know from the case of the Lago di Solfatara near Rome, that molluscous animals will not live in sulphureous springs.

It would be premature however to theorize about the origin of this formation, until its position be fully ascertained; and I must confess that I have not been able to satisfy myself with respect to the stratum on which it is incumbent.

There is one phænomenon connected with the natural history of the blue clay formation, which appears to be of pretty common occurrence, for it has been observed in several parts of Sicily, and likewise in the analogous rock that occurs at the foot of the Appennines.* From its supposed resemblance to the eruption of a burning mountain, it has obtained the name of a Mud or Air Volcano, though perhaps it has no more right to such a name, than the class of products called pseudo-volcanic have to an appellation, which places them in connection with the eruptions of a genuine volcano. I examined the most remarkable case of this kind, which occurs at the hill of Macaluba,† near Girgenti, in a country fully charged with sulphur and other inflammable minerals. Having reached the summit of the hill, I found the surface covered with dry clay, in which were a number of small crater-shaped cavities filled with water, mixed with mud and bitumen, somewhat above the natural temperature, and disengaging from time to time bubbles of gas, which I ascertained to consist of carbonic acid and carburetted hydrogen gases. These little craters,

---

* As near Modena.

† Plato in his Phædon, probably alludes to this place, when he speaks of " the torrent of mud which is in Sicily." Ferr. C. Phleg. 47. Von Hoff points out a curious analogy of name between this spot and one near the Dead Sea, where a similar phænomenon takes place, and which is called Ardh al Maclubah. Gesch. der Verand. p. 247.

when I visited the spot, were in a state of comparative quiescence, but it is said that at times the process goes on with considerable energy, for the mud has been known to rise to the height of 200 feet, accompanied with a strong odour of sulphur.

I cannot help imagining that the whole of these phænomena may be explained by the slow combustion of beds of sulphur, which is fully ascertained to be going on in many parts of the blue clay formation.

It is not long since the proprietor of some land in the interior congratulated himself on his good fortune, in being able to collect a large supply of sulphur already purified, by merely placing vessels to receive a stream of that substance, which was constantly issuing from the side of a hill. This was occasioned by a bed of sulphur in the interior of the mountain having caught fire, and the heat generated by the combustion of one portion serving to melt the remainder : nature having, in this instance, adopted the wasteful process employed from time immemorial by the Sicilians, for getting rid of the intermixed clay, which consists simply in collecting the materials into large heaps, and setting fire to them on the surface, thus causing the liquefaction of one portion by the combustion of another.

The sulphurous acid resulting from this process being retained by the moisture of the rock, and gradually converted into sulphuric acid, would act upon the calcareous particles, and give rise to the extrication of carbonic acid gas, whilst, if any bituminous matters were present, the heat generated might cause a slow decomposition, and resolve them into petroleum and carburetted hydrogen.

A continued stream of gas passing through the rock, would soon establish for itself a regular communication with the surface, and the same channel, when once formed, would afford the readiest means of escape for any water, which, from its existing at an higher level on the adjacent hills, might have a tendency to make its way upwards.

This water would carry with it the petroleum which resulted from the distillation of the bituminous matter, and would fill, in the manner which we now observe it to do, the little crater-shaped cavities caused originally by the escape of the gases.

In short the rise of the water to the summit of the argillaceous hill of Macaluba seems to depend upon precisely the same principles as the common rise of springs, when we bore through a bed of clay, and penetrate into a porous stratum underneath, saturated with humidity, and having its outgoings at an higher level than that of the upper surface of the aperture. In the case of the hill of Macaluba, the escape of aeriform fluids has done what art effects in the case of a well, and the position of its summit, overtopped, as it is, by the adjacent eminences, is such, that we may without difficulty suppose the relative height of the springs in that neighbourhood to be such as my hypothesis requires.

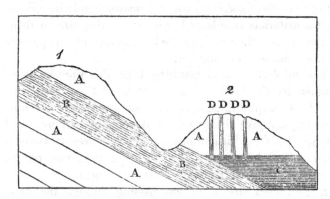

Thus let 2. be the hill of Macaluba, and 1. be a neighbouring hill of greater height. Let A. A. A. represent the bed of blue clay, and C. that portion of it in which the gaseous exhalations originate, having a vent through the channels D. D. D. D. Now it is evident that if the stratum B. B. be water-logged, a continual stream will pass upwards through the above channels, until the level of the springs be below the summit of the hill of Macaluba.

I confess that this plan is hypothetical, but as such alternations of argillaceous with porous beds occur everywhere in this formation, it is by no means improbable, that the above may be a true representation of the structure of the country.

Similar phænomena occur also at Terrapilata, near Caltanisetta,* and at Misterbianco, near Catania, and I should be disposed to view in the same light the extrications of unrespirable gas and of petroleum, which escape from a lake near Palagonia, about 20 miles west of Catania. This lake is situated in the midst of volcanic rocks, but the latter are not derived from modern eruptions, or accompanied, in their immediate neighbourhood at least, by any other indications of present volcanic action.

The gases given out from this lake appear, from Ferrara's report,† to be principally carbonic acid and carburetted hydrogen, which are the same as those of Macaluba; and when I visited the place, the water was in a state of violent ebullition in several places, but especially from two spots near the centre of the pool. The smell of Naptha, from which it derives its name of Lago Naftia, is perceived at a great distance, and this substance is seen floating on the surface. As at the Lacus Amsanti, the gas given out proves speedily fatal to animals that are thrown into the lake, but

---

* See Bulletin des Sciences for August, 1825, p. 406.

† Ferrara, Memoria sopra il Lago Naftia. Palermo, 1805.

LAGO NAFTIA.

as its great specific gravity prevents its rising to a height of
more than three feet, it is possible not only to stand on its
borders, but even to approach the spots from whence the
disengagement of gas takes place with the greatest vehemence.

The singular qualities possessed by the exhalations given
out from this spot rendered it at a very early period an
object of popular veneration, and we may perhaps recognize
in the fable attached to it some traces of its volcanic origin.
It was called the Fons, or Stagnum Palicorum, from two
sons of Jupiter, by the nymph Thalia, the daughter of
Vulcan,* who was concealed by the God from the ven-
geance of Juno, by being buried under ground, so that
when the time of her delivery was come, the earth opened
and brought into the world her two children, hence called
Palici ("απο τυ παλιν ικεσθαι") because they returned into the
world after being buried under it. This fable may perhaps
allude to the first origin of the gaseous exhalations from two
apertures, and the worship paid to these Deities, the human
sacrifices at first offered up, the temple built on the spot,
and the oracle that was consulted in the sanctuary, shew the
fear which had been inspired by the noxious qualities of
the vapour exhaled.

The description given by Ovid† of the Stagnum Palicorum so well corresponds with the Lago Naftia, near
Palagonia, that it seems most probable that he refers to this
very spot, though Virgil fixes the former rather nearer the
River Simethus, than is consistent with its actual position.‡

---

* See Diodorus Siculus, lib. xi. Stephanus in Epit. Servius, note on
Æneid, 9. v. 585, &c.

> † Perque lacus altos et olentia sulfure fertur
> Stagna Palicorum, ruptâ ferventia terrâ.
>
> METAMORPH. Lib. v.

> ‡ Eductum Matris luco, Symæthia circum
> Flumina: pinguis ubi et placabilis ara Palici.
>
> ÆNEID. Lib. ix. v. 584.

VAL DI NOTO.

South of Alicata we lose sight of the blue clay, and find ourselves upon a rock of a calcareous nature, sometimes homogeneous in structure, at others made up of limestone pebbles, imbedded in a calcareous basis. These beds continue as far as Cape Passero, the most southern point of the island, where they are seen to rest upon a volcanic tuff, containing fragments of a cellular lava.

It became therefore of great importance to determine by the petrifactions the age of this limestone. At Cape Passero the calcareous rock contains madrepores, nummulites, melonites, cypreæ, and that curious fossil called the hippurite, which has been described in an early volume of the Geological Transactions.

Now, although this latter petrifaction has been found in the Island of Aix, which the French geologists have set down, perhaps without sufficient reason, as belonging to the chalk formation, yet on the other hand, the remaining shells are such as plainly indicate a more recent origin. Admitting however that the phænomena presented by this single stratum are such as lead to no certain conclusion, the same cannot be said of the other calcareous beds belonging to the same series, which alternate with volcanic rocks between Noto and Lentini.

Among these I have found more unequivocal proofs of a recent origin, judging both from the characters of the stone itself, and from the shells which it so abundantly contains.

With this view of their origin, the cellular texture of the alternating beds of tuff and lava completely accords, and the tenor of the observations is altogether such, as enables us without hesitation to refer the whole to causes acting at nearly the same period, and under similar circumstances with those to which I have already ventured to attribute the rocks of Hungary, Styria, and the Vicentin.

It is almost needless to remark, that in a volcanic formation of such an age no craters can exist, although I found

such described by recent and respectable authorities. I
purposely visited the Monte Vennera, south of Lentini,
where there are said to be traces of a crater, and was re-
warded, not by the discovery of what I was in search of, but
by the opportunity which my excursion gave me of obtaining a good section of the volcanic and calcareous strata between that spot and Lentini.

From the antient volcanic rocks, let us now proceed to an
examination of the more modern. Although the greater
part of the country, included within the circumference of
Mount Etna, would appear to belong to a comparatively
recent epoch, yet there are some rocks in its vicinity that
were probably formed antecedently to the mountain at whose
foot they lie. I allude particularly to the Cyclopean Islands, with which every classical reader is acquainted, as
the rocks which Polyphemus is described by Homer as
hurling against the Bark, in which Ulysses and his crew
were taking their flight. These, though now detached,
must at one period have formed a connected stratum, for
they are covered with a bed of marl, which seems evidently
to have been continuous from the one to the other of these
islands. This circumstance, and their general compactness,
prove that these formations took place under the surface of
water.

The same remark will probably apply to the rock of
Castello d'Aci, on the coast near Catania. It consists of a
volcanic breccia, the cementing substance of a sandy nature;
the nodules, a cellular kind of lava. The latter however
are not rounded masses, but result from a sort of irregular
crystallization, most of them possessing a radiated structure,
so that they resemble a cluster of prisms meeting in a common centre. The above stellular arrangement is the most
common, but in other cases the prisms have more of a fan-
shaped structure; and in both instances, the point towards

MOUNT ETNA.

which they converge, as well as the interstices between them, consist of tuff.

Nothing of this kind is indicated by the structure of Etna. This mighty and imposing mountain, which rises in solitary grandeur to the height of above 10,000 feet, and embraces a circumference of 180 miles, is entirely composed of lavas, which, whatever subordinate differences may exist between them, all possess the appearance of having been ejected above the surface of water, and not under pressure.

In the structure of this mountain, every thing wears alike the character of vastness. The products of the eruptions of Vesuvius may be said almost to sink into insignificance, when compared with these coulées, some of which* are four or five miles in breadth, 15 in length, and from 50 to 100 feet in thickness, and the changes made on the coast by them is so considerable, that the natural boundaries between the sea and land seem almost to depend upon the movements of the volcano.

The height too of Etna is so great, that the lava frequently finds less resistance in piercing the flanks of the mountain, than in rising to its summit, and has in this manner formed a number of minor cones, many of which possess their respective craters, and have given rise to considerable streams of lava.

Hence an antient poet has very happily termed this volcano the Parent of Sicilian Mountains,† an expression strictly applicable to the relation which it bears to the hills in its immediate neighbourhood, all of which have been formed by successive ejections of matter from its interior.

The grandest and most original feature indeed in the physiognomy of Etna, is the zone of subordinate volcanic hills with which it is encompassed, and which look like a court of subaltern princes waiting upon their sovereign.

---

* Such, according to Ferrara, (Desc. dell Etna, p. 200) is the case with that of 1669, which Borelli calculated as containing 93,838,950 cubic feet.

† Σικελων ορεων ματηρ.' Euripides in Troadibus.

MOUNT ETNA.

Of these, some are covered with vegetation, others are bare and arid, their relative antiquity being probably denoted by the progress vegetation has made upon their surface, and the extraordinary difference that exists in this respect seems to indicate, that the mountain, to which they owe their origin, must have been in a state of activity, if not at a period antecedent to the commencement of the present order of things, at least at a distance of time exceedingly remote.

It must be remarked however, that the time which it takes to bring a volcanic mountain or a stream of lava into cultivation is very variable,* and that the progress is gene-

---

* This will appear from the following statement of the condition of a few of the lavas of Vesuvius, which I examined with reference to this question in 1823.

Lava of 1551.—Fossa de Gaetano. Much decomposed. Heaths grow upon it, and vines begin to be planted.

——— 1737.—But little decomposed. Moss alone grows on it.

——— 1760.—Near the hill of the Camalduli. Still unfit for vegetation. Surface however whitened and crumbly, owing to decomposition, which has proceeded farther than in that of 1737.

——— 1771.—Colour grey, moss grows upon it, but no heath.

——— 1785.—Fossa di Sventurato. Lava still quite hard and rough.

——— 1794.—Fossa di Cucazzello. Surface much decomposed, moss grows upon it, and a few heaths, but no trees or shrubs. It is to be observed, that even the latter are met with on the surface of the crater from which this lava flowed, and which was formed by heaps of scoriæ ejected at the same time. A proof of what I have asserted in the text with respect to the more rapid decomposition of loose ashes than of a bed of lava.

——— 1805.—Fossa del Noce. Colour very white; no moss appears to grow upon it, but, being covered with the loose scoriæ of later eruptions, it has trees growing upon it in a few parts.

——— 1810.—Colour grey, surface rough, though somewhat decomposed; moss grows upon it, but no heaths or trees are seen, except in one part where it is covered with cinders.

——— 1822.—Colour black, surface very rough and irregular, no moss as yet to be seen.

It will be seen that many of these lavas are in a more forward state than that of Ischia, which flowed in 1302, more than 200 years before.

rally more rapid in a cone composed of finely comminuted cinders, than in a stream of lava, which consists of an hard glossy substance, that yields but slowly to the causes of decomposition.

There is nothing I believe in the nature of lava, chemically speaking, injurious to vegetation; but mechanically the hard surface it presents proves an effectual preventative, as it gives no support or nidus to the tender shoots, and from its porosity often carries off all the moisture that descends upon its surface. Thus near Clermont, in Auvergne, scarcely a drop of water is to be found in the whole of the volcanic district which overhangs the town, but the waters collected at the bottom of the formation gush out in the valley of Royat beneath with such force and copiousness, as to turn immediately several mills.

From these causes the surface of a stream of lava must always require a long period to bring it into cultivation, unless as has been done by the Prince of Biscari in his garden at Catania, the surface is covered with artificial soil, or what comes to the same thing, a shower of volcanic ashes has overspread it of a nature sufficiently argillaceous to imbibe moisture, and to form a sort of mould.

Under these circumstances I naturally felt a desire to verify an observation reported by Brydone on the authority of the Canon Recupero, which might render us suspicious of the correctness of our received chronologies. This writer after giving an instance of a lava, the date of which he says goes back to the time of the second Punic war,[*] proceeds to state that at Aci Reale [†] we see seven such beds super-

---

[*] The real date of this current was more antient. It was that of the reign of the elder Dionysius tyrant of Syracuse.— Vide infra.

[†] The following is the passage to which I refer:—Near to a vault, which is now thirty feet below ground, and has probably been a burial place, there is a draw-well, where there are several strata of lavas, with earth to a considerable thickness over the surface of each stratum. Recupero has made use of this as an argument to prove the great antiquity of the eruptions of the volcano. For if it requires two thousand years, or upwards, to form but a

imposed one on the other, each of which has its surface throroughly decomposed, and converted into rich vegetable mould.

Now if a single bed of lava has continued for more than 2000 years without experiencing this alteration, what a lapse of time must it have required to reduce seven successive beds of the same material into a state of such complete decomposition.

Although I have no reason to doubt, that Brydone received from Recupero the observation on which he grounds his inferences, I think it most probable that the conclusion itself was in reality his own, though he perhaps thought it would sound more *piquant*, if put into the mouth of the Canon, whose scientific knowledge he seems willing to exalt at the expense of his orthodoxy. In reality however this good priest appears to have enjoyed in both respects a reputation which he very little deserved; the reports of Dolomieu* and other really scientific travellers make him out to have been a man of but slender philosophical attainments, but as one who at least was free from all imputation of scepticism. It is curious nevertheless, that another foreigner has stated, as an instance of the intolerant spirit prevailing

scanty soil on the surface of a lava, there must have been more than that space of time betwixt each of the eruptions which have formed these strata.

But what shall we say of a pit they sunk near to Jaci, of a great depth. They pierced through seven distinct lavas, one under the other, the surfaces of which were parallel, and most of them covered with a thick bed of rich earth. "Now," says he, "the eruption which formed the lowest of these lavas, if we may be allowed to reason from analogy, must have flowed from the mountain at least 14,000 years ago. Recupero tells me he is exceedingly embarrassed, by these discoveries, in writing the history of the mountain; that Moses hangs like a dead weight upon him, and blunts all his zeal for inquiry, for he really has not the conscience to make his mountain so young as that prophet makes the world. The bishop, who is strenuously orthodox—for its an excellent see—has already warned him to be upon his guard, and not pretend to be a better historian than Moses; nor to presume to urge any thing that may, in the smallest degree, be deemed contradictory to his sacred authority."

Brydone's Tour in Sicily. Vol. I. p. 140.

* See Dolomieu sur les Iles Ponces. p. 470.

in the country in which he lived, that the poor Abbe was
thrown into prison for his religious opinions, although the
truth appears to have been, that the reports circulated in his
favour by Brydone, Borch, and others, induced the Neapo-
litan government to grant him a pension on the score of his
scientific deserts. Indeed the only annoyance, it is said, he
ever experienced in consequence of his imagined discovery,
was the being informed that certain foreigners to whom he
communicated his observation, not content with wresting it
to a purpose of which he had never dreamt, had given him
credit for the inferences, which they had chosen to deduce
from it themselves.

The fact nevertheless reported by Brydone obtained a
currency proportionate to the popularity which his work
enjoyed, and the heterodox conclusion excited at the
time no slight degree of consternation among divines. It
was generally combated, by remarking the great variableness
as to the period which a bed of lava will take to undergo
decomposition, and even Spallanzani, though he visited
Sicily, seems to have contented himself with pointing out
instances, in which newer beds of lava have taken the start
of older ones in their progress towards cultivation.*

I was therefore not a little surprised, when on visiting the
celebrated spot of the Abbe's observation, I found that the
beds of vegetable mould, which proved according to Brydone the degree to which the decomposition of the lava had
extended, were in reality nothing more nor less than beds of
a ferruginous tuff, formed probably at the very period of the
flowing of the lava, and originating perhaps from a shower
of ashes that immediately succeeded its eruption. It is true
that the cliff, which exhibits a section of these lava beds
with interposed tuff, shews also the greater facility with
which the latter has yielded to the action of the elements,
as the bare and mural precipices presented by the lava are
in contrast with the gentler slope of the beds of tuff, which

* Voyage dans les deux Siciles. Vol. I. l. 7.

afford a soil sufficient for the hardy cactus, and in some places even for the vine. Still there is not the slightest evidence that the decomposition exists internally, or that it had taken place in any one instance before the superincumbent bed of lava was deposited.

Even had the tuff in question been in reality *vegetable mould,* the validity of Mr. Brydone's conclusion might very easily be disputed, for I think it cannot be shewn that any one of the beds, of which the cliff of Jaci Reale exposes a section, are of post-diluvial origin; so abrupt and lofty a face of rock would hardly have been cut by processes now in operation, but may be attributed with more probability to the cause which last reduced our continents to their existing form.

If we examine too the characters of these beds, we shall find them sufficiently distinguished by their greater compactness and stony aspect from modern lavas, whilst the general correspondence in mineralogical characters that exists between them all affords a strong presumption of their having been produced about the same period.

But it is useless to multiply proofs of the fallacy of Mr. Brydone's statement, and the only circumstance that needs surprise us is, that thirty years should have elapsed, without any traveller having visited the spot with the view of ascertaining the correctness of the observation.

Should the high antiquity I have assigned to this volcano be questioned, I may remark that there are vallies on the slope of the mountain which appeared to me too considerable to be the result of torrents, and that among the diluvial matter at its foot, I have found rolled masses of cellular as well as compact lava; the presence of the former seeming to prove that the volcano was in activity at some period intermediate between the general retreat of the ocean, and the event which formed the vallies, and reduced the fragments of rock detached to the rounded condition in which we observe them.

The silence of Homer on the subject of the eruptions of
Etna is indeed often quoted in proof of the more modern
date of this volcano;* but to such *negative* evidence we
have to oppose the *positive* statement of Diodorus Siculus,†
who notices an eruption long anterior to the age of this
poet, as he says that the Sicani, who with the exception of
the fabulous Cyclops ‡ and Lestrigons, were the first inhabitants of the island, and who are admitted on all sides to
have possessed it considerably before the Trojan war, deserted the neighbourhood of Mount Etna in consequence of
the terror caused by the eruptions of the volcano.

This is confirmed by Dionysius Halicarnassus, who states
that the Siculi, ‖ who passed over from Magna Græcia about
eighty years before the Trojan war, first took possession of
that part of the island which had been deserted by the
Sicanians, so that it is probable that the mountain was at
that period tolerably tranquil, and supposing no eruption
to have taken place from that time till the age of Homer, it
is by no means unlikely, that in a barbarous age, the tradition of events so remote may have been in great measure
effaced, and thus have never reached the ears of the Greek
poet.

The earliest historian by whom the volcano has been
noticed is Thucydides, who says, that up to the date of the

* It is true that the fire of Etna is alluded to in the poems attributed to
Orpheus, which some have supposed as antient as the time of Pisistratus, half
a century at least before the age of Pindar: these poems however are in general
referred to a much later period.

† Lib. v.

‡ Fazzello Decad, Lib. 1. l. vi. tells some wonderful stories of the bones
and teeth of these giants being discovered in caverns in the limestone of Tre-
pani, Palermo, &c. which, if examined by some skilful anatomist of the present
day, would probably be found to be the remains of elephants and other large
animals, and might supply my friend Professor Buckland with another chapter
for his Reliquiæ Diluvianæ.

‖ Dion. Hal. Lib. 1. There is great uncertainty however respecting the
date of this event. Cluverius places it 148 years after the taking of Troy
(Sicilia antiqua, p. 29.), but Thucydides expressly says, Lib. 6, that the
Siculi came over 300 years before the Greeks, who were driven to Sicily by a
tempest on their return from Troy.

Peloponnesian war, which commenced in the year 431 B.C.
three eruptions had taken place from Mount Etna, since
Sicily was peopled by the Greeks.   It is probably to one of
these that Pindar has alluded in his 1st Pythian Ode,*
written according to Heyné in consequence of the victory
obtained by Hiero in the year 470 B. C.   It may be remarked that this poet particularly speaks of the streams of
lava, which, if we may judge from Vesuvius, are less usual
concomitants of the first eruptions of a volcano.†

> Τας εϱευγονται μεν απλα—
> Του πυϱος αγνοταται
> Εκ μυχων παγαι ποταμοι
> Δάμεϱαισιν μεν πϱοχεοντι ϱοον καπνε
> Αιθων'.

Diodorus Siculus‡ mentions an eruption subsequent to
the above, namely in the 96th Olymp. or 396 years B. C.

* Æschylus as well as Pindar, alludes to the confinement of Typhon under
the Island of Sicily, and to the volcanic eruptions arising from his presence.

> Και νυν, αχϱειον και παϱηοϱον δεμας
> Κειται στενωπου πλησιον θαλασσιου
> Ιπνουμενος ϱιζησιν Αιτναιαις υπο.
> Κοϱυφαις δέν ακϱαις ημενος μυδϱοκτυπει
> Ηφαιστος, ενθεν εκϱαγησονται ποτε
> Ποταμοι πυϱος, δαπτοντες αγϱιαις γναθοις
> Της καλλικαϱπου Σικελιας λευϱας γυας·

Prometheus, line 363.

It may be remarked that this poet speaks of the volcanic phænomena as *to*
happen at some time subsequent to that at which the incidents of the play are
supposed to take place, and as being, at the period to which he refers, only in
preparation.

† In case any doubts should exist respecting the interpretation of this
passage, it may be remarked that the existence of a stream of lava is more
distinctly expressed by Thucydides, whose words are: εϱϱυη πεϱι αυτο το εαϱ
τουτο ο ϱυαξ τυ πυϱος εκ της Αιτνας. L. 3.— 116.

‡ Lib. 14.

which stopped the Carthaginian army in their march against Syracuse. The stream may be seen on the eastern slope of the mountain near Giarre, extending over a breadth of more than two miles, and having a length of twenty-four from the summit of the mountain to its final termination in the sea. The spot in question is called the Bosco di Aci; it contains many large trees, and has a partial coating of vegetable mould, and it is seen that this torrent covered lavas of an older date which existed on the spot.

Four eruptions are recorded to have happened between this period and the century immediately preceeding the Christian æra,* during which latter epoch the mountain seems to have been in a state of frequent agitation, so that it is noticed by the poets among the signs of the anger of the gods at the death of Cæsar.†

After this for about a thousand years its eruptions are but little noticed, but during the last eight centuries they have succeeded each other with considerable rapidity. Referring however to the chronological list of the eruptions of the mountain for a specification of these, I shall here merely allude to such as have produced some remarkable change in the character of the country.

I know not whether I ought to include among these events the supposed destruction of the port of Ulysses and the island adjoining, of which Homer and Virgil make mention, and which, from the position assigned to them by Pliny,‡ have been supposed to have stood near the village of Longnina a few miles north of Catania.

As the present size of the creek which is found there, adapted only for small fishing boats, is far too inconsiderable to correspond with the description given of it by

---

* Viz. B. C. 140—135— 126—122. Cluv. Sic. ant. p. 105.

† ———— Quoties Cyclopum effervere in agros
Vidimus undantem ruptis fornacibus Ætnam
Flammarumque globos, liquefactaque volvere saxa.
                                        Georgic, Lib. 1. 1.

‡ Plin. Nat. Hist. lib. 3. cap. 14.

the poets,* and as the island itself does not exist, it has
been imagined that there was once an harbour farther inland, and at the back of the present village, an idea to
which the configuration of the surrounding country seems
to lend some colour.   Bembo† even goes so far, as to attribute to an eruption that took place in the 14th century the
filling up of the harbour and the junction of the island with
the main land, and Fazzello follows him in this notion; but
Ferrara ‡ assures us that the lava of Longnina certainly belongs to the eruption recorded by Orosius as happening in the
year of Rome 631 or 122 B. C., so that the destruction of
the port must have occurred at that epoch, if at all.   It
must be remarked however that, with the single exception of
Pliny, no notice is taken of such a port by any of the prose
writers of antiquity, so that it is possible that the whole may
have been a figment of imagination, first introduced by
Homer, and copied with little variation by his Roman imi-
tator.

   The only semblance of an harbour, which the neighbourhood of Catania has to shew, it owes to the lava of 1669.
In this memorable eruption a rent twelve inches in length
took place on the flank of the mountain above Nicolosi,
about half way between Catania and the summit, and from
this fissure descended a torrent of melted matter, which continued flowing for several miles, destroyed a part of Catania,
and at length entering the sea formed a little promontory,
which serves to arrest the fury of the waves in that quarter.
At the same time the accumulation of matters ejected, raised
on the mountain two conical hills called the Monti Rossi,
which measure at their base about two Italian miles, and
are in height more than three hundred feet above the slope
of the mountain, on which they are placed.‡

   * Portus ab accessu ventorum immotus, et ingens
      Ipse, sed horrificis juxtà tonat Etna ruinis.

   † See P. Bembi Liber de Ætnâ, attached to Schelte's Edition of Corn. Sever.
Ætna.   Amstel. 1703.—p. 218.
   ‡ Ferrara. Descrizione dell' Etna.   Palermo. 1818.

The products of the volcanic action at these different periods hardly present sufficient variety to deserve a separate enumeration, although I have observed among the lavas that appear of the oldest date a nearer approach to the characters of trachyte and porphyry slate, than is ever observable among the more modern,* all of which, so far as I have examined, attract the magnetic needle, and therefore probably contain an admixture of titaniferous iron. The ejected masses are much more uniform in the composition than those found on Vesuvius, and I am not aware of the occurrence among them of any mineral that does not exist in the latter mountain. Signor Gemellaro has discovered a mass of granite, which seems to have been ejected, in the midst of antient lava.†

During the period at which I visited the mountain, sulphurous acid was given out in volumes from the crater, but the condensed vapours collected from the Famarcles on its exterior consisted simply of water, very slightly impregnated with muriatic acid. Sulphuretted hydrogen I did not discover near the summit, but at the bottom of the mountain it is given out from the spring of Santa Vennera near Jaci Reale.

* Giœni, who examined Mount Etna with much attention, has stated, that it consists of a nucleus of porphyry (trachyte) covered more or less by the lavas subsequently ejected. It is probable from analogy that this may be the case, but I could not satisfy myself on the point from actual examination.

† See his pamphlet " Sopra alcuni pezzi di Granito e di lava anticha trovati presso alla cima di Etna," del Dottor C. Gemmellaro. Catania. 1823.

TABLE,

Shewing the correspondence in point of time between the eruptions of Ætna, Vesuvius, and the other Volcanos connected with them.

(Extracted, with some few additions, from Hoff's Geschichte der veranderungen der Erdoberflache.)

ETNA.	VESUVIUS.
B. C.	
480 or thereabouts.	
427	
396	
	185 Eruption between the Eolian Islands, according to Pliny, 200 B. C.
140	
135	
126 or 125, in which year flames rose from the sea near Lipari.	
122	
	91, Eruption at Ischia.
56 or thereabouts.	
45 or 44.	
36 or thereabouts.	
A. D.	
40 or thereabouts.	**A. D.**
	79.
	203.
251	
	512.
	685.
812	
	983.
	993.
	1036.
	1049.
	1138 or 1139, 6 Kal. Iun.
1169, Feb. 4.	
Between 1198 and 1250.	
	1198, the Solfatara inflamed.
1284.	
	1302, Eruption of Mount Epomeo, in Ischia.

ETNA.	VESUVIUS.
A. D.	A. D.
	1306.
1329, June 28.	
1333.	
1408, November 9.	
1445.	
1446.	
1447, September.	
	1500.
1535, March till 1537.	

1538, 29th September, Formation of the Monte Nuovo, near Puzzuoli.

ETNA	VESUVIUS
1566. ⎫	
1578. ⎪	
1603, July. ⎪ Continuance of small	
1607. ⎬ eruptions during this	
1610, Feb. ⎪ interval.	
1614, July 2. ⎪	
1619. ⎪	
1624. ⎭	
	1631, December 16.
1633, February 22.	
1645, November.	
1654.	
	1660, July.
1669, March 8,	
1682, December.	1682, August 12.
1688	
1689, March 14.	
1694, March to December. (Eruption only of ashes.)	1694, March 12, with feeble recurrences of action till 1698.
	1701, July 2 till 15.
1702, March 8.	
	1707, May 20, with feeble recurrences of action till August, 1707.
	1712, Feb. 18, eruption continued till the following year.
	1717, June 6, continued as before.
1723, November, beginning of the month.	
	1727, July 26.
	1730, February 27.
1735, October, beginning of the month.	
	1737, May 14.
1747, September, volcanic action continued for some years.	
	1751, October 25.
	1754, December 2.
1755, March 2.	
1759.	
	1760, December 23.
1763, June 19.	
	1766, March 25.

ETNA.	VESUVIUS.
A. D.	A. D.
1766, April 27.	
	1767, October 23.
	1770.
	1778, September 22.
	1779, August 3.
1780, May 18.	
1781, April 24.	
	1783, August 18.
	1784, October 12, and December.
	1786, October 31.
1787, July 28	
	1787, December 21.
	1788, July 19.
	1789, September 6.
1792, March.	
	1794, June 15.
1798, June.	
1799, June.	1799, February.
1800, February 27.	
1802.	
	1804, August 12, and November 22.
	1805, July.
	1806, May.
1809, March 27.	
	1809, December 10.
1811, October 28.	1811, October 12.
	1811, December 31.
	1813, May to December.
	1817, December 22 to 26.
	1818.
	1819, April 17.
1819, May 29.	
	1819, November 25.
	1822, February 13 to 24,
	1822, October 22.

It appears from this Table that the nearest coincidence between the eruptions of the two Volcanos was in 1694 and in 1811, when they occurred within
a month of each other; and that on eight several occasions an interval of less
than half a year elapsed between them, viz. that of Vesuvius Dec. 2, 1754, was
followed by one of Etna on March 2, 1755; Vesuvius August 3, 1779, by Etna
May 18, 1780; Vesuvius October 31, by Etna July 28, 1787; Etna June,
1788, by Vesuvius February, 1799; again followed by one of Etna in June,
same year; Etna March 27, 1809, by Vesuvius 10 December, 1809; Vesuvius
October 12, 1811, by Etna October 25, 1811; again followed by Vesuvius
December 31, same year; Vesuvius May 27, 1819, by Etna, November 25,
same year.

LECTURE III.

ON VOLCANOS EXISTING IN COUNTRIES NOT VISITED BY THE AUTHOR.

———

EUROPE.

AFRICA.

*Lanzerote—volcanic phœnomena that occurred in 1730—36
—in 1825.*
*Madeira—Bowdich's account of the physical structure of
that Island—and of Porto Santo.*
Cape de Verde Islands.
*Azores—Dr. Webster's account of St. Michael— El Pico,
&c.—Island of Sabrina.*
*Question as to the former existence of the Island or Con-
tinent of Atlantis, in this situation.*
Island of Ascension—of Saint Helena.
*ISLANDS ON THE EASTERN COAST OF AFRICA—Madagas-
car, Isle of France—Isle of Bourbon.*
*Volcanic appearances on the CONTINENT OF AFRICA—
near Mount Atlas—Tripoli—Egypt.*

ASIA.

*Coasts of the RED SEA and PERSIAN GULPH—ARABIA—
PALESTINE—Dead Sea, the effect of a volcanic erup-
tion—ASIA MINOR—near Scandaroon—near Smyrna.—
In the CAUCASUS—near the CASPIAN SEA—in PER-
SIA—in MESOPOTAMIA—INDIA—THIBET—SIBERIA
—KAMSCATKA—JAPAN—LOO CHOO.*
*ISLANDS IN THE INDIAN ARCHIPELAGO.—Philippines—
Java—Sumatra, &c.*
ISLANDS IN THE SOUTH SEA.

AMERICA.

*Islands of the ANTILLES, divisible into four Classes, ac-
cording to their physical constitution.—Volcanic appear-
ances considered—Island of Trinidad.*
*NORTH AMERICA — California — Mexico — Guatimala—
Nicaragua.*
*SOUTH AMERICA — Columbia — Quito — Peru — Chili—
Humboldt's remarks on the Volcanos of the New World
generally.*

LECTURE III.

———

HAVING in my preceding Lectures confined myself, in great measure, to the consideration of those volcanic districts which I had been able personally to visit, it is my purpose in the present to lay before you such facts as I could collect respecting the existence of similar formations in other parts of the globe.

By so doing I shall the better enable you to judge of the soundness of the conclusions which I shall afterwards attempt to deduce, and be less likely to incur a censure similar to that which has been passed by Humboldt on the geologists of the last century, who, ignorant of the variety of aspects which these appearances assume in different parts of the world, considered Etna and Vesuvius the type of all existing volcanos, a proceeding no less absurd than that of the shepherd in Virgil, who expected his own little hamlet to contain within itself the type and image of imperial Rome.

In my first Lecture I alluded to the travels in Iceland, which seem to have been in some measure suggested by the discussions that were carried on with so much warmth at Edinburgh respecting the origin of basalt; in my present it is my intention to lay before you such of its contents as appear to illustrate the subjects more immediately under our consideration.

Sir G. Mackenzie, in the work to which I refer, notices two varieties of volcanic products in this island, one of which appeared to him of submarine, the other of terrestrial origin.

Among the rocks referred to the former period, the prevailing substance was a tuff containing fragments of cel-

lular lava, of pearlstone, and of amygdaloid, the cavities of
which were filled with calcareous spar. With this tuff
alternate beds of scoriform lava, and both are traversed by
dykes of greenstone, perfectly compact, and without any
vitreous aspect, thus serving to shew the manner in which
the characters of a rock depend upon the degree of pressure
exerted during its formation. In the case of the bed the
structure is cellular, because it probably flowed freely over
the surface, without being subjected to any pressure considerable enough to counterpoise the expansive force of the
elastic vapours disengaged; in the case of the greenstone
dykes, the rock itself, through which they forced their way,
may have opposed a resistance sufficiently considerable to
have prevented the formation of cells.

The circumstances, in short, of this case may be considered as corresponding with those which I have described
as occurring at the Monte Somma, where, though the rock
itself is analogous in its mineralogical characters to the
dykes that traverse it, yet the *mechanical* structure in the
former case is very different to what it is in the latter, owing
probably to the inferior degree of pressure exercised during
its formation.

In other cases the tuff contains horizontal beds of pearlstone, which I suspect to be analogous to the dykes noticed
in my former Lecture as occurring in the Lipari Islands, as
from the manner in which this stone disappeared among the
other rocks, becoming gradually mixed and blended with
them, Sir G. Mackenzie was led to think that the bed itself
was an interposed branch, or the trunk of a vein.

The cellular aspect of the constituents of these submarine
lavas seems to shew that their age is, comparatively speaking,
modern, and with this the almost total absence of any Neptunian products completely accords.

Sir G. Mackenzie has noticed an effect of volcanic action
of a kind rather different from that which has hitherto come
before us.

In many places, he says, an extensive stratum of volcanic matter has been heaved up into large bubbles or blisters, varying from a few feet to forty or fifty in diameter. It also contains numerous little craters, from which flames and scoriæ fhad issued, but no lava. These craters are often partially covered in by domes of the same material, as though the whole rock had been first softened by the operation of heat; and portions of it had then been made to swell outwards by the extrication of elastic vapours.

One author has chosen to distinguish this variety by the name of cavernous lava; its date is probably anterior to that of the commencement of the present order of things, for it is in many cases covered with gravel, and seems to extend under the sea.

When I come to speak of the Island of Lanzerote, I shall have occasion to point out appearances described by Von Buch, of a very analogous kind; and shall therefore defer any attempt to explain them for the present, proceeding in the mean time to some other phænomena connected with the same subject, which Iceland presents to our contemplation.

Amongst these are the Sulphur mountains of Krisiavik which, as described by Sir G. Mackenzie, present in some respects analogies to the pseudo-volcanic phænomena of Macaluba in Sicily, and in others to those of a genuine volcano, in continual but languid action.

The rock consists of alternating beds of white clay and sulphur, from all parts of which steam is given out. This was remarkably the case in a deep hollow into which the author descended, where a confused noise was heard of boiling and splashing, joined to the roaring of steam escaping from narrow crevices. At the bottom of this hollow was a cauldron of boiling mud about 15 feet in diameter. There was a constant sublimation of sulphur, which formed beautiful crystals round the sides of the cavity.

The celebrated springs of Geyser are, however, the phænomena which most forcibly arrest the attention of the

traveller in this country. The intermitting character of
these fountains may be, in some measure, imitated by pour-
ing a stream of water through a bent tube depressed about
the centre, and heated in that part alone.

Under these circumstances the steam suddenly generated
at bottom will force one portion of the water out in a jet
from the opposite extremity to that at which it entered,
driving back at the same time the current of water that
continued to flow in. In this manner the water might be
propelled in jerks, as happens in the case of the Geyser
springs.

Such an explanation however is far from being adequate
to account for the complicated phænomena of these fountains, which, after a pause of many hours, first threw up
water, and afterwards vast columns of steam, to the height
sometimes of 200 feet, and then immediately sunk into a
temporary repose; neither is it applicable to the singular
circumstance mentioned by Mr. Henderson, as to the possibility of bringing on the explosion at any given time by
merely throwing large stones into the orifice. The latter
fact indeed seems to prove that the generation of steam is
constant, and that nature has provided other *vents* sufficient
to carry off a certain portion of the elastic vapour, unless
when obstructed in the manner produced by Mr. Henderson, in which case its rapid accumulation gives rise to an
almost immediate explosion.

The presence of siliceous earth surrounding the waters of
the Geyser springs, is a phænomenon not altogether confined
to Iceland.

It exists as I have shewn in Ischia and in other places,
where it is less easy to attribute its occurrence, as has been
done in the present instance, to the chemical affinity exerted
for the earth by the mineral alkali present in the waters—
perhaps indeed in either case the pressure exerted contributes to the effect.

If we were to credit the accounts given by some travellers,

we must attribute still more extraordinary effects to the water of these Geysers than the mere solution of silica.

In the second volume of the Edinburgh Philosophical Journal, you will see an account of Iceland, by Menge, a German, in which he attributes the formation not only of siliceous sinter to these springs, but even in many instances of tuff, of basalt, of porphyry, and of obsidian. He even declares that he saw one hot spring producing lava, another forming basalt, and a third, trap porphyry, and notices a particular case where he extracted from a boiling marsh a muddy hot mass, which, when broken, exhibited the characters of basaltic lava in the centre, and towards the surface passed gradually into red and grey mud.

Such extraordinary facts however require for their belief the very best evidence, and I fear the testimony of Menge will hardly be considered sufficient to substantiate them, if at least I am rightly informed with respect to the estimation as a geologist in which he is held in his own country. That purely siliceous minerals, such as pearlstone, hyalite, and opal may be the productions of hot-springs is indeed less improbable, but all analogy is opposed to the idea that extensive strata of basalt or porphyry have ever been produced in the same manner.

Almost the only substance not connected with volcanic operations which occurs in Iceland is the surturbrand or bituminous wood, of the situation of which a recent traveller, Mr. Henderson, has given us a detailed account. The west side of a perpendicular cleft in the side of a mountain called Hagafiall exposes a section of ten or twelve horizontal strata, of which the surturbrand is undermost, occupying four layers, which are separated from each other by intermediate beds of soft sandstone and clay.

They vary in thickness from a foot and a half to three feet, and differ also in quality, the two lowest strata exhibiting the most perfect specimens mineralized wood, free from all foreign admixture and of a jet black, the numerous knots, roots, &c. leaving no doubt of its vegetable origin.

The two upper strata contain an admixture of earthy and
ferruginous matters, and in the midst of them occurs a thin
layer, four inches in thickness, consisting of a schistous
mass which appears to be made up entirely of leaves closely
pressed together, separated only by a little clay. These
leaves are chiefly of poplar, a tree, Mr. Henderson says, at
present not met with on the island. The beds of surturbrand support an alternation of basalt, tuff, and lava, which
extend to the summit of the hill.

With the sole exception perhaps of this substance, the
whole of the mineral structure of Iceland may be said to
have originated more or less directly from volcanos, and
there is probably no part of the globe in which operations
of this kind have been going on with so much activity, and
for so considerable a period. The existence of submarine
lavas proves the action to have commenced before the retreat
of the ocean, notwithstanding which eruptions occur here
more frequently at present than they do at Vesuvius or in
any other known case.

Besides Hecla, which has been twenty-two times in a
state of activity during the last eight hundred years, five
other volcanos are enumerated, from which the total number
of recorded eruptions during the same period is no less than
twenty. Some of these happened at the same time at which
the volcanos of the Mediterranean were in action, but the
instances of this coincidence are not sufficiently numerous
to lead to any certain conclusion.

In the year 1783 a submarine eruption took place six or
eight miles from Reykiavess, which gave birth to a new
island a mile in circumference, which however the following year again disappeared. A submarine eruption also
took place about the same time seventy miles from the same
cape, which is said to have thrown up pumice sufficient to
cover the sea for a space of 150 miles round.

A new island is also stated to have appeared opposite
Hecla in the year 1563, but there are some doubts as to this
latter fact.*

* See Raspe de novis Insulis. p. 126.

The only other volcano in the north of Europe is that in the island of Jan Mayen, off the coast of Greenland. This, when visited by Captain Scoresby* in the year 1817, exhibited the marks of a recent eruption, and was found to consist of cellular lava, of tuff, and of scoriæ; on its summit was a magnificent crater 500 feet in depth, and about 2000 in diameter.

Those who have made up their minds with regard to the volcanic origin of trap, will perhaps suppose a connexion between the rocks of Iceland, and those of Faro, the Hebrides, and the county of Antrim.

The line, however, which I have prescribed to myself in the prosecution of this inquiry, precludes me from entering into a description of these latter countries, which indeed are already made known to the public by some of the most distinguished geologists of this country, in publications at once easily accessible, and of recent date.†

It is true, there are certain phænomena noticed by Sir G. Mackenzie in his description of Faro, which remind one of those which I have noticed in my account of the volcanic rocks near Frankfort, especially the occurrence of a bed of amygdaloid having its upper surface filled with small insulated perpendicular cavities, as if caused by the escape of a gaseous fluid when the rock was in a soft pasty state.

As however I have not the same inducements, from having personally visited these spots, for entering upon the physical structure of Faro, as I had for noticing that of the trap formations in Germany, I shall return to the south of Europe, where there yet remain several tracts not less worthy of notice with reference to the present enquiry than those

---

* Scoresby in the Edinb. Philosophical Journal.

† Sir G. Mackenzie and Mr. Allan, Edinb. Phil. Trans. vol. 7.—Mr. Trevelyan, Ed. Phil. Trans. vol. 8.—Macculloch's Western Islands and Jameson's Mineralogical Travels.—Berger on the Geol. Features of the N. Eastern counties of Ireland. with an Introduction and Remarks by the Rev. W. Conybeare—and Descriptive Notes referring to an outline of Sections of the same Coast by Messrs. Conybeare and Buckland.

SANTORINO.

which have been treated of in the course of the preceding
Lectures.

Thus we have strong reasons, both from the accounts of
antient writers, and the observations of modern travellers,
to infer the prevalence of igneous action in many parts of the
Grecian Archipelago.

It is true, there may be room for questioning the suffi-
ciency of the former source of information, where not confirmed by the physical structure of the spots themselves;
for Pliny mentions Rhodes as having been thrown up from
the ocean* in a manner that would lead us to infer the
agency of some volcanic force, and yet it appears, that this
island consists of granite and other rocks, which present no
traces of any action of this kind.

In other cases, however, where the information obtained
from these two distinct sources concur in assigning to the
place this mode of formation, we can have no difficulty in
admitting the correctness of the statements given.

The Island of Santorino, known by the antients under
the name of Thera, as well as the smaller one near it called
Therasia, are mentioned by Pliny as having been thrown up
from the sea;† both which statements seem to have a
foundation in fact, though mixed up with much inaccuracy
of detail.

Thus in speaking of the larger Island Thera, the Roman
naturalist sets down the time of its appearance as happening
in the 135th Olympiad, or about 237 years B. C., a date
quite inconsistent with the mention made of the island by
Herodotus, who states that it was given by Cadmus to
Membliares, one of his followers ‡

If this historian indeed be depended on, we must likewise

* Plin. Hist. Nat. Lib. ii. CLXXXIX.
† Vide Plin. Hist. Nat. hoc. cit.
‡ Melpomene, c. 147.

regard as a poetical fiction the account, which Apollonius Rhodius* has given of the sudden appearance of the island, at the period of the return of the Argonautic expedition.

As to the origin however of the island, though not as to the date of the event, the reports of these writers are fully borne out by the accounts of Tournefort† and other moderns, who represent the island as made up entirely of volcanic matters, with the exception of one spot, the mountain of St. Stephen, where a block of white marble, of what size does not appear, occurs resting on the volcanic matter, or, to use the expression of the writer, *grafted on pumice.*‡

The island is of an horse-shoe form, and the bay inclosed within the horns of the crescent contains four small rocks or islets, the largest of which, according to Choiseul Gouffier,‖ though at present called Aspronesi, is the antient Therasia; whilst the modern Therasia or Theraiia, lies a little to the north of the bay, and consists of nothing more than a few barren rocks.

Aspronesi, on the other hand, nearly closes the mouth of the bay; whilst the three remaining islets, viz. Hiera, otherwise called Burnt Island, or Great Cammeni; New or Black Island; and Micronesi, or Little Cammeni, are situated within.

The subjoined map is taken from Tournefort's work, but as the representation of the smaller islands therein given does not correspond with the more modern sketch annexed to Choiseul Gouffier's Voyage Pittoresque de la Grèce, I have thought myself privileged to introduce such alterations as would make it accord more nearly with the latter.

---

* βωλον δε, θεοπρωπιησιν ιανθεις·
ηκεν υποβρυχιην. την δ'εκτοθι νησος αερθη
Καλλιστη, παιδων ιερη τροφος Ευφημοιο.
Argou. 4. 1755.

† See Tournefort's Travels in the Levant.

‡ Entè, pour ainsi dire, sur pierres ponces. Tournefort, vol. i. Letter vi.

‖ Voyage pittoresque de la Grèce, vol. i.

SANTORINO.

A. Antient Thera.   Modern Santorino.
    a. a. Black and calcined rocks, very abrupt.
    b.   Pyrgos.
    c.   Castle of Scaro.
    d.   San Nicolo.
    e.   Hill of St. Stephano.
    f.f. Sea unfathomable.
B. Little Cammeni or Micronesi, thrown up 1573.
C. New or Black Island, thrown up 1707.
D. Hiera, Burnt Island, or Great Cammeni.
E. Antient Therasia.   Modern Aspronesi, or White
    Island.
F. Rocks now called Therasia.

SANTORINO.

Of these islands, it seems pretty clear from a number of concurrent testimonies, that one must have been thrown up from the sea about three centuries before the Christian era, but it does not seem quite so certain to which of the four the authorities refer.

Pliny for instance declares, that Therasia, as well as Thera, made its appearance in the 135th Olympiad, but the inaccuracy of his statement with regard to the former island is alone sufficient to invalidate his testimony as regarding the latter.

He also mentions, that 130 years afterwards the island of Hiera was thrown up, a statement confirmed by Justin and Plutarch* as to the fact, though not as to the date, for these writers mention the appearance of Hiera as having occurred during the war between the Romans and Philip of Macedon, and if we adopt the date assigned by Justin,† which is that of the year in which the battle of Cynocephale was fought, we must fix it at 197 B. C., and substitute in the text of Pliny for the number cxxx, xxxx.

All these authorities however concur in placing Hiera between Thera and Therasia, and as of the three islands in that predicament, two of them, New Island, and Little Cammeni, are understood to have been produced in modern times, we must conclude that Hiera is the one now known by the name of Burnt Island, or Great Cammeni.

Pliny also speaks of another phænomenon of the same kind as happening in his own time,‡ for he tells us, that in the reign of Claudius, A. D. 46., a new island called Thia appeared near Thera. But as he mentions it as only two

---

* Justin, lib. 30. c. 4.—Plutarch De Pythiæ orac.

† Justin's words are : Eodem anno inter Theramenem et Therasiam, medio utriusque ripæ et maris spatio, terræ motus fuit; in quo, cum admiratione navigantium, repente ex profundo cum calidis aquis insula emersit. Lib. 30. c. 4.

‡ Plin. Hist. Nat. Lib. ii. c. lxxxix.

stadia distant from Hiera, it is possible that the island may have been joined to the latter by a subsequent revolution, as by that recorded to have taken place in the year 726, by which Hiera is said to have been greatly augmented in point of size.

In the centuries succeeding the latter epoch, other changes are noticed with respect to these islands, amongst which the production of a new rock, that of Little Cammeni, in 1573, was the most remarkable.

Thevenot mentions a great eruption of pumice as having taken place in the sea near Santorino in 1638, and Father Goree* in 1707 was eye witness of the appearance of a new rock between Little and Great Cammeni, which increased in size so rapidly, that in less than a month it became half a mile in circumference, and had risen twenty or thirty feet above the level of the waters.

The following is an extract from the account he has transmitted to us of the circumstances attending this event.

On the 23d of May, 1707, the commencement of a new island between Great and Little Cammeni, was perceived from Scaro, and from all that side of Santorino. It was at first taken for the wreck of a ship, but those who visited the spot under that impression, found that it was a mass of rocks, which rose from the bottom of the water. Some, whose curiosity got the better of their fear, had the hardihood to land upon it, and found the surface covered with a white and very soft stone; but, what was very remarkable, a large quantity of fresh oysters, which are rarely seen about Santorino, were found adhering to the rock newly thrown up. Whilst in the act of collecting them, they were frightened away by feeling the ground shake violently.

Between this and the month of July the island was observed to grow gradually larger, for though many of the

* See Phil. Transactions, vol. 26 and 27.

rocks which were added to it sunk again into the waters, a sufficient number remained to add considerably to its volume.

In July the appearances were more awful, as all at once there arose, at a distance of about sixty paces from the island already thrown up, a chain of black and calcined rocks, soon followed by a torrent of black smoke, which, from the odour that it spread around, from its effect on the natives in producing headache and vomiting, and from its blackening silver and copper vessels, seems to have consisted of sulphuretted hydrogen.

Some days afterwards the neighbouring waters grew hot, and many dead fish were thrown upon the shore. A frightful subterranean noise was at the same time heard, long streams of fire rose from the ground, and stones continued to be thrown out, until the rocks became joined to the White Island originally existing.

Showers of ashes and pumice extended over the sea, even to the coasts of Asia Minor and the Dardanelles, and destroyed all the productions of the earth in Santorino.

These, and similar frightful appearances continued round the island for nearly a year, after which nothing remained of them but a dense smoke.

On the 15th July, 1708, the same observer had the courage to attempt visiting the island, but when his boat approached within 500 paces of it, the boiling heat of the water deterred him from proceeding. He made another trial, but was driven back by a cloud of smoke and cinders that proceeded from the principal crater. This was followed by ejections of red-hot stones, from which he very narrowly escaped. The mariners remarked that the heat of the water had carried away all the pitch from their vessel.

During the ten subsequent years, the volcanic action had given rise to several other eruptions, but the same reporter states, that in 1712 all was quiet, and no other indication of

SANTORINO.

the kind existed, excepting a quantity of sulphur and bitumen, which floated on, without mixing with, the waters. Its circumference at that time was about four miles.

It is important, with reference to the natural history of volcanos, to remark that in this case, as in many others, the mountain appears to have been elevated, before the crater existed, or gaseous matters were given out. According to Bourguignon smoke was not observed till twenty-six days after the appearance of the raised rocks.

Not that I mean to represent this as universally true, for in many cases it is probable that the whole consists of a confused congeries of stones of various sizes, piled successively one upon another.

This indeed, if we believe the account transmitted to us by Seneca, appears to have been the case with the island of Hiera or Great Cammeni, as will be seen from the following passage.*

Majorum nostrorum memoriâ, ut Posidonius tradit, cum insula in Ægeo mari surgeret, spumabat interdiu mare, et fumus ex alto ferebatur. Nam primum producebat ignem, non continuum, sed ex intervallis emicantem, fulminum more, quotiens ardor inferiûs jacentis superum pondus evicerat. Deinde saxa revoluta, rupesque, partim illæsæ, partim exesæ, et in levitatem pumicis versæ; novissime cacumen montis emicuit. Posteâ altitudini adjectum et saxum illud in magnitudinem insulæ crevit.

It is curious, that the learned author of the Voyage Pittoresque de la Grèce, although he relates such an assemblage of facts, all tending to prove the agency of volcanos in building up tracts of land, prefaces his remarks by saying, that this cause, far from having been the means of elevating the Island of Santorino from the bosom of the waters, has, on the contrary, been only instrumental in swallowing up that part of it, which once occupied the space between it and Aspronesi.

* Seneca Nat. Quæst. l. 2. c. 26.

SANTORINO.

Whether indeed there be any truth in the idea that this has been separated from the larger island adjoining, and that the latter before this dismemberment went by the name, not of Thera, but of Callista, is more than we have the means of determining; but that none of the larger islands owe their origin to volcanos, or that they are merely the summits of the high mountains which existed on the surface of a tract of land which was overwhelmed by an imagined eruption of the waters, derived from the bursting of the Euxine Sea, is a notion too fanciful to deserve a serious confutation.

Whatever loss may have been at any time sustained by the islands in consequence of volcanic actions going on underneath them, it must be recollected, that it was only a recovery of a part of what had been originally wrested from the ocean by the very same cause, for the whole physical structure of Santorino and its appendages, is such, as proves that they have resulted from this, and from this cause alone.

The origin nevertheless of the two larger islands must be very remote, if the degree of their cultivation be any test of their antiquity; for the surface of Great Cammeni, though thrown up before the Christian æra, is still barren and un-decomposed.

The Island of Milo appears also to be volcanic.* All the vicinity of the port abounds in hot springs, which are sulphureous and chalybeate. Sulphur is sublimed in the crevices of the rock, and alum is everywhere met with. This alum Pliny† mentions as the best that could in his time be obtained, next to that of Egypt, and its origin is clearly explained by the sulphur, which, according to this naturalist, is found in the same island,‡ derived in all pro-

* See Choiseul-Gouffier—Tournefort—Richardson's Travels, &c.
† Plin. Hist. Nat. Lib. 35, c. lii.
‡ Plin. Hist. Nat. Lib. 35. c. l.

bability from exhalations consisting of sulphuretted hydrogen.

Besides these evidences of volcanic action, Tournefort[†] mentions the occurrence of " an uncommon kind of stone like pumice, but hard, blackish, light of weight, not susceptible of impressions of the air, and very fit for sharpening all sorts of iron tackle."

It is curious that here, as well as in the Phlegrean fields, near Naples, noxious miasmata so abound, that the few inhabitants of this once populous island are the very pictures of wretchedness and disease, insomuch that Choiseul Gouffier is disposed to attribute to the exhalations of the volcano, what is probably to be sought for in some other and more occult cause.

Dr. Richardson, the most recent traveller who has noticed this island, states, that the road from the shore to Castro the chief town lays up a steep ascent, in which he observed many masses of obsidian. The whole mountain was volcanic, and the effects of this action are seen in the theatre lying between the town and the entrance of the harbour, which has been completely buried under a shower of volcanic ashes. A few steps of it have been uncovered on one side, and are found to consist of the marble of the island fresh and uninjured.[‡]

Sonnini also states that porous lava occurs there, which is made into millstones.

Argentiere, antiently called Cimoli, is probably connected with Milo in its structure as well as its situation. It has obtained its modern name from the silver mines, which were formerly worked there, but are now abandoned; and which probably lie in a trachytic conglomerate, like those of Konigsburg in Hungary. Its only commerce at present is

† Voyage in the Levant, vol. i. p. 116.

‡ Richardson's Travels, vol. 2, p. 529. See also a notice by Sir Erasmus Darwin in the Ann. of Phil. vol. 22, p. 274. Dr. Thomson in vol. 14 of the same work, has given an analysis of the hot springs that occur in the island.

ARGENTIERE.

in a white heavy tasteless earth, much used in the Levant for cleaning woollen and other stuffs. It has a gritty feel, and effervesces according to Tournefort with acids, but this statement is not consistent with Klaproth's analysis.* Pliny† calls it Creta Cimolia, and mentions its uses, one of which appears to have been its fitness to. favour the generation of nitre.‡

It appears from Olivier, that the Cimolian Earth proceeds from the decomposition of trachyte, caused by sulphureous vapours. It exists in Milo as well as Argentiere, accompanied with gypsum and alum, which evidently have the same origin. The passage of the earth into trachyte is very evident.§

The only scientific collection that has ever been made to illustrate the physical structure of these islands, is the one given to the museum of Freyburg by Mr. Hawkins, a traveller, who, unfortunately for geology, could never be persuaded to publish any account of his researches.

On the authority of these specimens Beudant states, that the greater part of Santorino, Milo, and Argentiere, consists of semivitreous trachyte, accompanied with a trachytic conglomerate, containing organic remains.

It appears that Cerigo, the antient Cythera, is partly made up of the last mentioned material; for I find from an article in the Journal de Physique for 1799, that shells are found there inclosed in stones, having a volcanic appearance.‖

* Silica 63.00
Alumina............23.00
Iron..................1.25
Water...............12.00

† Pliny, Lib. 35. c. 18.

‡ See a note on the passage in Strabo lib. x. in which the island of Cimolus is mentioned.—Falconer's edition, vol. 2, p. 707.

§ Olivier's Travels, vol. 1, p. 323.

‖ Holland, Travels, p. 44, questions the volcanic origin of Cerigo, but he admits that he never visited it; Badhia, on the contrary, who resided some time there, mentions a chain of volcanic hills as traversing the island. See Ali Bey's Trav. v. 1. p. 205.

LEMNOS.

These stones would seem to consist of a sort of pumiceous conglomerate, like that of Hungary, and are accompanied with a red porphyry, probably trachyte.

It is stated in this memoir that the bones both of men and animals are met with in one of the mountains of this island inclosed in limestone, that is in all probability incrusted in stalactite.

Lemnos is another of the islands to which a volcanic origin may be assigned, and the fable of Vulcan, who, when expelled from heaven, first alighted on this spot, would lead us to suspect either that some tradition prevailed among the Greeks as to the existence of volcanic phæno-mena,[*] or that the rocks themselves had a burnt and sterile aspect.

The latter is the manner in which the fable is explained by Galen,[†] and it appears from modern travellers that many parts of the island are covered with pumice and ashes.[‡]

Choiseul Gouffier however is of opinion, that the volcano, from whence these products were derived, is now sunk in the sea, and that only certain portions of it may be recog-nized in some rocks near that island.[||]

It is probable, that the Lemnian Earth, famous from time immemorial in the cure of diseases, may be nothing more than a decomposed condition of trachyte, since it is found to

---

[*] This seems to be alluded to, when Philoctetes calls upon Neoptolemus to put an end to his sufferings with the Lemnian fire that was rolling round him.

τω Λημνιω τωδ' ανακυκλουμενω πυρι
εμπρησον, ω γενναιε.

Sophocles. Phil. v. 800.

[†] φαινεται γαρ ομοιοτατον κεκαυμενω κατα γε την χροαν, και δια το μηδεν εν αυτω φυεσθαι.

Galen, περι απλων φαρμακων.
Lib. ix. c. 2.

[‡] See Dr. Hunt's paper in Walpole's Turkey.

[||] Vol. i. p. 79. vol. ii. p. 130.

CHIMARIOT MOUNTAINS.

be associated with volcanic products, and would seem from analysis to consist of the same ingredients as that rock united in proportions not very different.*

It is probable that these are not the only spots in the same part of Europe which have been produced in the same manner, for pumice and similar productions are noticed by travellers as occurring in the Isle of Delos and other localities.

Further information is however wanted, to enable us to ascertain the *source* of these productions, for it may be said that they have been brought from a great distance, and constitute no part of the original soil of the place in which they are found.

Neither have we any very certain information with respect to the existence of volcanos on the Continent of Greece, though many phænomena are alluded to both by antient and modern writers that seem at first sight to favour such an opinion,

It is true that Lord Byron in his Childe Harolde, speaks of the mountains of Albania as

> Nature's volcanic amphitheatre
> Chimara's Alps.

---

* If long established belief be any test of truth, the Lemnian Earth ought to rank with our most approved remedies; its reputation has remained unimpaired in spite of all the changes of manners, government, and religion, that have since occurred; it is collected at present with the same superstitious ceremonies by the Christian as by the Heathen priest, and is given credit for equal virtues when it has received the impress of the Grand Signior's signet, as it was of old when it had obtained the seal of the chief magistrate of the place; its estimation has survived the very volcano to which it owes its existence, and has continued withont interruption from the time of Philoctetes to the present. (See Choiseul Gouffier. vol. 2.)

Yet on analysis it is found to consist merely of

Silex	66
Alumina	14.50
Oxide Iron	6.0
Water	8.50
Natron	3.50

Lime and magnesia an inappreciable quantity.

Its external characters may be seen described in Phillips' Mineralogy.

MOUNT PARNASSUS.

Adding in a note that the Chimariot mountains appear to have been volcanic.

On examining however the accounts of more scientific travellers, I find nothing to countenance such an opinion, and the phænomenon which probably led to this mistake, was an extrication of inflammable gases, which takes place on the Chimariot mountains, just as it does on the Appennines, between Bologna and Florence.

I have the authority of Dodwell for stating that such is the case at present, and that it was equally so in antient times we learn, not only from several antient writers, such as *Strabo, Pliny, Ælian, and Plutarch, who speak of the occurrence of a Nympheum on the coast of modern Albania near Apollonia, celebrated for the flames that rose continually from it, but likewise from the existence of a coin of the city of Apollonia, which represents on one side Apollo, and on the other three Nymphs dancing round a fire.

This however is plainly connected with the Asphaltum, which is so abundant in the mountains of Albania,† especially near Selenitza, and probably has no reference to any thing of a volcanic nature.

Megalopolis in Arcadia, is said by Pliny to have a burning mountain near it, but it is difficult to say, whether any thing more is meant by his description, than a flame emanating from the ground like that above mentioned.

I am not aware of any notice with respect to it existing among modern travels.

Such likewise appears to have been the spontaneous fire (" αυτοματον πυρ") that emanated from the summit of Mount

---

* Strabo vii. 316. Pliny ii. 106. Ælian xiii. 16. Plutarch in Sylla, p. 468. See also the singular poem on the Hot-baths of Pythia, which will be afterwards alluded to, vs. 40.

† See Strabo, Lib. 7, 458. Dion Cassius, Hist. 4. and Holland's Travels in Albania.

NEGROPONT.

Parnassus. It must be supposed to have proceeded from both its peaks, as the poets speak of the " διπορυφον σελας of the sacred mountain,* and though some have interpreted these passages as alluding to the sacrificial fires that were so frequent, it seems more simple to imagine, that some natural phænomenon of the kind alluded† to really existed.‡

As however it appears from Dr. Clarke's account‖ that all the higher parts of Parnassus are of limestone, containing Entrochi and other organic remains, we cannot for a moment entertain the idea, that the phænomenon itself was connected with any really volcanic cause.

I know not what to think of the statement of Strabo with respect to a phænomenon that occurred in the Lelantic fields, near Chalcis in Eubœa, and which Humboldt has cited in proof of volcanic action manifesting itself in that locality.

It is said, that in the Lelantic fields an earthquake took place,§ which did not cease until a chasm opened in the earth, from whence a stream of hot mud was vomited forth.

It is possible that the above phænomenon may have been of the same nature with that of Macaluba in Sicily, and therefore be connected, if at all, at least very remotely, with volcanic action; but this can be only ascertained when the physical structure of Eubœa has been determined by some geological traveller.

Mr. Hobhouse,¶ the only person who seems to have noticed the rocks of this island, remarks, that there is one next

---

* Phænissæ, v. 225.

† This seems to be the opinion of Elmsley, as appears from his note on vs. 306 of the Bacchæ of Euripides.

‡ It seems not unlikely, that certain gaseous exhalations that emanated from the ground on the site of the oracular shrine of Delphi, may have co-operated with the enthusiasm of the moment in bringing on the divine ecstacies of the priestess.

‖ Clarke, vol. 4, p. 207.

§ μη παυεσθαι δε σειομενην την νησον κατα τα μερη, πριν η χασμα γης ανοιχθεν εν τω Λελαντω πεδιω, πηλυ διαπυρε ποταμον εξεμασε. Strabo, l. l. P. 85. Ed. Falc.

¶ Travels in Albania.

ARGOLIS.

to the hills of Chalcis which presents a singular appearance, being of a red colour, and rent from top to bottom, with a huge chasm extending into the bowels of the mountain.

The neighbourhood of Trœzene in Argolis, would appear from Ovid to have been the seat of a volcanic eruption, which created an entire mountain, just in the same manner as in the last century the mountain of Jorullo was elevated in the midst of the Table Land of Mexico.

The description of Ovid is so applicable to both these events, that I have introduced an extract from it in the Frontispiece of this work, which represents the mountain Jorullo as described by Humboldt.

The following is the entire passage :

> Est prope Pithæam tumulus Trœzena, sine ullis
> Arduus arboribus, quondam planissima campi
> Area, nunc tumulus ; nam (res horrenda relatu)
> Vis fera ventorum, cæcis inclusa cavernis,
> Exspirare aliqua cupiens, luctataque frustra
> Liberiore frui coelo, cum carcere rima
> Nulla foret toto, nec pervia flatibus esset,
> Extentam tumefecit humum ; ceu spiritus oris
> Tendere vesicam solet, aut direpta bicornis
> Terga capri.　Tumor ille loci permansit; et alti
> Collis habet speciem, longoque induruit ævo.
>
> METAMORPH. l. 15.

It is probable that Strabo* may refer to the same event, where he speaks of a tract of land seven stadia high, being elevated round about Methone, owing to some exhalation of an igneous nature, for these two places are so near to each other, that they might very readily be confounded.

Diodorus Siculus† relates, that Phædra, when enamoured of Hippolytus, consecrated a temple to Venus upon the Acropolis of Athens, from whence she could distinguish Troezene, the residence of the object of her passion.

* Ed. Falc. vol. i. p. 87.

† Ιδρυσατο ιερον Αφροδιτης παρα την Ακροπολιν, οθεν ην καθοραν την Τροιζηνα.　B. 4. c. 62.

ARGOLIS.

Now Dodwell* remarks, " that the promontory of Me-thone, which at present obstructs the view, not only of Athens, but of its loftiest mountains, might possibly, in the time of Phædra, have been a flat surface, or not even have existed at all, as the whole of that, at present mountainous tract, has evidently been thrown up by the powerful opera-tion of a volcano, which, according to Pausanias, happened in the time of Antigonus. Were the promontory removed, Athens might be seen over the northern extremity of Ægina."

It would appear from Strabo, that even in his time the rage of the volcano was not exhausted, for he says that the mountain was sometimes inaccessible from the intensity of the heat which it occasioned, and the sulphureous odour which it diffused.

He adds, that it was visible at night from afar, and that the sea was hot for five stadia round.

Chandler, who visited the spot, merely mentions as still existing, the hot springs about 3¼ miles from Methone, which first appeared after the eruption in the reign of Antigonus. The springs are on the side of the mountain near a village, and tinge the soil near them with the colour of ochre.

The rocks before Methone, in the mouth of the bay, were called the Islets of Pelops. They were nine in number, produced, it is probable, by the volcano, and at one time bare. Some shrubs grew upon them at the time when Chandler visited the spot.†

I know not whether we are warranted in attributing a volcanic origin to any part of **Pallene** in Macedonia, a pe-ninsula bordering on the Gulph of Salonica, and terminating in Cape Paillouri. It is mentioned by Apollodorus as one of the spots assigned by the poets as the birth-place of the

* Dodwell's Classical Tour, vol. 2. p. 272.
† Travels in Greece, ch. li.

Giants, one of whom, Alcyoneus, is said to have been thrust
out of it by Hercules, and the Phlegrean fields, which in
later times were placed either in the vicinity of Mount Etna
or of Naples, appear antiently to have been fixed in this
peninsula, for Heyne observes,* that by Phlegra and Pallene is meant the same country, the latter peninsula being
remarkable for earthquakes and subterranean fires. The
same commentator further observes, I know not on what
authority, that the very aspect of this spot even at the present time proves the agency of earthquakes and subterranean
fires; but I do not find this statement confirmed by modern
travellers, on the contrary Dr. Holland states that the peninsula is in part at least of primitive formation.†

Dodwell conjectures that the mountains near the Pass of
Thermopylæ are volcanic, but he produces no better evidence than the story of Hercules and Deianira's tunic, which
seems somewhat far-fetched.‡

Omitting however these more questionable proofs of volcanic agency, let us proceed to the neighbourhood of Constantinople, where at least the most unequivocal indications
of subterranean fire are manifested in the nature of the rocks
themselves.

Dr. Clarke in his Travels has noticed, that the Cyanean

---

* See Heynè Annot. in Apollod. p. 29, and his Dissert. de Theog. Hes. in
Comm. Gott. v. 2. p. 151.

† It may be as well to notice a passage of Lucretius relating to another part
of the same country, by way of directing the attention of future travellers.
Scaptesula, the spot alluded to, was near Abdera in Thrace.

> Nonne vides, etiam terrâ quoque sulphur in ipsâ
> Gignier, et tetro concrescere odore bitumen?
> Denique ubi argenti venas aurique sequuntur
> Terraï penitus scrutantes abdita ferro,
> Qualeis exspirat *Scaptesula* subter odores?
>
>   LUCRET. Lib. 6. v. 540.

Scaptesula was the place mentioned as belonging to Thucydides, in Cimon's
Life, but I am inclined to think, that Lucretius alludes to sulphureous vapours arising rather from the metallic ores that occur there (of which Plutarch
speaks) than from any volcanic appearances.

‡ Classical Tour, vol. 2. p. 126.

rocks at the entrance of the Bosphorus consist of substances more or less modified by fire, containing fragments of lava, trap, basalt, and marble, together with veins of agate, chalcedony, and quartz. He also speaks of the ranges of basaltic pillars on the Asiatic side of the strait.

The most detailed account however of this formation is given by General Andreossi in his Remarks on the Lithology of the Bosphorus.

He states that the mountains of Bithynia, on the side of the Bosphorus, are composed of calcareous beds of a blue colour, and without shells.

From Bucuk-leman, however, on the European, and Kelscheli-leman on the Asiatic side, we observe a succession of irregular and black-looking rocks, made up of a congeries of angular fragments of a sombre grey colour, traversed by veins of chalcedony, and containing masses of compact felspar passing into basalt.

Cordier says that they consist either of a basaltic wacke, having a porphyritic character—of a porphyritic lava with a clinkstone basis—of an imperfect obsidian porphyry—or of a clay porphyry with green earth.

It is of a similar trachytic material (tephrine of Cordier) that the Cyanean rocks are composed, but the substratum on which they rest, is a series of horizontal beds consisting of wacke.

At Youm Bournou, on the Asiatic side, is a conglomerate traversed by prisms of basalt, the extremities of which cap the mountain.

The Crommyon, a rock about 80 feet high, is composed of clinkstone.

In the Gulph of Kabakos, the same observer found obsidian porphyry, clinkstone porphyry, passing into basalt, and of the common kind, and clay porphyry penetrated by a vein of saccharoid limestone. It appears that the borders of the Black Sea consist of a brecchia formed of marine shells, together with stalagmitical limestone and bituminous wood in beds.

Connected probably with this formation, are the hot baths of Pythia, in Bithynia, beyond Mount Olympus, and pretty far towards the eastern side, which are mentioned by several antient writers.

A friend has pointed out to me a curious passage referring to them in the 4th volume of Jacob's reprint of Brunck's Authology, in which is inserted a singular, though inelegant and almost barbarous poem, of Paulus Silentiarius, (Σιλιντιαριος) chief silence-keeper* to the Emperor Justinian. In speculating on the cause of the heat, the author chooses a theory very similar to that which prevails at present with respect to many hot springs not supposed to be connected with volcanic phænomena.

"It is conceived," he says,† "by some, that there are narrow

---

* An officer of high dignity in the Imperial palace of Constantinople.

† Ενερθε γης σηραγγας
εικαι ςεκας νοκσιν·
υδωρ εκειθεν ενθεν
αντιτρεχον πιλεισθαι
πιλυμενον δε, θερμην
κ την τυχκσαν πασχειν.

Αλλοι λεγκσι τκτο·
μεταλλα πκ θειωδη
γης εν μυχοις υπαρχειν·
το γειτονκν κν καμα
θερμης τυχον βιαιας
κατω μενειν κκ ισχον
ανω τρεχειν τω πληθει.

Ποιον δεχη; το πρωτον;
αλλ' κ δεδειγμαιτκτο·
τω δευτερω συμφημι.
οδμη γαρ εσιν, οιδας,
μυδωσα, δυσπνοκσα,
τρανον τε μαρτυρκσα.

Ουτω προηλθε πασι
το θερμοβλυσον ρειθρον,
Ιπποκρατης αψυχος
Τεχνης ατικ Γαληνος.

fissures below the earth; that opposing currents of water meeting from various quarters are compressed, and by that compression acquire no ordinary heat. Others on the contrary say that in the recesses of the earth there are somewhere sulphureous ores; that the neighbouring stream therefore, meeting with a violent heat, from its inability to remain below, rushes upwards in a mass.

Which opinion will my readers adopt? The former? I do not myself embrace this: I agree with the latter one. For there is a mephitic offensive stench clearly proving it.

'Twas thus the hot bubbling fluid issued for the benefit of mankind, an inanimate Hippocrates, a Galen untaught by art."

We hear also of volcanic phænomena in the little Island of Taman, which connects the chain of mountains traversing the Peninsula of Crimea with the Asiatic Continent.

From the best accounts however that have been transmitted to us respecting this phænomenon, such as those of Pallas and Heber, it seems not very probable that the cause is connected with real volcanic action.

Pallas represents the eruption as beginning with a thick smoke, followed by a column of flame fifty feet in height, which continued for eight hours and a half incessantly, during which time streams of mud flowed in all directions, but no lava or altered masses of stone were ejected.

The accounts given by Mr. Heber in his Mt. Journal attached to Dr. Clarke's Travels, is such as fully confirms this view, and renders it highly probable that the phænomenon is altogether analogous to that of Macaluba in Sicily. The same traveller informs us, that a sulphureous spring, like Harrowgate water, exists near the spot.

It is true that Dr. Clarke found on the coast loose fragments of lava, but these he supposes to have been brought from Italy as ballast, and to be quite unconnected with the processes before alluded to.*

Returning from this extreme point of Europe to the Mediterranean, we have yet to mention the volcanic rocks that occur on the western surface of Sardinia, of which an account has recently appeared in the Annales du Musée† for 1824.

It may be collected from this statement, that the volcanos occur in almost every case in groupes of greater or less extent; and that they in general repose on rocks belonging to the most recent order of formations.

Part of these products are of a date posterior to the excavation of the vallies, but others are distinctly recognized as anterior to that event. Thus in the south of the island, between the village of Nurri and the plain called Campidano, the calcareous rocks of the country are capped by a platform of well-characterized lava, which follows the general inclination of the country from east to west.

The name given to these platforms is Giarra, and there are several of them, such as the Giarra de Serri, de Gestori, &c.

The inclination of the beds, the direction of the cells, and the abundance of the lava which is found alike on the summits of all the calcareous and marly hills of this neighbourhood, lead to the belief that their origin is in all cases the same, or that they belong, to speak more correctly, to one and the same current that proceeded from a crater near Nurri, at an epoque antecedent to the period at which the vallies were excavated.

The craters are in great measure effaced, and it is only with hesitation that our author admits that there exist traces of any. In his search he was directed rather by the shape and direction of the cells found in the lavas, than by the actual form of the masses themselves.

* For a further description of this phænomenon, see Engelhardt and Parrot *Reise* in dem Krym and dem Kaukasus, page 69.

† The author's name is Mons. de la Marmora.

Among the volcanic formations of this island, the pre-
dominant rock is a felspathic (petrosiliceuse) porphyry. It
constitutes two-thirds of the *lithoide* lavas of the country.
It occurs in great masses on the two islands already cited;
among the mountains of Ales, Bortigali, and the environs
of Macomer; forms a large portion of the mountains called
Villa Nuova, Monte Leone, and Bosa; and is found at last
at Ploaghe, near Osilo and Castel Sardo, where it passes into
obsidian.

The most remarkable variety which this porphyry presents,
is a rock of a prismatic form, with a fine rosy hue, often
ramified with dendrites. It was observed at the islands of
St. Pierre, St. Antioco, and Isola Prima. Those portions
which are preserved from the action of air and light, retain
a very bright colour, the lustre of which is relieved by fine
ramifications of very large and varied dendrites.

This is the only prismatic rock observed in the island.

The Island of St. Antioco is also very rich in pearlstone,
which constitutes part of a species of conglomerate or
breccia, inclosing likewise other substances.

This pearlstone seems to have been rolled, and occurs in
masses from the size of a nut to that of twice or thrice a man's
head. It is always accompanied with Puzzolana.

Glassy obsidian with a conchoidal fracture, occurs in
the Island of St. Peter's, on the summit of Trebina, near
Ales, &c.

True pumice has not been met with. Red jasper abounds,
especially in the Isle of St. Peter, and in the volcanic rocks
of Alghero, Eteri, and Bosa.

Basaltic lavas, often scoriform, form the greatest part of
Mount St. Lussurgio and Caglieri. On the eastern flank of
this mountain we see the lava that flowed from the extinct
crater of Lussurgio.

There appear to be some indications of volcanic action
likewise in the Spanish Peninsula.

Thus Dr. Maclure, the American Geologist, mentions in

a letter to Delametherie, published in the Journal de Physique,* and dated 12th February, 1808, that having passed the Pyrenees to go to Barcelona, he found in the bed of the *fluvia* lavas and scoriæ. He ascended towards the scource of the river, traversed four leagues of a volcanic country round Ollot, and observed there several streams of lava, volcanic cinders or Puzzolana, and lastly craters not yet effaced. This volcanic district extends from six to eight leagues to the south beyond Amera, where in 1428 there was an eruption which destroyed Ollot, and left only one house standing. He found much lava in the bed of the River Tor, and traversed near Massanite a current of antient lava, almost a league in breadth, in a state of decomposition, and covered by an alluvial soil. From Massanite to Ollot is a distance of fifteen leagues, so that the theatre of volcanic action in these countries is much more extended than that around Vesuvius.

In the American Journal of Science,† there is also a notice respecting some specimens of volcanic rocks from Murcia in Spain, four miles west of Jumella. They are antient lavas, having their cells filled with phosphate of lime. They are thrown up from beneath a compact shelly limestone. Being covered also by calcareous deposits, it is to be presumed that the volcano itself was submarine.

There appear likewise to be indications of volcanic action, in Valentia at Cape di Gaieta near Almeira, and on the opposite side of the Peninsula, the chain of mountains that separates Portugal on its southern extremity from the province of Algarve, and terminates in Cape St. Vincent, is said to be in many places traversed by volcanos. The unhappy condition of the Peninsula has however for many years past thrown such obstacles in the way of travelling, that we are

* Vol. 66, p. 266.
† Vol. 5. No. 1. p. 187.

even less informed with regard to its geological structure, than we are respecting many parts of America, or even of Asia.

Dolomieu* has however noticed alternations of basalt and limestone, as occurring near Lisbon. North of that capital the limestone rests on the basalt, but on the road from Cintra to Maffra, the two rocks are intermixed.

In a mountain a league from Maffra, at the foot of which the Lisbon road passes, basalt is seen on the summit resting on limestone. This basalt is of a semivitreous character, and is coated with a sort of enamel-like porcelain. The calcareous rock which alternates with it is pronounced by Bowdich† to belong to the tertiary class.

Dolomieu notices also in the province of Biera in Portugal, a mountain called Siera de l'Estrella, the Mons Herminius of the antients, which is very lofty, is of a conical form, and emits an hollow sound when we tread upon it, as though it contained caverns. On the summit is a large excavation with a lake at bottom, through which bubbles of air arise. At its base are columns of basalt.

ON THE VOLCANOS OF AFRICA.

The most unequivocal proofs of igneous agency, that occur in this portion of the globe, are exhibited in the islands that may be considered as dependencies of the great African Continent.

The whole group of the Canaries, for example, seems to be placed, as it were, within the sphere of the same sub-

* See Dolomieu's Letter to Faujas St. Fond in his work "Sur les Volcans du Vivarais."

† See Bowdich's Posthumous Work.

TENERIFFE.

marine volcano, for although vestiges of other rocks are to
be met with, as granite and mica slate in Gomera, and
limestone in the Great Canary, Fortaventura, and Lanzerote, yet none of these islands are exempt from occasional
manifestations of the same igneous action.

The most remarkable phænomena arising from the above
cause occur in the Island of Teneriffe, where the lofty peak
of Teyde, though tranquil at present on its summit, still
exhibits on its flanks occasional evidences of the same
volcanic action, from which the rocks composing its colossal
structure seem wholly to be derived.

In considering this island, we must in the first place distinguish between the productions of the actual volcano, and
the range of basaltic rocks surrounding it. The latter do
not rise to an height of more than five or six hundred toises,
whilst the elevation of the peak itself is, according to Humboldt, 1909 toises.*

It is through the midst of this basaltic formation that the
rocks constituting the principal mass of this volcano have
been protruded, and hence we may characterize the two
classes under the name of antient and modern lavas, just as
has been done in the case of those which are found at the
foot, and which compose the mass of Mount Etna.

The modern lavas however of the peak admit likewise of
a two-fold division, 1st, into those composing the nucleus of
the mountain, which are of a trachytic character, and appear
to have been forced up through the midst of the older
basalts, and 2dly, into the products of the volcanic action to
which this central mass furnished an appropriate vent.

The latter are very various in their nature and characters :
we may distinguish, first, the lavas, which have sometimes a
stony, and sometimes a vitreous aspect ; and secondly, the
loose ejected masses, such as pumice, obsidian, and lapilli.

* That is (reckoning the toise at 6 feet 1 inches English) 12,090 feet.

TENERIFFE.

Of the lavas, such as have a stony aspect, appear to be confined to a comparatively low elevation, and to have proceeded exclusively from the flanks of the volcano—whilst the vitreous are found only near the summit, the lowest point at which they occur being 8900 feet above the level of the sea.

The source of the latter description of lavas appears to have been the adjoining mountain Chahorra, which holds the same relation to the peak, that the Monte Rossi do to Mount Etna, being a sort of appendage to the principal volcano, and produced by one of its lateral eruptions.

Humboldt nevertheless mentions one stream of vitreous lava as having been traced to the very summit of the peak, where there exists a circular cavity, which must be considered at present in the light of a Solfatara rather than of a Crater, as it is never known to emit flames, though sulphurous acid vapours constantly arise from it. It would appear however, that it has in former times given vent not only to the stream of lava above noticed, but likewise to showers of pumice and obsidian, loose masses of which strew all the upper part of the mountain.

The latter description of ejected masses does not appear to extend to the lower parts of the mountain, the surface there being mostly covered by rapilli, consisting of black lava, possessing more of a stony aspect, not mixed with either obsidian or pumice.

This latter distribution, says Humboldt, seems to confirm the observation made a long time ago at Vesuvius, that the white ashes are thrown out last, and indicate that the eruption is at an end. In proportion as the elasticity of the vapours diminishes, the matter is thrown to a less distance ; and the black rapilli, which issue the first, when the lava has ceased flowing, must necessarily reach farther than the white rapilli. The last appear to have undergone the action of a more intense fire.

TENERIFFE.

The size of the crater that exists in the summit of the Peak is diminutive compared with that of Etna or of Vesuvius, being only 300 feet in its greatest, and 200 in its lesser diameter, whilst its depth does not exceed 100 feet.

Indeed it may be remarked in general, although the rule is liable to exceptions, that the dimensions of a crater are in an inverse ratio to the elevation of the mountain; for in proportion to the height which the ejected masses must attain before they reach the orifice, will be the resistance to be overcome in forcing a passage by this channel, so that in a mountain like the Peak of Teneriffe, the force applied will in most instances be instrumental in creating apertures in the flanks of the mountain, rather than in enlarging the cavity on its summit.

The existence nevertheless of this chimney preserves the island in Von Buch's opinion from those destructive eruptions which convulse some of those adjoining it, since elastic vapours, the immediate and necessary concomitants of volcanic action, thus find a readier vent, and confine their violence to the immediate precincts of the volcano.

We must not however go so far as to suppose, that Teneriffe itself is altogether exempted from those convulsions of nature which are so common in the neighbouring islands.

Its lofty peak, although it may act as a safety valve, and moderate the violence of the volcanic action by determining it to a point at which it can obtain a vent, proves nevertheless from this very circumstance a dangerous neighbour to the towns that lie underneath it. In the years 1704 and 1706 lateral eruptions took place from the Peak, the latter of which destroyed the port of Garachico, the finest and most frequented harbour in the island. In 1798 too, the mountain Chahorra threw out lavas and scoriæ for the space of more than three months, and the violence of the eruption may be judged of by the fact mentioned by Humboldt on the authority of a eye witness; namely that considerable

TENERIFFE.

fragments of stones were thrown to such an height, that from twelve to fifteen seconds were reckoned during their descent. This curious observation proves, that rocks were projected from this crater to a height of 3000 feet and upwards.

Before I conclude the subject we are upon, I may remark, how strikingly the difference between the volcanic products of Teneriffe illustrates the manner, in which the effects of heat are modified in such cases by the influence of pressure.

At the bottom of the mountain are the basaltic lavas or tuffs, which being produced probably under the ocean, and at a very remote period, are compact and possess a stony fracture. Through these have been protruded the trachytes of the peak, which, having had the resistance of so large a body of rock to overcome, also possess a considerable degree of compactness.

This conical and upheaved mass having become the centre of the volcanic operations subsequently carried on, is surrounded by products of later formation, some of which were ejected from the summit at a time when a free channel of communication existed between it and the interior of the volcano, others from the flanks at a later period, when the aperture had become obstructed by the falling in of its sides, or the accumulation of ejected substances. It is clear, that in either of these cases, the pressure exerted upon the substance whilst in a melted state was less considerable than that which prevailed during the formation of the submarine lavas, or even of the trachyte, and hence it is found to possess more of a vitreous aspect, and is more completely penetrated with cells.*

The remainder of this group, as described by Von Buch,†

* The pumice never covers any of the currents of lava, a proof of its greater antiquity. See Von Buch. In Leonh. Min. Tasch. 4 part. 1823.

† See the Transactions of the Royal Academy of Berlin, for the valuable memoirs of Von Buch—on Craters of Elevation—and on the Island of Lanzerote. 1818—19.

appears to consist of submarine lavas, similar to those which I have described as forming the basis of the Island of Teneriffe. The strata of which they are constituted lie in such a position, that they would seem to have been elevated from the bottom of the ocean by the force of elastic vapours; for they dip away in all directions from some central point, where a crater still exists to attest the former agency of aeriform fluids.

This peculiar structure is best illustrated in the Island of Palma, where one of those deep vallies called Barancos exposes an excellent section of the alternating beds. Amongst them Von Buch distinguished one of basalt containing augite and olivine, covered by a stratum of rolled masses chiefly of the same material. Repeated alternations occurred to him as he proceeded, between beds of this conglomerate and continuous strata of compact or amygdaloidal basalt, and below them all was a single bed of trachyte, the only rock of a clearly felspathic nature that is found. Its ground is of a dark grey colour, and is made up of an infinite number of very small tessular concretions, arising from the separation effected in the mass of the stone by a multitude of minute drusy cavities distributed every where over it. These hollows are in general only partially filled, but contain chabasite, analcime and other crystals. Glassy felspar is met with in the rock in long narrow crystals, which in general are in parallel lines, unless when the drusy cavities before mentioned interfere with their direction.

These beds are all intersected by dykes of granular basalt, which become more and more abundant as we proceed along the valley, until at length the lofty wall of rock which bounds it is covered with a net work of them.

These beds all rise towards the crater, or, as it is called by the people, the Great Caldera, a circular opening in the centre of the island, the depth of which is stated by Von Buch as exceeding 5000 feet. From its brim we are en-

PALMA.

abled to look down upon the abyss, and observe underneath us the terminations of the strata, which we have successively passed in our way to it. Viewed from this point they all appear horizontal, but this, as I observed in speaking of the Monte Somma, is an illusion, and arises from their terminations only being visible, and from their ranging at an equal elevation in every part of the circular wall, which bounds the internal cavity of the crater.

The Caldera of the Isle of Palma, says Von Buch, differs much from the crater of an ordinary volcano. Here are no streams of lava, no slags, no rapilli or ashes. Nor do we ever find the latter of such a circumference, or so profound and abrupt. Its general aspect seems to shew that it was formed by the pressure of those elastic fluids which raised the whole island above the level of the ocean, and changed the strata composing it from an horizontal to their present highly inclined position. The aspect of the Barancos is such as favors this hypothesis; these vallies are too narrow and abrupt to be attributed to diluvial action, and are so devoid of water, that they cannot be referred to torrents; but if we suppose a succession of solid and unelastic strata to be suddenly lifted up in the manner of those in the Island of Palma, it is evident that not merely would a central aperture be formed where the crater now exists, but that the strain would occasion a number of lateral fissures corresponding with those called in the island Barancos.

Considering therefore that the crater in this instance is unattended with the usual phænomena of a volcano, and is even distinguished from the latter by the preceding characters, Von Buch has chosen to denote it by the name of " Erhebungs Crater" or Crater of Elevation, and he proceeds to shew that the same distinctive title is applicable to many craters both among these islands and in other parts of the globe.

The structure of the Island of Great Canary is very similar to that of Palma,—the same heaving up of the strata round a central point, the same deep and abrupt Barancos, the same description of crater exhibiting the successive outcrops of the adjoining beds.

The order of superposition in the latter is such as to illustrate apparently the gradation that often occurs in the character of volcanic products, and perhaps the manner in which they have been derived by successive changes from the fundamental granite. Lowest of all Von Buch descried the primitive rocks ; then masses of trachyte ; afterwards an aggregate consisting of angular fragments of the latter rock, forming either a conglomerate or a tuff, which alternate with one another several successive times ; still higher an augite rock (dolerite) with felspar, interstratified with beds of rolled masses of the same composition, but of a cellular structure ; then an amygdaloid ; and last of all basalt.

The structure of Fortaventura is also similar, but Lanzerote, though originally raised up in the same manner as the other islands, has since been augmented by the eruptions of volcanic matter that have subsequently taken place upon its surface. Lanzerote is distinguished from the other Canary Islands by its comparative flatness, possessing none of those lofty precipices, or abrupt conical hills, that occur in the rest. There is nevertheless in one corner of the island a vestige of the same kind of crater which I have just been noticing, but, a part being in all probability sunk in the ocean, the strata are only seen rising on one side, which faces the water.

Von Buch has given a striking description of the aspect of that portion of the island, from whence proceeded the lava which in the year 1730 caused so much devastation.

After a painful walk, he says, over a tract of harsh, un-decomposed lava, I reached at length an eminence composed entirely of an accumulation of slag and lapilli, which were heaped in successive layers upon each other. In the centre was a crater walled in by precipitous rocks, of which one side was broken away by a lava which had proceeded from its interior. Within the compass of this hollow two other minor craters appear, which emitted at the time volumes of aqueous vapour mixed with sulphureous exhalations. Hence it is that the hill has obtained the name of Montagna di Fuego.

It is impossible, continues Von Buch, to describe the scene of desolation, which presents itself from the summit of this crater. A surface of more than three square miles in a westernly direction is covered with black lava, in the whole of which space nothing occurs to break the uniformity of the prospect, but occasional small cones of basalt scattered over the plain.

It is clear that this vast mass of lava is not derived from any one point; even the Montagna di Fuego appearing to have contributed but little to its formation, for the lava actually proceeding from the latter is found to take an easternly instead of a westernly direction. During my ascent, I felt very anxious to ascertain, what the other sources might be which assisted in emitting so vast a mass of lava. How much was I astonished, when on reaching the summit I perceived an entire series of cones, all nearly as lofty as the Montagna de Fuego, placed so exactly in a line, that the nearest covered the farther ones in such a manner, that their summits alone were seen peeping from behind.

Between the western coast and the little village of Florida, I counted twelve cones of larger size, of which the Montagna di Fuego was the sixth in the series, besides a considerable number of smaller cones, partly between and partly on the side of the larger ones. It was an exact repetition of the phænomena of Jorullo, or of the Puys in Auvergne.

LANZEROTE.

The whole of this eruption proceeded in all probability from a large fissure, the existence of which is in all cases found to produce effects of the more alarming kind, the more distant it is from any volcano, the latter serving as a sort of chimney for the escape of the matter within.

On my road to Florida, I visited several such cones. They all alike consist of heaps, three or four hundred feet in height, of harsh, porous, sharp, lapilli, of the size of a bean, which cause a grating sound when they roll upon each other.

These craters open for the most part towards the interior of the island, where the streams of lava unite to from one vast continuous bed, which, the farther we trace it from its source, is found to be less and less charged with olivine.

The larger part of these effects is to be attributed to the great eruption, or rather series of eruptions, which took place in this island between the 1st of September, 1730, and the 16th of April, 1736. The details are given in the interesting memoir by Von Buch, which has been already referred to, but it would be inconsistent with my plan to do more than particularize some of the leading features. A number of rents took place successively in the island, generally occurring in the same direction.[*] From these issued in all cases flames and smoke, and in the majority loose fragments of volcanic matter and streams of melted lava. The former, accumulating round the apertures from which they were ejected, often formed conical hills of considerable height; the latter, taking different directions, ravaged various parts of the island, and in general continued to flow on until stopped by the sea. In one case the lava was diverted from its original direction by an·huge rock which suddenly rose

---

[*] The reader will immediately be led to consider the effect itself as analogous to that noticed under the name of cavernous lava, in Sir G. Mackenzie's Description of Iceland.

in the midst of it, but of which no vestiges are to be seen at present. Gaseous exhalations likewise were emitted, which proved fatal to the cattle. At length the inhabitants wearied out by such a series of misfortunes, seeing the most fertile parts of their country successively reduced to irretrievable ruin, and despairing that the eruptions would ever cease, determined on leaving their homes, and took refuge in the neighbouring island of Great Canary.

One of the most curious phænomena attendant on this eruption, though one not altogether peculiar to it, was the rise of flames from the midst of the sea. The nature of these is worth enquiring into, as it may hereafter assist us towards a theory of volcanos.

The only gas at present known, which inflames spontaneously at ordinary temperatures on the surface of water, is phosphoretted hydrogen, and this can hardly be suspected, as the product of its combustion is phosphorous acid, a substance which is possessed of several striking properties, and therefore could hardly fail to be detected, if it were ever produced by volcanic action.

It is true that phosphoric acid exists in the cavities of certain volcanic rocks, as in those of Estremadura, combined with lime and other earths, but I do not know that it has ever been detected in a free or uncombined state among the products of any active volcano.

Are we at liberty to suppose, that some less inflammable gas, such as simple hydrogen, or its combinations with carbon and sulphur, might, during their rapid ascent through the water, retain a sufficiently exalted temperature, to inflame spontaneously on coming in contact with the air. Such an hypothesis might explain the circumstances related respecting the constant combustion of the gas at the Pietra Mala, without recurring to the somewhat forced solution, that it was inflamed in the first instance by the application of a light. That this is often the cause of the flame, is pretty clear; but that it always is so, does not seem altogether established.

Since the eruption of 1730—36, the Island of Lanzerote enjoyed a state of tranquillity until the 29th of August in last year, when at the port of Rescif and its environs earthquakes occurred, which became more terrible at night. They increased in violence the next day, and on the 31st, at seven in the evening, a volcano broke out a league from the harbour of Rescif, and half a league from the mountain called Famia. It vomited from its crater terrible flames, which lighted up the whole island, and stones of an enormous size reddened by the fire, and in such large quantity, that in less than twenty-four hours they formed a mountain of considerable size. This eruption continued till ten in the morning of the 1st of September; when the volcano seemed to close, and to leave only fissures, from whence escaped a thick smoke, which covered all the neighbourhood. On the 2d there formed three great columns of smoke, each of a different colour, one perfectly white, the other black, the third, which was the farthest off, red.

This volcano, says the account, still burns over a space of half a league in length, and a quarter in breadth, and the mountain newly formed appears to be inaccessible, and does not exhibit lavas in any direction.

On the 4th a large column of smoke rose from the volcano, and on the 22nd of September it became again active, and poured forth a quantity of water so considerable as to form a large stream, which diminished on the 23d, and on the 26th had nearly disappeared.—Bulletin de Sciences. May, 1825, copied from the Constitutionel for 23d October, 1824.

If we believe Von Buch, the Island of Madeira is formed after the same manner as the Canaries, consisting of beds which have been elevated above the level of the ocean by elastic fluids, but destitute of any crater from whence smoke and lava have been ejected.

The account however, which Mr. Bowdich, in his posthumous work just published,* has given of the physical structure of the island, is somewhat different. That it is composed of volcanic matter cannot indeed be doubted, for the rocks principally consist of tuff, scoriæ, or basalt. The former sometimes contains bands, as it were, of pumice, traversing it in the direction of the bed; it is sometimes yellowish and sometimes red; it alternates repeatedly with the scoriæ, which again are continuous with a very cellular variety of the basalt, forming a connected bed or current, the direction of which may be presumed to be indicated by that of the cells, which are oval and elongated all in one direction.

Besides this form of basalt, a compact and columnar variety is also seen, which in many places appears to cap all the other strata, but in others is itself covered by a thin stratum, consisting of fragments of porous basalt, cemented by a yellow tufa. This conglomerate likewise insinuates itself into the crevices between the other strata. Compact basalt however is not confined to the uppermost beds, for in the centre of the island it is found immediately incumbent on a limestone, which Mr. Bowdich considers to belong to the transition series, and which is the fundamental rock hitherto discovered, being 700 feet in thickness.

All these beds, even the limestone, are traversed by very numerous dykes of basalt, which appear to be always compact, and have a natural cleavage at right angles to their walls.

Mr. Bowdich describes a small conical hill situated on the high table land in the centre of the island, which has on its summit a small elliptical cavity, about 200 feet in diameter, and 54 in depth.

No stream of lava however has flowed from it, nor has it

* Account of Madeira and Porto Santo, London, 1825.

any of the usual appearances of a crater. He also notices
an elliptical funnel-shaped depression about 80 feet above
the sea, between the fort and Praya Bay, which seems to
answer to the description given by Von Buch of a crater
of elevation. It presents every evidence, says Bowdich, of
having been formed by a minor volcanic heave, which threw
up vast blocks of the rock it rent from beneath the ocean
to form a passage, but did not eject any lava or contents
of its own.

The strata here are stated to be inclined, but it would
have been more satisfactory if Mr. Bowdich had informed us
whether they dip in all directions away from the crater, as
we should expect to be the case.

Granting however these cavities to have been produced
in the manner Von Buch represents, it is still impossible to
reconcile his view of the general elevation of the island by
volcanic agency to the statements given by this last traveller.

It is true that there exists in the interior of the island a
vast fissure, 1634 feet deep, and 3700 feet above the sea,
" where," says Bowdich, " the basaltic rocks that compose
the mountains, seem to have been blasted and shivered by
the great convulsion which rent the foundation strata, so as
to create at once this stupendous valley, enlarged and
deepened by the action of torrents which have battered it
for ages."

But besides that the strata are not said to have that disposition, which is necessary in all cases to establish the fact
of their having been heaved up, the circumstance of their
resting upon an horizontal bed of limestone 700 feet thick,
seems fatal to such an hypothesis. This, as Mr. Bowdich
very justly remarks, would demonstrate that Madeira pre-
existed as a mass of transition, or probably of primitive and
transition rocks, which were covered and elevated by successive streams and ejections of basalt and tufa, derived from
some submarine volcano.

I should therefore hope that the next person who visits

the Madeiras with scientific views, will particularly attend
to the stratification of the volcanic rocks of Coural, and of
the limestone underneath, in order to ascertain whether they
possess the inclination which seems implied in Von Buch's
statement, or owe their irregularity to the hills and vallies
existing on the surface of the limestone on which they were
deposited, as Mr. Bowdich has represented.

Another interesting enquiry relates to the manner in which
these volcanic strata have been formed, whether from a
crater as Mr. Bowdich supposes, without however being able
to discover any traces of its existence, or through the
medium of dykes, as might be presumed to be the case, from
the vast number which penetrate all the rocks even down to
the fundamental limestone. At present we must be guided
on this point by analogy, rather than by actual evidence, so
that the subject will come more naturally under consideration in the latter part of my undertaking, in which I shall
attempt to deduce some general inferences from the facts
previously detailed.

The Island of Porto Santo differs from that of Madeira,
in the occurrence of tertiary sandstone and limestone alternating with the volcanic strata.* The lowest visible deposit
is a calcareous tufa of a greenish grey colour, which extends
on the N.E. parts of the island to the height of 1600 feet,
and is ribbed throughout with vertical dykes of a reddish
brown basalt.

In the middle of the island there is a depression in the
tuff, and in this basin a bed of sandstone has been deposited, the lowest part of which is hard and solid, of a reddish
buff colour, and slaty structure, with indurated veins effervescing pretty vigorously, and presenting small black spots,
apparently ferruginous, the upper of a looser consistence,
and containing helices, bulimi, ampullariæ, and other shells,
belonging to a recent epoch.

This sandstone seems to be of more recent origin than the

* See Bowdich's work.

volcanic matter, for it is not traversed by any dykes. In the neighbouring island of Basco, the calcareous tufa is covered with beds of limestone 100 feet above the level of the sea, alternating with a conglomerate, in which nodules of basalt and wacke are inserted in a ferruginous sandy earth of a brick and dull orange red, traversed by veins of mammillated carbonate of lime. This limestone presented immense masses of Lamarck's cateniporæ (tubifera catinulata) and contained specimens of the cardium, mytilus, venus, voluta, turritella, conus, and other shells. These beds are traversed in the same manner with dykes.

The tuff of Porto Santo is sometimes covered with a compact basaltic rock approaching to phonolite, which forms the peaks in the north-eastern part of the island. It is distinguished by its vitreous colour, numerous crystals of glassy felspar, and lamellar structure; immediately beneath it are the dykes which descend through the tufa to the sea. Mr. Bowdich observed in one of them a deposit of native alum, derived probably from the decomposition of sulphuret of iron, and another of a bright orange ochre, which also appeared to have proceeded from an altered condition of the basalt.

There are no traces of primitive or secondary formations in Porto Santo, as there are in Madeira, but the calcareous tufa, which, from its association with the limestone, Mr. Bowdich is disposed to consider tertiary, seems to form its base, and at all events descends lower than the present level of the sea. The horizontality of the beds is such, that it would be even more difficult to admit the elevation of the strata in this island, than in the neighbouring one of Madeira.

The structure of the Cape de Verde Islands is too imperfectly known to detain us long; they are said to consist principally of volcanic matter, and the Island of Fogo contains an active volcano.*

---

* See a notice in the 1st volume of the 2nd series of the Geological Transactions, p. 120, and Bowdich's Posthumous Work.

Of the Azores we have lately obtained a geological account from my friend and fellow-student Dr. Webster, of Boston, in the United States.*

The Island of St. Michael, the largest of them, is described by him as entirely volcanic, and containing a number of conical hills of trachyte, several of which have craters, and appear at some former period to have been the mouths of so many volcanos. The trachyte is completely covered by the ejections of pumice and obsidian that have proceeded from these sources, and it is seldom that we are able to discover it, except in the ravines, or on the summit of the hills.

This rock however, so far as I can collect from Dr. Webster's description, does not constitute the fundamental stratum, it has rather been forced through strata of tuff and of basalt, which extend over the island, covered however in most places by the ejections of pumice and obsidian. Dr. Webster notices likewise a description of lava which he compares to the cavernous variety mentioned by Sir G. Mackenzie; it is swollen into large blisters and other irregularities, and contains several remarkable caves, from the roof of which hang stalactites, as they may be termed, of lava, which assume a variety of curious arborescent figures.

The Island of St. Michael is famous for its hot springs, which are impregnated with sulphuretted hydrogen and carbonic acid gases, thus seeming to attest that the volcanic action is still going on. Of this however, evidences have, within the present century, been afforded of a far more direct and positive nature. In the year 1811 a phænomenon occurred, similar in kind to that, which I have already described as having happened in the Grecian Archipelago. After a succession of earthquakes experienced more or less sensibly in all the neighbouring parts, a new island arose in the midst of the sea, of a conical form, and with a crater on its summit, from which flame and smoke continually issued. The island, when visited soon after its appearance by the crew of the frigate Sabrina, was about a mile in circum-

* Webster. Description of the Island of St. Michael, &c. Boston, 1822.

ference, and two or three hundred feet above the level of
the ocean, it continued for some weeks, and then sunk again
into the sea.*

A singular phænomenon occurs at a short distance from
the coast near the town of Villa Franca, which has all the
appearance of the crater of a volcano half sunk below the
level of the water. It consists of a circular basin, the interior of which contains enough water to float a small vessel,
but the sides are elevated 400 feet above the sea. The basin
is perfect except on one side, where the depth of water is
sufficient to allow a ship to enter, and it would ride securely
in the interior, were it not for a sinking of the basin on one
side, which allows of the sea breaking over it, when the wind
is in a south-easterly direction.

The sides of the basin consist of tuff, presenting an
abrupt precipice externally, but dipping in a more gradual
manner towards the interior, and Dr. Webster observes that
the strata all slope towards the interior and not towards the
circumference, so that the structure is the very reverse of
that described by Von Buch as belonging to the waters in
the Canary Islands, and the circular cavity must in this case
have been formed, not by an uplifting, but by a subsidence
of the strata.

The following sketch may represent the appearance of the basin.

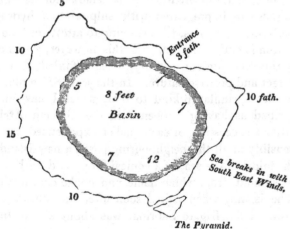

The Pyramid.

* Humboldt (Pers. Narr. vol. i. p. 243) remarks, that the formation of the
island was anterior to that of the crater.

There also occurs in the midst of the sea a pyramidal mass of tuff only 30 or 40 feet in diameter, consisting of horizontal strata from one to two feet in thickness. It is evidently the remnant of a formation, the greater part of which has been washed away, and seems to be analogous to that which I have noticed, as occurring in the valley of the Puy en Velay, in the south of France.

The other islands belonging to the same groupe appear to be similarly constituted. That of El Pico is the only one which contains a volcano at present in activity, for the great currents of lava which flowed in 1812 from the adjoining island of St. George, are considered as the results of a lateral eruption from this volcano. The summit of El Pico is no less than 9000 feet above the sea; it consists of a conical mass of trachyte, and is constantly emitting smoke.

It may perhaps be inquired, what degree of light geology is capable of affording with respect to the existence of a former continent or large island, serving to connect Europe with America, which the antients sometimes allude to under the name of Atlantis.

It might appear at first sight, that the knowledge we have of late obtained with respect to the physical structure of the islands, which in such case must be regarded as the relics of this submerged country, lends some weight to the historical evidence in favour of its existence, since the volcanos that are proved to be in action in so many parts of the intermediate tract of ocean, might afford an adequate explanation of the *effect* supposed to have taken place.

But on the other hand, when we examine more narrowly into the appearances presented in those islands which I have been describing, we shall find it greatly more probable that they have been separately raised from the bottom of the sea by volcanic agency, than that they have been severed apart, after having once constituted a single continuous tract of land. The details which I have extracted from Von Buch's interesting memoirs lead inevitably to this conclusion, as they tend to demonstrate, that the strata which form the basis of these islands, and through which the volcanic cones, where such exist, have been protruded, were formed originally at the bottom of the sea, and have been afterwards heaved up by the elastic vapours acting from beneath.

It may be alledged indeed, that the whole of these islands do not consist of volcanic matter, for Gomera, Fortaventura, and others, contain rocks either consisting of primitive materials or of limestone, but the rare occurrence of the latter over a tract so vast, as that which the Continent of Atlantis must be supposed to have occupied, certainly lends but little countenance to the hypothesis. It may be remarked too, that volcanos seem much more active in building up strata than in destroying them, and that, if they had operated on so extensive a scale as must be assumed by those who attribute to their agency the destruction of a continent, they would on the other hand have raised up more extensive tracts of volcanic materials in the place of those they had been the means of subverting. It is sufficient to cast a glance over the conjectural map of the Atlantis, by Bory St. Vincent,* to be convinced of the absurdity of any such hypothesis, for who can imagine a tract of land extending from 40 to 15 N. latitude, having for its northern boundary the Azores, for its southern the Cape de Verde Islands, and for its eastern promontory the Peak of Teneriffe, to have been swallowed up in the ocean by any causes now in operation, or at any period since the creation of man.

For my own part, were I persuaded that the Atlantis of the Greeks referred to any thing real, and was not throughout a figment of the imagination, the hypothesis I should be most inclined to adopt, is that proposed by Ali Bey in his Travels.† We find it there stated, that the whole of the flat coast towards the Atlantic Ocean is caused by the wash-

---

* See Bory St. Vincent sur les Iles Fortunées.
† Tom. 1. p, 36—37.

ings of the sea, by the ground sand carried into the ocean, and by masses of clay which he considers as the product of some submarine volcano. These materials together form a sort of tufa, upon which beds of marl and of animal exuviæ have since been deposited. He finds these desert strands on the whole of the southern border of the table-land of Mount Atlas, towards the deep-laying flat of the Sahara, and extending as far as the Syrtis, and therefore supposes, that at some former period the elevated ground of Mount Atlas may have constituted a sort of island—perhaps that of Atlantis.

This notion as to the volcanic origin of certain parts of Mount Atlas, is borne out by the existence of some mountains in the same chain, called Black Harusch, which are conjectured to be of an igneous nature.

It would also appear from the author of the work— " Περι θαυμασιων ακουσματων" vulgarly attributed to Aristotle, that the Greeks regarded the whole of the coast of Africa beyond the Pillars of Hercules as thrown into disorder by the fire of volcanos, and Solinus expressly states that the snowy summit of Mount Atlas glitters with nightly flames ; so that we might be led to infer, either that this mountain was situated ·in one of the islands of the Hesperides, and was probably identical with the Peak of Teneriffe, or that some volcanic appearances exist in the chain which the moderns speak of as Mount Atlas.

In the account of the Periplus of Hanno, as well as in that of Eudoxus, as quoted by Mela, (which is considered a fabrication, and a direct copy of the former,) we read, that as these navigators were coasting in the above direction along this part of Africa, torrents of light were seen to fall on the sea, that every night the shore was covered with fires, that the Great Mountain, called the Car of the Gods, (θεων οχημα) had appeared to throw up sheets of fire that rose

even to the clouds, and that the sand of the sea was into-
lerably hot.

Now the mountain that went by this name is placed by
Polybius south of the town of Lixus, a Carthaginian colony
beyond the pillars of Hercules, and in the midst of the chain
of Mount Atlas. Nor is it probable, that this antient navigator should have deviated so far from the coast, as would
have been necessary, in order to enable him to catch a
glimpse of the volcanic fires of Teneriffe.

It is remarkable indeed that the antients, although they
have described the Canaries, under the names of Canariæ,
Purpurariæ,* and Fortunate Islands, seem to have taken no
notice of the volcanos that occur there. Even the Peak of
Teneriffe is only alluded to in the mention of the perpetual
snow found upon it, from whence the island obtained the
name of Nivaria.

It is therefore most probable, that the description of Hanno
refers to the Continent of Africa, and not to its islands; but
before we decide that any thing of a volcanic nature is
hinted at, we ought to recollect, that the custom† which
exists there, as in many other hot countries, of setting fire at
certain seasons to the forests and dry grass, might have given
rise to the statements of the Carthaginian navigator Even
in our own times the Island of Amsterdam was set down as
volcanic from the very same mistake.

ASCENSION ISLAND.

The Island of Ascension‡ is composed entirely of trachyte, and the rocks allied to it, which appear to have been

* Bory St Vincent thinks, that by the Purpurariæ, Madeira, and Porto
Santo, were intended.

† This is the opinion of the Abbe Gosselin in his Geographie des Anciens.

‡ See Ed. Phil. Journ. for January, 1826.

ST. HELENA.

thrown up in a dome-like form, and to have subsequently afforded a vent to vesicular, spumous, and corded lavas, by which they are frequently covered. On one of its hills, about 700 feet above the level of the sea, is a circular hollow, which has since been in great measure filled up. The beach is formed of a sand composed of comminuted shells, with fragments of echini and corals, which in parts near the sea is sometimes agglutinated by a calcareous cement into a pretty solid mass. The whole island has a most rugged and forbidding aspect. Its highest mountain is 2818 feet above the level of the sea, and is covered at top with pumice, having upon it a scanty covering of soil.

ISLE OF ST. HELENA.

It has been supposed by Forster, that the present island is only the wreck of a larger tract of land, submerged by the action of volcanos.

This however does not seem consistent with more recent statements, which lead rather to the idea, that the whole island has been raised above the surface of the sea by the same agency.

It would appear from Governor Beatson's account,[*] that the volcanic matter rests on a substratum of sand, clay, and coral, which extends under the sea over a much larger surface.

The same remark applies to the Islands of Ascension and Tristan d'Acunha, and to Gough's Island.

Neither is there that general appearance of fracture and disturbance in the strata, which would have been observed had there been any truth in Forster's idea. The volcanic matter chiefly consists of a rock of a basaltic character,

[*] See Beatson's St. Helena, London, 1816.

either cellular or compact, together with a species of tuff or puzzolana. Near the level of the sea is a conglomerate of recent formation, consisting of shells and sand agglutinated together. There exists a crater, called the Devil's Punch Bowl, in the interior of the island, 1000 yards from east to west, 700 across, and about 250 yards in depth; but it does not appear that any lava has been given out from it.

The islands on the opposite coast of Africa afford abundant evidences of igneous action, though one only is known to possess an active volcano.

Madagascar, the largest of them, is as yet but little explored, and we only know that its structure is very different from that of the adjacent continent; but of the Isles of France and of Bourbon there is an account by Bory St. Vincent, of a more circumstantial nature.*

The Isle of France is of an oval shape, about 11 leagues in its greatest length, and 8 in its greatest breadth; it rises on all sides from the circumference to the centre, so as to form a conical mountain, called Le Piton.

The other mountains, of which the most elevated is Peter Boot, constitute a chain extending across the island.

All these mountains consist of volcanic matter, either lava or basalt, which is sometimes prismatic, and intersected by dykes. The strata have a slight inclination towards the sea.

The low region to the north, and the isolated rocks off the coast, consist of a coralline limestone, produced at a very recent epoch. It lies in a sort of basin, enclosed in the midst of volcanic rocks, and intermixed with lava, so that it would appear that it was of posterior formation to certain of the latter products.

* Voyage ause Iles d'Afrique.

ISLE OF BOURBON.

The Island of Bourbon, like the Island of France, slopes
on all sides upwards towards its centre.

It may be viewed as consisting of two volcanic mountains,
of different dates—the southern, which is the smallest, still
in a state of activity; the western, extinct.

The point of contact between the two coincides with a
line drawn from N.E. to S.W., along which the country is of
inferior elevation.

The mountain of Salazes, which is the active volcano,
contains two craters; the first of which, the crater Com-
merson, is composed of basaltic lava having a prismatic form,
and therefore seeming to indicate, that the expansive force
had in this instance made its way through volcanic matter of
more antient date.

The extinct volcano, called Les Fournaises, exhibits a
number of basaltic dykes, which, it is to be remarked, are
never observed in the one now burning.

Some of these dykes seem no larger than cords, whilst
others are of the same thickness as the beds which they
traverse. They penetrate indiscriminately all the rocks,
sometimes in a direction perpendicular to their planes, at
others in one more or less oblique. They also intersect
each other in the figure of the letter X. They are usually
divided into prisms, which, whatever may be the inclination
of the dyke, range at right angles to it. Being more compact than the lavas they penetrate, they frequently stand
out in bold relief.

It is worth observing, that earthquakes are only experienced in that part of the island, which is farthest from
the site of the active volcano, and even there but rarely.

The most curious production of this island is the variety
of pumice like spun glass, which is thrown out occasionally
by its volcano.

Bory St. Vincent describes the quantity of these films as

being on one occasion sufficient to form a cloud, covering the entire summit of the volcano.

Scarcely had he observed it, when the whole party found themselves covered with small shining and capillary flakes, possessing the flexibility and appearance of silk, or of a spider's web. This substance was accompanied with showers, of light, vitreous, spongy scoriæ, in fragments varying from the size of a cherry to that of an apple. It fell into powder with the slightest force.

The threads, of which we have just been speaking, appeared to him nothing but a modification of the scoriform lava peculiar to the Isle of Bourbon. He supposes, they may have been formed, owing to the extrication of elastic matter from this substance, whilst in a state of partial fusion, on the same principle that the threads are formed in sealing-wax, when the stick is suddenly withdrawn from the surface of a portion, dropped upon paper, and not completely cooled. He was confirmed in this opinion by observing, attached to these threads, little pear-shaped globules, which were found on examination to be identical with the vitreous scoriæ before alluded to.*

———

Volcanos therefore seem pretty abundantly distributed on all sides of the African coast, but whether they exist also on the continent, we have not the means of determining.

I have hinted above at the possibility, that volcanic rocks may be met with in the chain of Mount Atlas, on the western extremity of Africa; on the northern coast we know of no active volcano, nor have any rocks been noticed even of a basaltic nature, excepting in Fezzan and Tripoli, where Captain Lyon speaks of a formation of that sort associated

———

* Similar specimens have been noticed as occurring in the Island of Guadaloupe, and by Mr. Ellis in Owhyhee. See his late Tour in the Sandwich Islands. London, 1826.

EGYPT.

with calcareous beds, which I have the authority of Professor Buckland for pronouncing tertiary.*

There is also an account in the 11th volume of the correspondence of Baron Zach (No. 3.) respecting some rocks in the interior of Egypt, between the Nile and the Red Sea, in the province called Kordoufan.

Rüppell, a German traveller, on whose authority these facts are given, states merely, that there exists in that country a chain of some half-extinguished volcanos of great interest, at a place called Gebel-Koldagi, where a conical mountain of great height, smokes continually on the summit, and throws out hot cinders without intermission.

Mons. Jomard, who accompanied Buonaparte in his invasion of Egypt, states, that in the centre of that country, between the Nile and the Red Sea, in the midst of the Alabaster Quarries, there exists a mountain called Djebel Dokhan, which means the Mountain of Smoke. The Arabs speak of the Petroleum observed to flow some distance off. Djebel Kebryt, or the Sulphur Mountain, lies more to the south, in the 24th degree of latitude, and on the borders of the sea. According to the reports of the Arabs, Djebel Dokhan smokes continually.

It is probable that Rüppell and Jomard both refer to the same chain of mountains, but whether the facts they notice relate to some volcano, or to phænomena similar to those of Macaluba in Sicily, or the Pietra Mala in the Appennines, may admit of doubt.

It must be remarked that neither reporter has visited the spot to which he refers, and that it is not stated even on the authority of the Arab informants, that lava had been ejected from any of the mountains specified.

* As quoted in Brongniart's work, " Sur les environs de Paris." Dr. Oudney also speaks of these rocks between Tripoli and Mourzouk. See Denham and Clapperton's Travels in Africa, p. xxix.

ON THE VOLCANOS OF ASIA.

WHATEVER opinion may be entertained with respect to
the existence of volcanos in the interior or coast of Egypt,
there is no question as to their presence on the opposite
shore, as well as in the Red Sea itself.

The Island of Zibbel-Teir, in north latitude 16°. appears,
from Bruce's statement, to contain an active volcano, and
rocks possessing the same characters are mentioned, as occurring in a groupe of smaller islets in the same part of the
Red Sea, off Loheia.*

Niebuhr likewise has given some accounts of phænomena
allied to those arising from volcanos, and states in particular
that in the valley of Girondel, near Suez, he met with some
hot sulphureous springs, on the spot, near which, according to vulgar tradition, Pharoah and his host were swallowed up.†

Lord Valentia‡ likewise remarks, that the rocky Peninsula, on which the town of Aden, at the extremity of the
Arabian Gulph, is situated, has all the appearance of being
the half of a volcano, the crater of which is covered by the
sea, whilst the edge of it is occupied by the present town;
from which the rocks rise to a considerable height.

The unfortunate Seetzen,|| who is supposed to have been
poisoned during his travels over this part of the Arabian
Peninsula, notices particularly the mountain near Aden, as
containing a kind of volcanic rock, nearly as light as pumice,
and mentions the account which he received of an eruption

* Bruce's Travels, vol. i. p. 330 & 340.
† Niebuhr, vol. i. p. 184.
‡ Travels, vol. 2. p. 86.
|| Some account of Seetzen's researches may be seen in Baron Zach's
Correspondence— in Hammer's publication, entitled Fundgrube des Orients,—
and in a small volume entitled, " A brief account of the countries adjoining
the Lake Tiberias, the Jordan, and the Dead Sea. London, 1810."

which took place on the site of its present harbour. The same traveller ascertained the existence of porous lava at Damar, about 15° of N. latitude, and traced the same in various situations between that place and Mecca.

Similar appearances are exhibited about Medina, which is perhaps not more than 200 miles south of Sherm, where we have the authority of the accurate Burckhardt for the existence of volcanos.

Neither have we any reason to doubt the facts detailed by Seetzen, for Von Hoff, a writer, whom I have already had occasion to notice with approbation, assures us from personal knowledge, that this traveller was fully competent to render a correct account of the physical structure of the places he visited.

At Sherm, in the Peninsula of Mount Sinai, the hills for a distance of two miles presented, says Burckhardt, perpendicular cliffs, formed in half circles, none more than 60 or 80 feet in height; whilst in other places there was the appearance of volcanic craters. The rock, of which these mountains are composed, is black, with a slight tinge of red, full of cavities, and with a rough surface; fragments that had been detached from them were seen lying on the road. The cliffs were covered by deep layers of sand, which also overspread the vallies.

Burckhardt thinks it probable that other rocks of the same kind may be found near Ras Abou Mohammed, and that the name of Black Mountains (μελανα ορη) applied to them by the Greeks, may have arisen from this cause.* It should be observed however, that low sand hills intervene between the volcanic rocks and the sea, and that above them towards the higher mountains no traces of lava are found, which circumstance seems to prove that the volcanic matter is confined to this spot. Burckhardt adds, in a letter to the Association,†

---

* Burckhardt's Syria, p. 529.

† Burckhardt's Nubia, Life, &c. p. lxviii.

that the Arabs, as well as the priests of the convent, mention, that loud explosions are sometimes heard, accompanied with smoke, proceeding from a mountain called Om Shommar, eight hours S. S. W. of Djebel Mousa, where however he searched in vain for any traces of the kind.

Humboldt observes, that he is satisfied from the numerous specimens which have fallen under his observation, that the rock from which these volcanos have proceeded, is a transition porphyry, like that of Mexico.

If we proceed northwards from this Peninsula to Palestine, we shall meet with abundant evidences of igneous action to corroborate the accounts that have been handed down to us by antient writers, whether sacred and profane, from both which it might be inferred, that volcanos were in activity, at a period even so late as to be included within the limits of history.

From their familiarity with such phænomena, the Prophets seem often to have derived some of their most splendid imagery. Thus Nahum, describing the majesty of God, says, that, *the mountains quake at him, and the hills melt, and that the earth is burned at his presence. His fury is poured out like fire, and the rocks are thrown down by him.* (Nah. i. 5. 6.)

Behold, says Micah, *the Lord cometh forth out of his place, and will come down, and tread upon the high places of the earth. And the mountains shall be molten under him, and the vallies shall be cleft as wax before the fire, and as the waters that are poured down a steep place.* (Mic. i. 3. 4.)

O that thou wouldest rend the heavens, says Isaiah, *that thou wouldest come down, that the mountains might flow down at thy presence. As when the melting fire burneth, the fire causeth the waters to boil, to make thy name known to thine adversaries, that the nations may tremble at thy presence. When thou didst terrible things which we looked not for, thou comest down, the mountains flowed down at thy presence.* (Isaiah, lxiv. 1. 3.)

And Jeremiah, evidently alluding to a volcano, says— *Behold, I am against thee, O destroying mountain, saith the Lord, which destroyest all the earth, and I will stretch out mine hand upon thee, and roll thee down from the rocks, and will make thee a burnt mountain. And they shall not take of thee a stone for a corner, nor a stone for foundations ; but thou shalt be desolate for ever.* (Jer. li. 25. 26.)

The destruction of the five cities on the borders of the Lake Asphaltitis or Dead Sea, can be attributed, I conceive, to nothing else than a volcanic eruption, judging both from the description given by Moses of the manner in which it took place,* and from the present aspect of the country itself.

I presume it is unnecessary to urge, that the reason assigned in Holy Writ for the destruction of the cities alluded to, does not exclude the operation of natural causes in bringing it about, and that there can be no greater impropriety in supposing a volcano to have executed the will of the Deity against the cities of Sodom and Gomorrah, than it would be to imagine, if such an idea were on other grounds admissible, that the sea might have been the instrument in the hands of the same Being for effecting the general destruction of the human race in the case of the Deluge.

---

* The following are the words of Scripture :
Gen. Ch. xix.
Vs. 24. Then the Lord rained upon Sodom and Gomorrah brimstone and fire out of heaven.
   25. And he overthrew these cities, and all the plain, and all the inhabitants of these cities, and that which grew upon the ground.
   26. And he (Abraham) looked toward Sodom and Gomorrah, and toward all the land of the plain, and behold, and lo, the smoke of the country went up as the smoke of the furnace.

In Deut. ch. 29, vs. 23, the neighbourhood of the Dead Sea is described as a country, the land of which is brimstone, and salt, and burning, which is not sown nor beareth, nor has any grass growing therein.

DEAD SEA.

Whether indeed we chuse to suppose the fire which laid waste these places, to have originated from *above* or from *below*, the employment of secondary causes seems equally implied; and if it be urged, that the words of Genesis denote that it proceeded from the former quarter, it may, I think, be replied, that a volcanic eruption seen from a distance might be naturally mistaken for a shower of stones, and that we cannot expect from the sacred historian in the case before us, any greater insight into the real nature of such phænomena, than we attribute to him in the analogous instance, in which the Sun is said to have stood still at the command of Joshua.

That the individuals who witnessed the destruction of these places might have been impressed with this notion, will be more readily believed, when we reflect, that in most eruptions the greater part of the mischief occasioned proceeds from the matters ejected, which are often *perceived* only to fall from above; and those who recollect the description given by the younger Pliny of that from Vesuvius, will admit, that a person who had fled from the neighbourhood of that volcano, as Lot is stated to have done from the one near the Dead Sea, at the commencement of the eruption, would probably have formed the same idea of what was taking place; for it appears from the Roman writer, that it was long before he was enabled, even at Misenum, to determine in the midst of the general obscurity, that the cloud of unusual appearance, which was the precursor of the volcanic phænomena, proceeded from the mountain itself.

When Livy mentions the shower of stones, which, according to common report, fell from heaven on Mount Albano,* there can be little doubt, that the phænomenon that gave rise to such an idea was of an analogous description, and we shall see hereafter, that the volcanic action, of which there are such decided evidences in Phrygia, was attributed

* See page 130.

by some to heavenly meteors : " εικαζουσι τινις," says Strabo, " εκ κεραυνοβολιων και πρησηρων συμβηναι τυτο."

As therefore we have no authority for supposing Moses a natural historian, or for imagining that he possessed a knowledge of physics beyond that of the age in which he lived, we may venture to apply to his narrative of the destruction of these cities the same remark, which Strabo has made respecting the indications of igneous action presented by the country round Laodicea. " ουκ ευλογον υπο τοιυτων παθων την τοιαυτην χωραν εκπρησθηναι αθροως, αλλα μαλλον υπο γηγενυς πυρος."

Volney's description of the present state of this country, fully coincides with this view.*

The south of Syria, (he remarks) that is, the hollow through which the Jordan flows, is a country of volcanos : the bituminous and sulphureous sources of the lake Asphaltitis, the lava, the pumice-stones thrown upon its banks, and the hot-baths of Tabaria, demonstrate, that this valley has been the seat of a subterraneous fire, which is not yet extinguished.

Clouds of smoke are often observed to issue from the lake, and new crevices to be formed upon its banks. If conjectures in such cases were not too liable to error, we might suspect, that the whole valley has been formed only by a violent sinking of a country which formerly poured the Jordan into the Mediterranean. It appears certain, at least, that the catastrophe of five cities destroyed by fire, must have been occasioned by the eruption of a volcano then burning. Strabo expressly says, " that the tradition of the inhabitants of the country (that is of the Jews themselves) was, that formerly the valley of the Lake was peopled by thirteen flourishing cities, and that they were swallowed

* Travels in Egypt and Syria, vol. i. p. 281, 282. See likewise in the commencement of the new novel of the Talisman, a very picturesque and apparently exact description of the neighbourhood of the Dead Sea.

up by a volcano." This account seems to be confirmed by the quantities of ruins still found by travellers on the western border."

"The eruptions themselves have ceased long since, but the effects, which usually succeed them, still continue to be felt at intervals in this country. The coast in general is subject to earthquakes, and history notices several, which have changed the face of Antioch, Laodicea, Tripoli, Berytus, Tyre, and Sidon. In our time, in the year 1759, there happened one which caused the greatest ravages. It is said to have destroyed, in the valley of Balbec, upwards of twenty thousand persons; a loss which has never been repaired. For three months the shock of it terrified the inhabitants of Lebanon so much, as to make them abandon their houses, and dwell under tents."

In addition to these remarks of Volney's, a recent traveller, Mr. Legh,* states, that on the south-east side of the Dead Sea, on the right of the road that leads to Karrac, red and brown hornstone porphyry, in the latter of which the felspar is much decomposed, syenite, breccia, and a heavy black amygdaloid, containing white specks, apparently of zeolite, are the prevailing rocks. Not far from Shubac, (near the spot marked in D'Anville's Map, Patriarchatus Hierosolymitanus), where there were formerly copper mines, he observed portions of scoriæ. Near the fortress of Shubac, on the left, are two volcanic craters; on the right, one.

The Roman road on the same side is formed of pieces of lava. Masses of volcanic rock also occur in the valley of Ellasar.

The chemical properties of the waters of the Dead Sea, rather lend countenance to the volcanic origin of the surrounding country, as they contain scarcely any thing except

---

* See his account of Syria, attached to Macmichael's Journey from Moscow to Constantinople.

muriatic salts, Dr. Marcet's analysis* giving in 100 parts of the water

Muriate of lime 3.920
Muriate of magnesia 10.246
Muriate of soda 10.360
Sulphate of lime............ 0.054
                            ————
                            24.580

Now we not only know that muriatic acid is commonly exhaled from volcanos in a state of activity, but that muriatic salts are also frequent products of their eruption.

The other substances met with are no less corroborative of the cause assigned. Great quantities of asphaltum appear floating on the surface of the sea, and are driven by the winds to the east and west bank, where they remain fixed. Antient writers inform us, that the neighbouring inhabitants went out in boats to collect this substance, and that it constituted a considerable branch of commerce. On the southwest bank are hot springs and deep gullies, dangerous to the traveller, were not their position indicated by small pyramidic edifices on the sides. Sulphur and bitumen are also met with on the mountains round.

On the shore of the lake Mr. Maundrel found a kind of bituminous stone, which I infer from his description to be analogous to that of Ragusa in Sicily, noticed in my memoir on the Geology of that Island.†—" It is a black sort of pebble, which being held to the flame of a candle, soon burns, and yields a smoke of a most intolerable stench. It has this property, that it loses a part of its weight, but not of its bulk, by burning. The hills bordering on the lake are said to abound with this sort of sulphureous (qu. bituminous) stone. I saw pieces of it, adds our author, at the convent of St. John in the Wilderness, two feet square. They were carved in basso relievo, and polished to so high a lustre, as black marble is

* See Phil. Trans. vol. xcvii. p. 269.

† I have since received a specimen of this stone, which turns out to be precisely similar to that of Ragusa.

capable of, and were designed for the ornament of the new church in the convent."

It would appear, that even antecedently to the eruption mentioned in Scripture, bitumen-pits abounded in the plain of Siddim. Thus in the account of the battle between the kings of Sodom and Gomorrah, and some of the neighbouring princes (Gen. ch. 14.) it is said, *And the vale of Siddim was full of slime-pits*—which a learned friend assures me ought to be translated *fountains of bitumen*.

Mr. Henderson in his Travels in Iceland, will have it, that phænomena similar to those of the Geysers of Iceland, existed likewise in this neighbourhood.* The word Siddim, he says, is derived from a Hebrew root, signifying " to gush out," which is the identical meaning of the Icelandic word Geyser, and it is remarkable, that there exists in Iceland a valley called Geysadal, which signifies the valley of Geysers, and consequently corresponds with the " valley of Siddim."

The latter therefore he thinks should be translated the valley of the Gushing Mountains.

Mr. Henderson further believes, that Sheddim, the object of the idolatrous worship of the Israelites, (Deuter. xxxii. 17. Psalms cvi. 37) translated in our version " Devils," were boiling springs derived from volcanos, and I may add, as some little corroboration of this opinion, that somewhat similar phænomena at the Lacus Palicorum in Sicily, were the objects among the Greeks of peculiar and equally sanguinary superstition.

Mr. Henderson thinks, that it was in imitation of these natural fountains, that Solomon caused to be constructed a number of Jetting Fountains, (as he translates the passage,) of which we read in Ecclesiasticus, cap. ii. 8. My ignorance of the Hebrew language precludes me from forming any opinion as to the probability of these conjectures, but the existence of hot springs in the valley, at a much later period than that to which he refers, is fully established.

---

* Vol. i. p. 154.

DEAD SEA.

But besides this volcanic eruption, which brought about the destruction of these cities, it would appear that the very plain itself, in which they stood, was obliterated, and that a lake was formed in its stead. This is collected, not only from the apparent non-existence of the valley in which these cities were placed, but likewise from the express words of Scripture, where, in speaking of the wars which took place between the kings of Sodom and Gomorrah and certain adjoining tribes, it is added, that the latter assembled in the valley of Siddim, which *is* the Salt (i. e. the Dead) Sea. It is therefore supposed that the Lake itself occupies the site of this once fertile valley, and in order to account for the change, Volney and others have imagined, that the destruction of the cities was followed by a tremendous earthquake, which sunk the whole country considerably below its former level.

But the sinking of a valley, besides that it is quite an unprecedented phænomenon in the extent assumed, would hardly account for the obliteration of the antient bed of the Jordan, a river, which, though now absorbed in the Dead Sea, from whence it is carried off by the mere influence of evaporation, must, before that Lake existed, have continued its course either to the Red Sea or the Mediterranean.

Now if the Dead Sea had been formed by the cause assigned, the waters I conceive would still continue to have discharged themselves by their old channel, unless indeed the subsidence had been very considerable, and then the course of the Jordan, just north of the Dead Sea, would have presented, what I believe no traveller, antient or modern, has remarked, a succession of rapids and cataracts, proportionate to the greatness of the descent.

That the Jordan really did discharge its waters at one period into the Red Sea, is rendered extremely probable by the late interesting researches of Mr. Burckhardt, who has been the first to discover the existence of a great longitu-

DEAD SEA.

dinal valley, extending, in nearly a straight line south west, from the Dead Sea as far as Akaba, at the extremity of the eastern branch of the Red Sea, and continuous with that, in which the Jordan flows from its origin in the mountains near Damascus.* It was probably through this very valley, that the trade between Jerusalem and the Red Sea was in former times carried on. The caravan, loaded at Ezengebe with the treasures of Ophir, might after a march of six or seven days deposit their loads in the warehouses of Solomon.

This important discovery seems to place it beyond question, that if there ever was a time at which the Jordan was not received into a lake, which presented a surface considerable enough to carry off its waters by evaporation,† the latter would have been discharged by this valley into the Red Sea, and hence every theory of the origin of the Lake Asphaltitis must be regarded as imperfect, which does not account for the obliteration of this channel.

For my own part were I to offer a conjecture on the subject, I should suppose, that the same volcano which overwhelmed with its ejected materials the cities of the plain, threw out at the same time a current of lava sufficiently considerable to stop the course of the Jordan, the waters of which, unable to overcome this barrier, accumulated in the plain of Siddim until they converted it into the present lake.‡ I do not know that any traveller has observed what is the ordinary depth of the Dead Sea, but if we only imagine a current of lava, like that which in 1667 proceeded from Etna, and flowed into the sea above Catania, to have

---

* See the Map of the Dead Sea attached to this volume.

† Several lakes are mentioned in Persia similarly circumstanced, and particularly one near Tabriz. See Morier's 2nd Journey in Persia. It is evident that every lake without an outlet, which is supplied with water from rivers flowing through a stratum impregnated with salt, must necessarily be strongly saline, so that there is no absolute necessity for supposing the ingredients contained in the waters of the Dead Sea to have arisen from volcanic exhalations.

‡ This must necessarily have been the case, as (according to Maundrell and other travellers), the Dead Sea is enclosed on the east and west with exceeding high mountains.

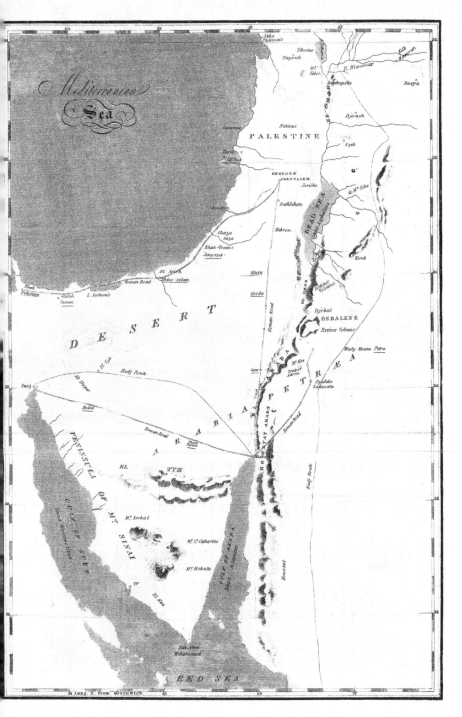

The material originally positioned here is too large for reproduction in this reissue. A PDF can be downloaded from the web address given on page iv of this book, by clicking on 'Resources Available'.

descended at right angles to the bed of the River Jordan, the lake need not be supposed very shallow.*

Nor need we be startled at the magnitude of the effect, that we find to have resulted from a cause which, comparatively speaking, appears so insignificant; for, if the little rivulet, that flows at the foot of the Puy de la Vache in Auvergne, was adequate to produce the lake of Aidat, there seems no disproportion, in attributing to a river of the size of Jordan, to say nothing of the other streams, nowise inconsiderable which must have been affected by the same cause, the formation of a piece of water, which according to the best authorities, is, after all, not more than twenty-four leagues in length by six or seven in breadth.

That the volcanic eruption which destroyed the cities of the Pentapolis was accompanied by the flowing of a stream of lava, may be inferred, I think, from the very words of scripture. Thus when Eliphaz reminds Job of this catastrophe, he makes use of the following expressions, according to Henderson's translation of the passage:

> Hast thou observed the ancient tract
> That was trodden by wicked mortals?
> Who were arrested on a sudden,
> Whose foundation is a *molten flood.*
> Who said to God: depart from us,
> What can Shaddai do to us?
> Though he had filled their houses with wealth;
> (Far from me be the counsel of the wicked!)
> The righteous beheld and rejoiced,
> The innocent laughed them to scorn;
> Surely their substance was carried away,
> And their riches devoured by fire.
>
> Job xxii. 15—20.

The same fact, Mr. Henderson thinks, is implied in the

* A recent traveller, Mr. Carne, speaks of the Dead Sea as so shallow, at least for some distance from its banks, that he was unable to swim in it. See his Letters from the East. London, 1826.

description of the circumstances connected with Lot's escape.*

" Why was he prohibited from lingering in any part of the low land, if not because he would be there exposed to the pestilential volcanic effluvia and to the lava? And what reason can be assigned for his obtaining leave to stop in Zoar, but its lying at some distance from the spot where the lava began to act, as likewise on an elevation whence he could survey the approaching ruin, and retire before the stream reached that place? We accordingly find, that however desirous he was to stay there at first, he quitted it before night, for a still more elevated and safe retreat. *" And Lot went up out of Zoar, and dwelt in the mountain, for he feared to dwell in Zoar."* (Gen. xix. 30.)

" How natural is the incrustation of his wife on this hypothesis! Remaining in a lower part of the valley, and looking with a wistful eye towards Sodom, she was surrounded, ere she was aware, by the lava, which rising and swelling, at length reached her, and (whilst the volcanic effluvia deprived her of life,) incrusted her where she stood, so that being, as it were, embalmed by the salso-bituminous mass, she became a conspicuous beacon, and admonitory example to future generations. The power of this asphaltic substance in preserving from corruption is evident, from its being employed by the Egyptians for embalming their mummies."†

" She is said to have been converted into a pillar of *salt*, on account of the quantity of that substance which appeared on the crust, and its abundance in those countries is notorious, both from sacred and profane history; so much so, that the lake, which now fills the caverns made by the earthquake, has, among other names, that of the Salt Sea."

* Henderson's Iceland, Vol 1. p. 153, 154.

† Diod. Sic. lib. xix. c. 109. Mr. Carne in his Letters from the East, states, that about 15 years ago a human body, or what had the form of one, was discovered floating not far from the shore, and on taking it out it was found to be incrusted all over with a thick and hard coating of bitumen and salt.

DAMASCUS.

I know not what opinion may be entertained with regard to this explanation of the disaster that awaited Lot's wife, but it will at least be allowed, that the eruption of a stream of lava, which might have interfered with the course of the River Jordan, is not only in itself a probable circumstance, but one that derives some support from the sacred writings themselves.

Much however, it is confessed, remains to be explored, before this or any other theory can be finally adopted, and it is to be hoped, that the first individual, who has spirit and resolution enough to venture into these inhospitable regions, will pay attention to the physical structure of the country.

He should in particular search the rocks which bound the Dead Sea, in order to discover, if possible, the crater of the volcano which was in a state of eruption at the period alluded to; he should ascertain, whether there are any proofs of that sinking of the ground, which, notwithstanding Volney's authority, I have regarded as so problematical; whether traces of the antient bed of the river can be discovered south of the lake, or of a barrier of lava stretching across it; nor should he omit to examine, whether the vestiges of these devoted cities have been submerged, as some have stated, beneath the waters, or are buried, like Pompeii, under heaps of the ejected materials.

It appears from Dr. Clarke that there are traces of volcanic rocks between Jerusalem and the sea, especially about the Lake of Gennesereth, and Badhia describes a volcanic desert lying between the River Jordan and Damascus, which was traversed by him in his way to that city.[*]

" The Phlegrean fields, and all that can present an idea of volcanic destruction, form but a feeble image of the

---

[*] It is probable that the Iron mountain (Σιδηρουν ορος) mentioned by Josephus, as extending from Julias on the Lake Gennesereth to the land of the Moabites, was of a volcanic nature. See Lib. 5. c. 27.

T

frightful country through which I passed. From the bridge
of Jacob to Sassa, the whole ground is composed of nothing
but *lava, basalt, and other* volcanic productions: all is
black, porous, or carious; it was like travelling in the infernal regions. Besides these productions, which cover the
country, either in detached masses, or in loose strata, the
surface of the ground is entirely covered with loose *volcanic
stones,* from three to four inches in circumference to a foot
in diameter, all equally black, porous, or carious; as if they
had just come out of the crater. But it is particularly at the
approaches to Sassa, that the traveller meets with groupes
of crevices, and volcanic mounds, of so frightful a size, that
he is seized with horror, which is increased if he allows his
imagination to wander to the period, when these masses were
hurled forth with violence from the bowels of the earth.
There are evident signs that all this country was formerly
filled with volcanos, for we beheld several small craters in
traversing the plain."*

From this, coupled with Mr. Buckingham's description of
the lake of Tiberias,† which he represents as situated in a
deep basin, surrounded on all sides with lofty hills, excepting only the narrow entrance and outlet of the Jordan at
each extreme, we might almost be induced to extend to this
piece of water the same explanation, which I have given
with respect to the formation of the Dead Sea. The stream
of the Hieromax at least, which flows into the Jordan, just
where that river issues from the Lake of Tiberias, has cut
its course through lava, and though the same is not expressly
stated to be the case with the Jordan, yet it is evident, that
the impression upon Mr. Buckingham's mind was, that the
whole country had a volcanic appearance. It is possible
then, that a still more antient eruption may, by blocking up
the course of the Jordan near this spot, have caused the

* Ali Bey's Travels. Vol. II. p. 261, 262.
† See Buckingham's Travels in Palestine. p. 468 and 448.

lake, and that the water may afterwards have cut itself a
passage through the volcanic matter, and thus have regained
its old channel.

The imperfect acquaintance, which we possess with respect to the physical structure of this part of Asia, does not
admit of our tracing in a connected manner the chain of
volcanic appearances, which may be suspected to extend
from the Red Sea to Asia Minor.

One of the links of this chain may probably be seen near
Scandaroon, as Mr. Otter, when at Bylan in 1737, was told
of a mountain called Araz Dagiri, about three miles from
that place, from whence fire had for some time issued.
Pococke mentions, his having heard of this volcano, on his
road to Seleucia (now Souvadi), from an English gentleman,
but that he had not seen it himself. He also specifies a
small hill north of the town of Kepse, between the mouth
of the Orontes and the Bay of Scandaroon, from which
smoke and occasionally flames are seen to proceed. No
recent traveller however has confirmed this statement,
though Olivier observed at a place not far distant traces of
antient eruptions.

Russel, in his Natural History of Aleppo,* speaks of a
crater-shaped cavity as existing near Scandaroon, which
is called by the natives the Sunken Village. Although
it has probably no reference to any thing of a volcanic nature, as the same author informs us, that it is
composed of coral, and various sea shells incrusted and
consolidated by calcareous matter almost as white as snow,
yet I may as well introduce the account of it given by
Pococke in his Description of the East.† According to
this traveller, it is a round oval pit, about 100 yards in
diameter and 40 in depth, the upper half perpendicular,
the lower exceedingly steep. There is only one way down
to it, which is passable for beasts. Half way down there

* Vol. 1, p. 57.
† Vol. 1. p. 169

is a grotto worked in the earth, about five feet high and 20 broad.

It seems probable I think, that some tradition at least of the volcanic eruptions, that appear to have taken place in Syria and Cilicia, had reached the ears of the Greeks. Thus Homer, in the only passage in which he mentions the giant Typhœus, whom after poets place underneath the volcano of Sicily or of Naples, says, that the resting place of the monster is reputed to be among the Arimeans, or in the Arimean mountains * (εⅳ Αριμοις). Now the Arimeans were the antient inhabitants of Syria,† and therefore we might infer, that this was the country to which Homer alludes ‡ It is however a curious coincidence, that the Arimeans peopled that part of Mysia near Smyrna in which volcanic appearances are stated by Strabo to exist, and hence that geographer supposes Homer to refer to the latter country, and not to

---

* Οι δ'αρ'ισαν, ωσει τε πυρι χθων πασα νεμοιτο.

Γαια δ'υπεστοναχιζε, Διι ως τερπικεραυνω

Χωομενω, οτε τ'αμφι Τυφωει γαιαν ιμασση

Εⅳ Αριμοις, οθι φασι Τυφωεος εμμεναι ευνας.

Iliad. B. 781.

† Syria is called in Scripture Aram, and Padan-Aram (Gen. xxv. 20. and xxviii. 2.) from Aram, the fifth son of Shem, from whom the Syrian people were reputed descendants.

‡ I do not call to my assistance the disputed verse, which is added in some copies to the lines above from Homer, viz.

Χωρωεⅴι δρυοεⅴτι, Υδης εⅴ πιοⅴι δημω.

as Heynè and the best commentators view it as an interpolation, and are still more opposed to the substitution of the word Ιⅴδης for Υδης, which was originally proposed by Dickinson in his Delphi Phœnicissantes, and afterwards adopted by Wood in his Essay on the original Genius of Homer. As the word Ιⅴδα does not occur in any classical writer, it is hardly probable that Homer could have employed it.

It has been supposed by Leclerc with somewhat more probability, that the fable of the Giant Typhœus, so well described by Hesiod in his Theogenia Vers. 820 to 870, arose from a vague tradition which had reached the Greeks from Phœnicia of the destruction of the cities of the valley of Siddim : he remarks that the lines beginning

" Σκληρον δ'εβροⅴτησε και οβριμον"

allude to the manner in which the event took place.

Syria.* Thus after speaking of the country round Phila-
delphia in Asia Minor (now Ali Shahir near Smyrna) which
I shall presently allude to, he adds that it is here the old
writers lay the scene of the circumstances related of Typhon, and place there the Arimi and the Catacecaumene,
under which names indeed they venture to rank the whole
country between Lydia and the Meander.

But whatever may be thought with respect to the spot
referred to by Homer, there can be no doubt, that the Greeks
were aware of the existence of volcanic phænomena in some
part of Cilicia. Thus Pindar, when he speaks of Typhœus
being crushed under Cumæ and Sicily, says that he was
before confined in Cilicia.†

It appears from Olivier, that there are in Lydia, on the
borders of Caramania, conical hills, on one of which is situated an old castle called " the Black Rock."

Riedesel has discovered lava nearly opposite in Samos, and
Clarke mentions a fire arising from this island.

In Lycia the ancient writers notice a burning mountain,
the top of which was the resort of lions, the middle of goats,
whilst the bottom abounded in serpents, whence arose the
fable of the monster Chimæra, the head of which was that
of a lion, the middle of a goat, and the hinder parts those
of a dragon, and which continually vomited forth flames.

Beaufort, in his Caramania, describes his visit to this mountain, which as of old sent forth fire, not like that of Etna,

---

* Και δη και τα περι τον Τυφωνα παθη ενταυθα μυθευεσι, και τες Αριμες,
και την Κατακεκαυμενην ταυτην ειναι φασιν· ουκ οκνεσι δε και τα μεταξυ
Μαικνδεα και Λυδων απανθ' υπονοειν τοιαυτα——.

Strabo. Lib. 12.

† Ος εν αινα Ταρταρω κει—
τ̇αι Θεων πολεμιος
Τυφως εκατοντακαρανος, τον ποτε
Κιλικιων θρεψεν πολυω—
νυμον αντρον· νυν γε μαν
ται θ'υπερ Κυμας αλιερκεες οχθαι,
Σικελια τ'αυτου πιεζει
Στερνα λαχνοεντα

Pyth. Od. 1.

but quiet and regular. The antients state that this fire did
not even destroy the plants that grew around, and Beaufort
remarks the very same thing.

" Trees, brushwood, and weeds," he says, " grow close
round the cratei, never accompanied with noise or earthquakes; nor does it ever eject stones, smoke, or noxious
vapours."*

It is evident from this account, that the above phænomenon ought not to rank with those of volcanos, but like
that of Apollonia in the mountains of Albania, or the Pietra mala in the Appennines, owes its origin to a slow distillation of bituminous matters.

In Mysia however we have accounts from Strabo of appearances more to our present purpose. Near Smyrna is a
tract of country, known in the time of that geographer by
the name of Catacecaumene from its burnt and arid appearance. It is without trees, he says,† with the exception of
the vine, from which wine is made, that yields to none of
the most celebrated. The surface of the ground is cindery,
and the mountains and rocks are black, as if they had been
calcined. Some, he adds, have supposed the country to
have been affected by fire from heaven, but it is most probable, that so large a tract as this should have been burnt
by fire proceeding from the earth,

* See the Description of Caramania, by Capt. Beaufort, R.N. 1 Vol. 8vo.

† Μετα δε ταυτ'εστιν ή Κατακεκαυμενη λεγομενη χωρα, μηκος μεν και
πεντακοσιων σταδιων, πλατος δε τετρακοσιων, ειτε Μυσιαν χρη καλειν, ειτε
Μηονιαν· λεγεται γαρ αμφοτερως· απαντα αδενδρος, πλην αμπελη της τον
Κατακεκαυμενιτην φερουσης οινον, ουδενος των ελλογιμων αρετη λειπομενον.
Εστι δε ή επιφανεια τεφρωδης των πεδιων· ή δ'ορεινη και πετρωδης, μελαινα,
ως αν εξ επικαυσεως. Εικαζουσι μεν ουν τινες εκ κεραυνοβολιων και πρηστηρων
συμβηναι τουτο, και εκ οκνουσι τα περι τον Τυφωνα ενταυθα μυθολογειν.
Ουκ ευλογον δε υπο τοιουτων παθων την τοσαυτην χωραν εκπρησθηναι αθροως,
αλλα μαλλον υπο γηγενους πυρος· εκλιπειν δε νυν τας πηγας· δεικνυνται δε και
βοθροι τρεις, ους φυσας καλουσιν, οσον τετταρακοντα αλληλων διεστωτας
σαδιας· υπερκεινται δε λοφοι τραχεις, ους εικος εκ των αναφυσηθεντων
εισωρευσθαι μυδρων. Strabo. Ed. Fal. p. 900.

It is true, that its sources are now exhausted, but three hollows, called in the country (φυσαι) still exist, as much as 40 stadia apart from one another, which are overhung by rugged rocks, formed in all probability from melted masses of stone, heaved up like bladders. Strabo also makes mention of volcanic appearances in the neighbourhood of Laodicea, and the cave called Plutonium, near the town of Hierapolis in the same district, seems to have been another Grotto del Cane, being filled with a noxious vapour speedily fatal to animal life.* The carbonic acid given out probably imparted to the springs the property of dissolving calcareous earth, which from Strabo's account they possessed in so remarkable a degree, that, when the water was conducted along the vineyards and gardens, the channels became long fences, each a single stone. Yet this strong petrifying property did not prevent them from being potable.†

The great difficulties that attend travelling in these countries, and the circumstance, that hitherto the attention of European enquirers has been directed, rather to the antiquities, than to the natural history of these classical spots, will readily account for the want of more precise information on the subject of their volcanos, and ought to prevent us from rejecting the accounts of ancient writers, because they have not as yet been fully confirmed by the researches of the moderns.

---

* It seems that the priests of Cybele claimed a singular exemption from the influence of this vapour, and Dion Cassius, in his account of the cavern, extends it to *all* Eunuchs. It is not difficult to understand, how this fraud might be maintained, as the specific gravity of the gaseous fluid is such, that it only occupies the bottom of the cavern, so that a bird, or even a quadruped, whose head was low, would be immediately suffocated; whilst the priest, walking more upright, might easily proceed as far as the entrance (which is all that Strabo says he saw him do) without feeling the effects. οι δ'αποκοποι Γαλλοι πχριατιν απαθεις, ωστε και μεχρι τ8 στομι8 πλησιαζειν, και εγκυπτειν, και καταδυνειν μεχρι ποσ8, συνεχοντας ως επι το πολυ το πνευμα. Strabo. p. 903.

I find that (handler has already made a remark to nearly the same purport. See his Travels in Asia Minor. ch. 48. & 49.

† Το δε της απολιθωσεως και επι των εν Λαοσικεια ποταμων φασι συμβαινειν, καιπερ οντων ποτιμων. Strabo. p. 903

We must never forget, that in Asia things appear to be taking quite an opposite course to that which they pursue in Europe, and that whilst among ourselves the advancement of knowledge is so uninterrupted, that we are in the habit, not altogether without reason, of measuring the value of information on subjects of this description, in some degree by the modernness of their date, the very contrary rule must hold good in the East, where any supposed superiority in the observers is more than counterbalanced by the retrograde condition of the people, and the increased obstacles which are thus presented to the traveller.

Nevertheless we have in this instance the evidence of a modern observer for the existence of volcanos not far from Smyrna, as well as in other farther in the interior. Mr. Brown, the African traveller, in a letter addressed to Mr. Tennant during his last journey, and dated Tefliz, thus expresses himself:—

" My eyes have been very much opened in this journey, to the volcanic nature of certain parts of Asia Minor and its confines. At Kôlah, near the Hermus, only three days from Smyrna, may be seen an unquestionable site of volcanic eruption. It is one of the most recent, though still probably of a very remote period. Carabignar is another, but this probably may have been noticed by others. Kôlah, I imagine, has not hitherto been observed. I shall have something to say of Afium Karahissar. The neighbourhood of Konié, and still more of Kaisarié, is overspread with fragments of lava, some of it almost in the state of scoriæ. The quantity of lava in the district of Erzerûm is immense, and the whole country about Mount Ararat is volcanic. The eruptions in these places seem to be of the great antiquity.

Volcanic matter about Erzerum is so widely diffused, that I am disposed to acquit Sestini of exaggeration in his route to Diarbekir."—Walpole's Travels in the East, p. 178.

Maltbrun * likewise quotes an Arabian geographer,

---

* Maltbrun's Geography, vol. 3, p. 102.

MOUNT ARARAT.

Hadji Khalfah, in proof of the existence of an extinct volcano in Armenia, in a hill called Djebel-Nimroud, or the Hill of Nimrod, which has on its summit a lake, probably once the crater.

Olivier also notices an extinct volcano between Birt and Orfa, two leagues from the latter place, and likewise between Orfa and Mosul, the capital of Curdistan.* He suspects the great mountain Sindsjaar, to be volcanic.

It would seem that some indications of the same description existed formerly in the country of Adiabene,† between the Tigris and Euphrates, near the city of Niniveh, for Szetzes, in his Scholia on Lycophron, v. 704, mentions, that the Lake of Avernus, (λιμνη αορνος) which most persons refer to the neighbourhood of Naples, stood, according to some, near that place. In this country there occurred in the year 1822, the earthquake which destroyed the town of Alep, and with it no less than 8000 persons, and which extended its effects over a radius of 50 leagues.‡

Kinneir, in his memoir on the Persian Empire, conjectures, that the Island of Ormus, at the entrance of the Persian Gulph, may also be volcanic.

The chain of the Caucasus exhibits, according to Reinegg, rocks containing glassy felspar, and columnar basalt, and Mount Ararat has long possessed the character of a volcanic mountain. It is remarkable however, that

* Ker Porter also speaks of what he calls a Sulphur Desert in Lower Curdistan near Sulimania, and mentions very abundant Naphtha springs at Kirkook, a town south of Mosul.

† A German writer, Sickler, (Ideen zu einer vulcanischen Erd-globus) has *somewhat boldly* imagined, that the fiery sword, by which our first parents were driven from Paradise, was a volcanic eruption, and he appeals to the existence of volcanos near the Euphrates, in confirmation of this opinion.—For my own part I shall feel satisfied, if I have succeeded in tracing historical records of volcanic eruptions to the period of the destruction of the cities in the vale of Siddim, and am content to search the *book of nature* for those of an earlier date.

‡ Ferussac, Bulletin des Sciences for May, 1825.

MOUNT ARARAT.

Morier, who spent some time at its foot, makes no mention of any appearances of this kind, and that nothing which favours such an hypothesis can be extracted from Tournefort, a traveller, who, though principally attentive to botany, appears, from the notice he has taken of the Island of Santorino and other places in the Archipelago, to have been alive to the existence of volcanic phænomena when they came before him. The most recent account of this mountain is that given by Ker Porter,* which, as it includes all the information that is to be gleaned from other sources, I shall extract entire.

" On viewing Mount Ararat from the northern side of the plain, its two heads are seen separated by a wide cleft, or rather glen, in the body of the mountain. The rocky side of the greater head runs almost perpendicularly down to the north-east, whilst the lesser head rises from the sloping bosom of the cleft in a perfectly conical shape. Both heads are covered with snow. The form of the greater is similar to the less, only broader and rounder at the top, and shows, to the north-west, a broken and abrupt front, opening, about half-way down, into a stupendous chasm, deep, rocky, and peculiarly black. At that part of the mountain, the hollow of the chasm receives an interruption from the projections of minor mountains, which start from the sides of Ararat, like branches from the root of a tree, and run along in undulating progression, till lost in the distant vapours of the plain.

The dark chasm, which I have mentioned as being on the side of the great head of the mountain, is supposed by some travellers to have been the exhausted crater of Ararat. Dr. Reineggs even affirms it, by stating, that in the year 1783, during certain days in the months of January and February, an eruption took place in the mountain, and he suggests the probability, of the burning ashes ejected thence, at that time

* Sir Ker Porter's Travels in Georgia, &c. vol. i. p. 182.

MOUNT ARARAT.

reaching to the south side of Caucasus, (a distance in a direct line of 220 wersts) and so depositing the volcanic productions, which are found there. The reason he gives for this latter supposition is, that the trap seen there did not originate in these mountains, and must consequently have been sent thither by volcanic explosions elsewhere. And that this elsewhere, which he concludes to have been Ararat, may have been that mountain, I do not pretend to deny; but those events must have taken place many centuries ago, even before history took note of the spot, for since that period we have no intimation whatever of any part of Ararat having been seen in a burning state. This part of Asia was well known to the antient historians, from being the seat of certain wars they describe; and it cannot be supposed, that had so conspicuous a mountain been often or ever (within the knowledge of man) in a state of volcanic eruption, we should not have heard of it from Strabo, Pliny, Ptolemy, or others: but on the contrary, all these writers are silent on such a subject with regard to Ararat; while every one who wrote in the vicinities of Etna or of Vesuvius, had something to say of the thunders and molten fires of these mountains. That there are volcanic remains to a vast extent round Ararat, every one who visits that neighbourhood must testify, and giving credit to Dr. Reinegg's assertion, that an explosion of the mountain had happened in his time, I determined to support so interesting a fact with the evidence of every observation on, my part, when I could reach the spot

But on arriving at the monastery of Eitch-mai-adzen, where my remarks must chiefly be made, and discoursing with the fathers on the idea of Ararat having been a volcano, I found that a register of the general appearances of the mountain had been regularly kept by their predecessors and themselves for upwards of 800 years; and that nothing of an eruption, or any thing tending to such an event, was to be found in any of their notices.

When I spoke of an explosion of the mountain having taken place in 1783, and which had been made known in Europe by a traveller declaring himself an eye-witness, they were all in surprise; and besides the written documents to the contrary, I was assured by several of the holy brethren who had been resident on the plain for upwards of 40 years, that during the whole of that period they had never seen even a smoke from the mountain. Therefore how the author fell into so very erroneous a misstatement, I can form no guess.

Kinneir also notices a mountain called Sepan Dag which, he says, is conical and has every appearance of being volcanic. It is situated near the Lake of Van, in Armenia, between Erzerum and Betlis. Quantities of obsidian are found along the borders of the lake.*

Ker Porter† also describes and gives a plan of a valley near Erivan, which he passed on his way to Mount Ararat, the rocks of which consist of columnar basalt.

The principal mountain is called from its turretted form, the Castle of Tiridates.

Still farther north, near the River Kuban, between the Black Sea and the Caspian, Engelhardt and Parrot‡ have noticed several hills composed of similar materials to the Wolkenberg, in the Siebengebirge, near Bonn.

The rock consists of needle-shaped crystals of basaltic hornblende, small ditto of augite, glassy felspar, common hornblende in nests, and a little quartz. Its structure is sometimes lamellar, sometimes columnar. Its colour red.

This rock is found associated with Muschelkalk, but the boundary is not well defined

* Kinneir's Travels in Asia Minor. p. 374
† Travels in Georgia, vol. 2. p. 624.
‡ Engelhardt and Parrot!. Reise in dem Krym und dem Caucasus Berlin 1825.

Neither lava nor craters occur among these mountains.

Mount Kasbeck, the highest eminence in this chain, estimated by these travellers at no less than 14,400 feet, is composed of porphyry (qu. trachyte) which is columnar.

The valley of Terek, a river which runs from west to east,
and empties itself into the Caspian, seems to exhibit a repetition of the same rock formations which I have noticed as
occurring in Hungary. We have here not only an alternation of clay-slate and greenstone with a porphyry, which
appears to correspond with the older porphyry of Schemnitz,
with sienite or granite, with gneiss, and even with a bituminous limestone; but we find in the vallies, and resting on
the other rock formations, a clay porphyry, containing fragments of the older porphyry, a conglomerate with a porphyritic basis, and what is more curious, beds consisting
chiefly of pumice. The latter is greyish, reddish, or yellowish white, porous, with a fracture partly uneven, partly
fibrous—the fibres often curved. It contains plates of mica,
and crystals of glassy felspar, sometimes even a little hornblende.

The pumice constitutes detached conical hills, which rise
up from the slope of the valley of Terek.

The rock is in part like a pumiceous conglomerate, in
which each separate fragment is clothed with a blackish-
brown metallic coating, owing to which the parts appear
with black and white stripes.

Under the blocks of pumice are fragments of clayslate,
which are in like manner coated.

The lower stratum of these detached conical hills, is a bed
of conglomerate, consisting of a yellowish-white argillaceous
mass, interspersed with blocks of pumiceous porphyry and
clay-slate of various sizes.

A similar formation occurs at the foot of Kasbeck.

It would seem from this description, that evidences of volcanic action extend over this part of the Caucasus as they do

over Hungary, for though Engelhardt is disposed to consider the pumice a porphyry altered by weathering, yet I cannot believe that this intelligent naturalist means to question the igneous origin of such a deposit. The description, which he has himself given of the characters of this substance, is decisive on the point.

At Baku, a town situated on a small peninsula which stands out on the western side of the Caspian Sea, forming a sort of advanced post of the Russian dominions, phænomena occur which have been hastily set down as volcanic. The rocks consist chiefly of a bituminous shale, which is in some places so impregnated with petroleum, that wells are sunk for the purpose of obtaining it, from some of which as much as 1000 or 1500 pounds are daily obtained. It is natural that a formation of this kind should give rise to certain pseudo-volcanic phænomena, and I believe that all the facts related by travellers, which have led to the notion of volcanic action existing there, may be referred to this cause.

It would be interesting to ascertain, whether the same formation which contains the petroleum at Baku extends to the Island of Taman, near Crim Tartary, where similar phænomena have been observed, and whether it has any connexion with the bituminous limestone noticed by Engelhardt as existing in the valley of Terek.

It should be mentioned however, that the sudden rise of an island in the Sea of Azof, which is stated to have happened in the year 1814, seems to lend countenance to the idea of the continuance of true volcanic action in that district even at the present period.

The account is, that on the 10th of May in that year, a frightful noise was heard in the sea, round a distance of 200 toises. Flames rose from the water, accompanied by explosions as loud as those of a cannon. A thick smoke was blown about by the violence of the wind, and enormous masses of earth were seen thrown up in the air, together with large stones.

Ten eruptions of this kind took place at intervals of a quarter of a hour. Similar phænomena continued during the night. There then rose out of the sea an island, which threw out from several apertures a muddy substance, that acquired by degrees some consistency.

During this time, a remarkable smell, which had nothing of a sulphureous nature, was perceived over a space of ten wersts. On the 20th of April, a nearer examination of the island was undertaken, and it was found almost inaccessible, being surrounded on all sides with hardened mud. When they had at last succeeded in reaching the interior of the island, its height above the level of the sea was found to be a toise and a half, and its surface was seen to be every where covered with a stony material of a whitish colour.*

In the chain of Elburs, which bounds the Caspian Sea on the south, there occurs a lofty mountain called Demavend, which has long been noted as a volcano; and as the Greeks attributed the agitations of Mount Etna to the Giant Typhœus burned under it, so the Persians believe that Zohag, one of their sovereigns, remarkable for his tyrannies, after being conquered by Feridoun, the ancestor of Zoroaster, was imprisoned in this mountain.†

Feridoun appears to have been a personification of the good, as Zohag was of the evil principle, being confounded with Ahriman; and if we were disposed to follow Buffon in the fanciful picture he has drawn in his Epoques de la

* Leonhard. Taschenb. der Mineral. X. p. 476.

† Zend-Avesta, translated by Anquetil du Perron, vol. i. part 3. p. 422, and vol. ii. p. 410.—Morier, in his Travels in Persia, vol. ii. p. 355. gives rather a different version to the tale. My readers will recollect other particulars respecting Zohag, in the new Novel of the Talisman.

ELBURS.

Nature* of the early state of the world, which he supposes
to have been at first so subject to the ravages of fire, as to be
unfit for the maintenance of animal life, we might then
imagine the confinement of Zohag within the entrails of the
mountain, to have been typical of the gradual diminution of
the volcanic action, which rendered the country more and
more fitted for the habitation of man.

The analogy between the Persian Zohag and the Greek
Typhœus, holds good in other respects; as Zohag was
figured with a serpent growing out of either shoulder, which
he was obliged to feed with human blood, so Typhœus is
described with a hundred snakes, or a snake with a hundred
heads, proceeding from the same part.

εκ δε οι ωμων
Ην εκατον κεφαλαι οφιος, δεινοιο δρακοντος
Γλωσσησι δνοφερησι λελειχμοτες

Hesiod Θ. 824.

Conformably with this legend, Olivier found on Demavend
lava and columnar basalt; Morier states, that it is reported
sometimes to emit smoke, and that the circumstance of finding sulphur in small craters near the base of the mountain
may lead to the conclusion, that the cone is itself volcanic.
It appears that a considerable commerce in sulphur, as well
as saltpetre, is carried on from it. Morier further mentions
traces of the action of fire, extending south of this point,
between Teheran and Ispahan, and near Tabriz, and
Olivier makes the same observation respecting this neighbourhood, noticing particularly the country about Sava
and Cashan.

Among our possessions in India, the accounts of volcanic
appearances are vague and scanty.

* Buffon Epoques de la Nature, p. 355.

We know indeed that the Island of Salsette, near Bombay, is basaltic, but it does not appear to be of recent formation ; and Sir John Malcolm in his Memoir on Central India, has inserted a letter from Captain Dangerfield, giving an account of the geological structure of the province of Malwa, an elevated table land, which stretches across the centre of the peninsula between 21°. 30. and 24°. of north latitude, and which is chiefly composed of trap rocks. These however he considers to be secondary, and recognizes among them nothing of a volcanic nature.*

There is likewise in the Eastern Magazine for September, 1823, a notice of a journey up the Ganges, in which it is stated, that the rocks of Peerpoint and of Sacrigully are black and porous, like lavas.

Amongst the Himalaya range there is said to have been discovered a volcanic mountain, which is at present emitting much smoke, but no flame. As no one however has yet examined the actual spot, it is impossible to determine at present what the real nature of the phænomenon may be, and the probabilities are certainly against its connexion with a genuine volcano, both from its situation in the centre of a vast continent, and from the apparent absence of any lavas or ejected masses in its vicinity.†

In central Tartary we have long had obscure notions as to the existence of volcanos, and Mons. Ferussac has taken the trouble of collecting, in a late number of the Bulletin des Sciences,‡ the principal particulars that have been transmitted to us respecting them.

It would appear, that at the north of Khouei-thsu, and on the southern frontier occupied at the close of the 1st century of the Christian æra by the Hioungnou Turks, driven westward by the Chinese, there rises a burning mountain called Ho-chan. On one side of this mountain, add

* Malcolm's Central India, vol. 2. *Appendix.*
† See Brewster's Philos. Journal. April, 1826.
‡ Vol. 3, for 1824.

the accounts, all the stones are in a burning and melted state, and flow to a distance of some tens of *li* (i. e. leagues). This melted mass afterwards becomes cold and hard. The inhabitants of the country use it for medicine. Sulphur is also met with.

A Chinese writer of the 7th century, in speaking of Khouei-thsu, says : At 200 *li* (20 leagues) north of this town, there is a white mountain, which is called Aghie. Fire and smoke continually proceed from it ; it is from thence that the sal ammoniac comes.

The antient town of Khouei-thsu is the town of Khoutché of the present time, situated in 41°. 37. north latitude, 80° 35. east longitude, according to the observations of the missionaries sent towards the middle of the last century to prepare a map of it. This volcano, which forms a part of the snowy chain of the celestial mountains (Thian-Chan) must therefore be found nearly in 42.35 of N. latitude. It is probably the same which has at present the name of Khalar. According to the account of the Boukharies, who bring the Sal Ammoniac to Siberia and Russia, the latter is found south of Korgos, a town situated on the Ili. So large a quantity of this salt is collected there, that the inhabitants of Khoutché employ it to pay their tribute to China.

The new description of central Asia, published at Pekin in 1777, contains the following notice :

" The territory of Khoutche produces copper, saltpetre, sulphur, and sal ammoniac. The latter proceeds from a mountain called *the Mountain of Sal Ammoniac*, which is found on the north of the town. It has many caverns and crevices, which, in spring, summer, and autumn, are filled with fire, so that during the night the mountain seems illuminated by thousands of lamps. No one then can come near it. It is only in winter, during the coldest season, and when the great quantity of snow has stifled the fire, that the people of the country approach ; they strip themselves quite naked, in order to collect the sal ammoniac, which is found in the caverns in the form of very hard stalactites; it is for this reason very difficult to detach it.

" Twelve days journey by the caravan, north of Korgos, is found another town, commonly called Tchougoultchak. It is situated at the foot of Mount Tarbagatai, 46°. 5. N. lat. and 80°. 45. E. long. Four stations to the east of this town we arrive at the canton of Khoboksar, near Khobok, which falls into the Lake Darlai; there is there a small mountain full of fissures, which are excessively hot, but do not exhale any smoke. In these fissures sal ammoniac sub-limes, and attaches itself so strongly to the walls, that it is necessary to break the rock in order to collect it."——— *Klaproth.*

The Abbe Remusat, in a letter to Cordier,* states his having found in the Japanese edition of the Chinese ency-clopedia other particulars respecting this volcano.

" The sal ammoniac in persian " Nouchader," in chinese " Naocha," &c. is drawn from two volcanic mountains of Central Tartary; one is the volcano of Tourfan, lat. 43. 30. long. 87. 11. which has given to the town near it, the name of Ho-Tcheou, town of fire ; the other is the White Moun-tain, in the country of Bisch-Balikh ; these two mountains throw out continually flames and smoke. There are cavities there in which they collect a greenish liquid ; exposed to the air this liquid is changed into salt, which is the sal ammoniac ; the people of the country use it in the prepa-ration of leather. As to the Mountain of Tourfan, we ob-serve a column of smoke rise from it perpetually ; this smoke is replaced in the evening by a flame like that of a torch ; birds and other animals lighted up by it appear of a red colour. The mountain is called the Mount of Fire. In order to search for the salt, they put on wooden shoes, for leather ones would soon be burnt. The people of the country likewise collect the mother waters, which they boil in cauldrons, and obtain from thence the sal ammoniac, under the form of loaves, like those of common salt.

* Annales des Mines, v. 1820—p. 135 & 137.

These volcanos are 400 leagues from the Caspian Sea.

Cordier observes that the existence of two burning mountains in the midst of the immense table land bounded by the
Oural, the Altaic Mountains, the frontiers of China, and the
Himalaya Chain, is very worthy of attention. Sal ammoniac is never found in Europe in any but a volcanic rock;
it is therefore probable, a priori, that the origin of it in
Asia is that assigned by the Abbé Remusat, and the pro-
fessed learning of that scholar gives an authority to the facts
detailed.

Ferussac however remarks, that the volcano of Tourfan is
situated, according to Pere Gaubil, in lat. 43. 30. long.
87. 11. so that it lies in the midst of some considerable lakes.
That of the White Mountain, in lat. 46., is also between two
large lakes, that of Balgasch and Alakougoul.

The Mount of Fire or Ho-chan, now Khoutche, seems to be
the same as that noticed by the Chinese author of the 7th
century, under the name of Pechan, 20 leagues north of
Khouei-thou. According to this author the mountain is
called Aghie, a word which, according to M. Klaproth,
signifies the same as Hochan in Chinese. Klaproth thinks
that the volcano is situated in 42°. 35. N. lat. probably the
same as Mount Khalar, which, according to the-Boukharies,
is found south of Korgas, on the R. Ili. Thus all this part
of Klaproth's statement, seems to relate to a single volcano;
and the different names of Mountain of Fire and White
Mountain, by which it is known, seem to correspond with
the two volcanos of M. Remusat.

The volcano of the environs of Tchougoultchak, must be
found at the foot of M Chamar, near the Lake Zaisan, and
does not appear to correspond with either of those of M.
Remusat. The number consequently of the actual volcanos and solfataras in central Asia does not appear fully
made out.

In the translation made by Hylander the Father* of a

* See Bulletin des Sciences for January, 1825.

geographical work by Ibn-el-Wardi, mention is made of a mountain in the interior of Asia, from which smoke is seen to arise in the day, and flames by night. This mountain is situated in the country called Tim, which Hylander and his son think are the same as the Botom of Edrisi and Abulfeda, and the Bastam of Bakoui.

This country is situated between the Oxus and the Iaxartes. The mountains give rise to the Sogd (the Polytimetus of ancient geographers) a river which waters the Sogdiana. It is worth while to observe also, that according to the same Arabian writer, this country produces native sal ammoniac, and the substance called Zadj, which is either alum or an aluminous slate. The burning mountain made known to us in this passage, must be to the east of the Lake Aral, and of the Caspian Sea.

Pallas also in his Travels mentions a burning mountain, which he visited in the spring of the year 1770, in the government of Oremburg, near the village of Soulpa occupied by the Baschkires, and the river Jourjousen. He speaks of vapours like smoke in the day, and of slight flames at night, but he does not view them as volcanic.*

Other notices might be collected of similar phænomena occurring in parts of this vast tract, but they principally depend upon the authority of the Tartar tribes, and may perhaps turn out to be the same as those already alluded to.†

The only active volcanos on the Continent of Asia, that appear to be fully ascertained, are those on the Peninsula of Kamtschatka.

Kraskeninikoff, in his history of that province, translated into French in 1767, makes mention of three ; viz.

1. Awachinski, north of the Bay Awatscha, which had an eruption in 1737, followed by a tremendous earthquake,

---

* Vol. 2. p. 76.

† See Schlangin, in Pallas. Nordischen Beitragen, vi. p. 117, and do. vii. p. 327.—Also Falk Topog. des Russ. Reichs, vol. 1. p. 380.

during which the sea overflowed the land, and afterwards receded so far, as to leave its bed, between the first and second of the Kurule Islands, dry.

2. Tulbatchinski, situated on a tongue of land between the rivers of Kamtschatka and Tulbatchik. Its first eruption took place in 1739, and caused the country for 50 wersts to be covered with ashes.

3. Kamtschatka Mountain, the loftiest in the country. It was in a state of eruption in 1737, the same year in which the mountain Awachinski was in activity. It continued for a week to throw out streams of lava with great vehemence. Since that time it usually ejects ashes and scoriæ three or four times a year.

Besides the above, there are said to be two which only emit smoke, and two which are extinct. They appear to be all situated about the southern part of the Peninsula.*

From Kamtschatka we may, to all appearance, trace a line of volcanic operations along the chain of the Aleutian Islands to the Peninsula of Alaschka in North America, where indications of the same kind are said to occur. Among the Aleutian groupe, Langsdorf has described a rock near the Island of Unalaschka, 3000 feet in height, consisting of trachyte, which made its appearance in 1795, and seems to have been thrown up all at once from the bottom of the ocean, and not formed by successive accumulations of ejected materials.† It appears from Otto Von Kotzebue's Travels, that the geological structure of the surrounding rocks is basaltic, though they possess a greater degree of hardness, and resist the action of the weather more completely than basalt usually does. This he attributes to the quartz and augite they contain.

Trachyte and porphyry slate, however, appear from Moritz Von Engelhardt to occur in Unalaschka.‡

* Kraskeninikof, vol i. p. 148.
† See Langsdorff's Voyages, vol. ii.
‡ See Kotzebue's Voyage, vol. iii. p. 337.

The volcanos of Kamtschatka are connected again in the south with those of Japan, by means of the Kurule Islands, where no less than nine active volcanos occur according to Krascheninikou.

In the Islands of Japan ten volcanos have been enumerated, but little is known concerning them. In the southern part of Jesso, the most northern of these islands, near the town of Matsmai, is a harbour, having on its south side three volcanos, to which Broughton‡ gave the name of Volcano Bay. Near the same spot are two small detached islands discovered by Capt. Krusenstern, but more particularly described by Dr. Tilesius,‖ who accompanied that navigator in the quality of naturalist. They are called Oosima and Coosima, and are remarkable for their diminutive size, that of Coosima indeed rising only to the height of 150 fathoms above the level of the water. Both islands appeared to smoke incessantly, and the volcanic appearances at Oosima emanated from a crater, one side of which had fallen in, and was filled with red puzzolana. The hollow between the separated rocks was penetrated with spiracles, which continued in activity under the sea, and the old lava, from which the present cone rises, is seen distinctly near the level of the water.

The Island of Jesso is supposed by Krusenstern to have been once connected with that of Niphon, where the straights of Matsmai now exist, and the volcanic action going on in all directions around certainly supplies a sufficient cause for their subsequent disruption.

That Jesso and Niphon were once continuous, may be inferred from the similar composition of the opposite coasts, the basaltic character of their rocks, and their broken and shattered condition. Krusenstern however admits the necessity of examining the coast more accurately, before the conjecture can be regarded as established on solid grounds.

‡ Broughton, Voyage of Discovery in the Northern Pacific Ocean, London, 1804. See also Langsdorff, vol. i. p. 325.

‖ Tilesius in Edin. Philos. Journal, vol. 3.

Langsdorff particularly mentions a volcano in Satzuma Bay, into the crater of which the Christian proselytes were thrown, if they would not renounce their faith, during the severe persecution carried on against them in the last century.

From the southern point of Japan, a chain of headlands is continued along the groupe of Loo Choo to Formosa, and thence to the Philippine Islands.

Off Loo Choo* Captain Hall discovered an isolated rock, on which was the crater of a volcano reduced apparently to the condition of a solfatara. Its sides were stratified, as were also the rocks on the south side of the Island, which are penetrated with great dykes of a material more durable than the stone they intersect, and therefore standing out to a considerable distance in relief from the face of the rock.

The Island of Formosa is described by Klaproth in Maltbrun's Voyages, and in the Asiatic Journal for December, 1824. A high chain of mountains, which is covered with snow in November and December, stretches across the country. Abundance of salt and sulphur is met with, and flames are said to rise occasionally from the waters of the lakes and from the ground. There is a tradition as to the summit of one of these mountains having become the seat of a volcano. There is said to be on the top of the mountain, called Pa-lee-fen-shan, a block of iron of the highest antiquity, to which the natives attribute many extraordinary qualities.

Luçon, one of the Philippine Islands, contains three active volcanos, one of which, Taal, south of Manilla, had an eruption in 1754.†

---

* See Captain Basil Hall's Voyage to Loo Choo.

† See Chamisso in Kotzebue's Travels, vol. iii. p. 52. It was visited by that Naturalist, who describes it as situated in a small island, on a lake communicating with the sea. It does not appear that lava has ever flowed from it, but the whole island consists of a mass of ashes and scoriæ. He could not succeed in reaching the bottom of the crater. Sulphureous vapours were given off, and plumose alum was collected.

The Islands of Fugo and Magindanao likewise, each contain a burning mountain.

We know nothing of the volcanos said to exist in Borneo, but it appears that the Andaman Islands, west of Pegu, and north of Sumatra, contain one in activity, called Barren Island,* nearly 4000 feet in height, which frequently emits vast columns of smoke, and red hot stones three or four tons in weight.

In Sumatra, Marsden has described four as existing, but the following are all the particulars known concerning them :

Lava has been seen to flow from a considerable volcano near *Priamang*, but the only volcano this observer† had an opportunity of visiting, opened on the side of a mountain about 20 miles inland of Bencoolen, one fourth way from the top, so far as he could judge. It scarcely ever failed to emit smoke, but the column was only visible for two or three hours in the morning, seldom rising and preserving its form above the upper edge of the hill, which is not of a conical shape, but slopes gradually upwards. He never observed any connexion between the state of the mountain and the earthquake, but it was stated to him, that a few years before his arrival it was remarked to send forth flame during an earthquake, which it does not usually do.

The inhabitants are however alarmed, when these vents all remain tranquil for a considerable time together, as they find by experience, that they then become more liable to earthquakes.

Dr. Jack in a short notice of the Geology of Sumatra, (Geol. Trans. vol. i. p. 398, new series) observes, that the volcanos of this island have somewhat of a different character from those of Java : the former generally terminating at the summit on a ridge or crest ; while the latter are more exactly conical, and have for the most part a much broader

---

* See Colebrooke in the Asiatic Transactions.

† Marsden's Sumatra, p. 29.

basis. It appears, that the country near Bencoolen before alluded to (between Indapore and Bencoolen) was visited by the order of Raffles in 1818. There is a lake with a cultivated valley to the west, watered by a small river, that descends from a high volcanic mountain called Gunong Api, which is always smoking. The rocks in this neighbourhood are of trap, either compact or amygdaloidal, sometimes tufaceous; the most remarkable hill is called Gunong Bungko, or Sugar Loaf Hill; it is composed of irregular masses of trap.

In crossing the island from Bencoolen to Palembang, we traverse a plain, having in the midst of it the mountain called Gunong Dempo, which is the highest in the island, being 12,000 feet above the level of the sea. It is almost always emitting smoke. Hot springs and other volcanic phænomena are common in the neighbourhood. Basalt and trap compose the range of hills between Bencoolen and Cawoor.

The basis of Sumatra is however, in all probability, primitive: Granite being found near Menang Kabau, and at Ayer Bangy. Much limestone derived from Coral Reefs likewise occurs.

Very different to this is the structure of Java, where the researches of Dr. Horsfield* have made known to us one of the most extended tracts of volcanic formations that perhaps any where exist.—An uninterrupted series of large mountains, varying in elevation from 5 to 11 and even 12,000 feet, and exhibiting by their round or pointed tops, their volcanic origin, extend through the whole of the island. The several large mountains composed in this series, in number 38, though different from each other in external figure, agree in the general attribute of volcanos, having a broad base gradually verging towards the summit in the form of a cone.

---

* See his remarks, as inserted in Sir Stamford Raffles' History of Java.

They all rise from a plain but little elevated above the level of the sea, and each must, with very few exceptions, be considered as a separate mountain, raised by a cause independent of that which produced the others. Most of these have been formed at a very remote period, and are covered by the vegetation of many ages : but the indications and remains of their former eruptions are numerous and unequivocal. The craters of several are completely extinct; those of others contain small apertures, which continually discharge sulphureous vapours and smoke. Many of them have had eruptions during late years. Almost all the mountains or volcanos in the large series before noticed, are found on examination to have the same general constitution; they are striped vertically by sharp ridges, which, as they approach the foot of the mountain, take a more winding course.

These ridges alternate with valleys, whose sides are of a very various declivity.

Large blocks of basalt occasionally project, and in several instances the vallies form the beds of rivers towards the tops of the volcanos; in the rainy season they all convey large volumes of water.

There are also various ridges of smaller mountains, which, though evidently volcanic, may be termed *secondary*, as they appear to have originated from the *primary* volcanos before noticed. They generally extend in long narrow stripes, with but a moderate elevation, and their sides are less regularly composed of the vertical ridges above mentioned. In most cases, a stratified structure and submarine origin may be discovered. They are generally covered with large rocks of basalt; and in some cases, they consist of wacke and hornblende, which is found along their base in immense piles.

Hills of a mixed nature, partly calcareous, partly volcanic are also found. The southern coast of the island consists almost entirely of them, rising in many places to the perpendicular height of 80 or 100 feet, and sometimes much higher. These, as they branch inward, and approach the central or higher districts, gradually disappear, and give place to the volcanic series, or alternate with huge masses of basaltic hornblende that appear to assume a regular stratification.

A Dutch writer * has since communicated some further particulars respecting these basaltic rocks. He informs us that almost all the lofty summits of this basaltic chain are truncated cones; that they consist in general of irregular columns, presenting every variety in point of length, thickness, direction, and other circumstances. They are sometimes placed immediately one on the other, sometimes divided by beds of a different material. This arrangement is only exposed in a few places where the streams have laid bare the strata. The action of the elements, assisted perhaps by that of sulphureous vapours, has effected a great destruction of the rock, large masses of which are washed down into the plains after heavy showers. The basalt is of two qualities,—1st, that composing the lower mountains, which is less compact and porphyritic, but contains large concretions of felspar, quartz? crystallized hornblende, and olivine; 2dly, that constituting the loftier ranges, which is more compact and uniform in structure, and is so impregnated with iron, as to attract the magnetic needle, and even to be obedient to its influence.

With regard to the modern lavas of Java, having been favoured with a sight of the specimens brought by Dr. Horsfield himself from the island, and now at the India House, I may remark that they struck me as being very similar to those of Vesuvius, and in several instances appeared to contain much leucite. I observed several specimens of pitchstone, which I was assured constituted dykes. It is much to be wished, that so interesting a suite were properly arranged, and made a part of some public and general collection of volcanic products.

Dr. Horsfield has related an effect of volcanic action in

* Reinweldt in the Batavian Transactions, published at Batavia in 1823.

this island, which for its extent, seems to exceed almost any that has been hitherto noticed.

The Papandayang, situated on the south western part of the island, was formerly one of the largest volcanos, but the greater part of it was swallowed up in the earth, after a short but severe combustion in the year 1772. The account which has remained of this event asserts, that near midnight, between the 11th and 12th of August, there was observed about the mountain an uncommonly luminous cloud, by which it appeared to be completely enveloped. The inhabitants, as well about the foot, as on the declivities of the mountain, alarmed by the appearance, betook themselves to flight; but before they could all save themselves, the mountain began to give way, and the greatest part of it actually *fell in,* and disappeared in the earth. At the same time a tremendous noise was heard, resembling the discharge of the heaviest cannon. Immense quantities of volcanic substances, which were thrown out at the same time and spread in every direction, propagated the effects of the explosion through the space of many miles.

It is estimated, that an extent of ground, of the mountain itself, and its immediate environs, fifteen miles long and six broad, was by this commotion swallowed up in the bowels of the earth. Several persons, sent to examine the condition of the neighbourhood, made report, that they found it impossible to approach the mountain, on account of the heat of the substances which covered its circumference, and which were piled on each other to the height of three feet, although this was the 24th of September, and thus full six weeks after the catastrophe. It is also mentioned that forty villages, partly swallowed up by the ground, and partly covered by the substances thrown out, were destroyed on this occasion, and that 2,957 of the inhabitants perished.

The mountain of Galoen-gong, which had never been reckoned among the volcanos of the island, broke out with terrific violence in 1822. The eruption began by a tremendous explosion of stones and ashes, followed by a stream of lava which covered a large tract. Four thousand persons were destroyed.*

It would appear likewise from Dr. Horsfields' description, that Java exhibits phænomena of a similar kind to those noticed in Sicily, and at the foot of the Appennines, and there known under the name of " Salses." In the calcareous district (which I suspect to belong to the same class of formations as the blue clay and tertiary limestone of Sicily,) occur a number of hot springs, containing a large quantity of calcareous earth in solution, which incrusts the surface of the ground near it. Some of them are much mixed with petroleum, and others highly saline.

The latter are dispersed through a district of country consisting of limestone, several miles in circumference. They are of considerable number, and force themselves upwards through apertures in the rocks with some violence and ebullition. The waters are strongly impregnated with muriate of soda, and yield upon evaporation very good salt for culinary purposes (not less than 200 tons in the year).

About the centre of this limestone district, is found an extraordinary volcanic phænomenon. On approaching it from a distance, it is first discovered by a large volume of smoke rising and disappearing at intervals of a few seconds, resembling the vapours arising from a violent surf: a dull noise is heard, like that of distant thunder. Having advanced so near, that the vision was no longer impeded by the smoke, a large hemispherical mass was observed, consisting of black earth, mixed with water, about sixteen feet in diameter, rising to the height of 20 or 30 feet in a perfectly regular manner, and, as it were, pushed up by a force beneath ; which suddenly exploded with a dull noise, and scattered about a volume of black mud in every direction. After an interval of two or three, or sometimes four or five seconds, the hemispherical body of mud or earth rose and exploded again.

* Phil. Mag. August, 1823.

In the same manner this volcanic ebullition goes on without interruption, throwing up a globular body of mud, and dispersing it with violence through the neighbouring places. The spot where the ebullition occurs is nearly circular and perfectly level, it is covered only with the earthy particles impregnated with salt water, which are thrown up from below; the circumference may be estimated at about half an English mile. In order to conduct the salt water to the circumference, small passages, or gutters, are made in the loose muddy earth, which lead it to the borders, where it is collected in holes dug in the ground for the purpose of evaporation.

A strong, pungent, sulphureous smell, somewhat resembling that of earth-oil (naphtha), is perceived on standing near the explosion; and the mud recently thrown up possesses a degree of heat greater than that of the surrounding atmosphere. During the rainy season these explosions are more violent, the mud is thrown up much higher, and the noise is heard at a greater distance.

This volcanic phænomenon is situated near the centre of the large plain, which interrupts the large series of volcanoes, and owes its origin to the general cause of the numerous volcanic eruptions which occur on the island.*

A tradition prevails that Java, Sumatra, Bali and Sumbawa were once united, and formed with Hindostan one unbroken continent. Even the dates of the separation of these islands are given in the Javanese records.—Thus the separation of:

	In the Javan year.†
Sumatra and Java took place	1114
Bali, Balembangan, and Tava	1204
Gilling, Trawangan, and Bali	1260
Selo-Parang, and Sambawa	1280

* Raffles' Java, chap. x.—See also Breislac. Institut. Geolog. vol. 3. p. 47, for an extract from the Bib. Univ. Juillet, 1817, giving an account of the same phænomenon.

† The Javan Æra begins 73 years later than the Christian.

Raffles however justly remarks, that the physical structure of Sumatra is not such as to give countenance to this opinion.

The volcanos of Java and of the Philippine groupe appear almost connected with one another, through the medium of those which exist in Sumbawa, Flores, Daumer, Banda, and the Moluccas.

That of Tomboro in the Island of Sambawa is perhaps one of the most considerable in the world, according to the description given of it by Sir Stamford Raffles.*

Almost every one, says this writer, is acquainted with the intermitting convulsions of Etna and Vesuvius, as they appear in the descriptions of the poet, and the authentic accounts of the naturalist, but the most extraordinary of them can bear no comparison, in point of duration and force, with that of Mount Tomboro in the Island of Sambawa. This eruption extended perceptible evidences of its existence, over the whole of the Molucca Islands, over Java, a considerable portion of Celebes, Sumatra, and Borneo, to a circumference of a thousand statute miles from its centre, by tremulous motions and the report of explosions; while, within the range of its more immediate activity, embracing a space of 300 miles around it, it produced the most astonishing effects, and excited the most alarming apprehensions. In Java, at the distance of 300 miles, it seemed to be awfully present. The sky was overcast at midday with clouds of ashes, the sun was enveloped in an atmosphere, whose " palpable density " he was unable to penetrate; a shower of ashes covered the houses, the streets, and the fields, to the depth of several inches, and amid this darkness explosions were heard at intervals, like the report of artillery, or the noise of distant thunder.

At Sambawa itself three distinct columns of flame appeared to burst forth, near the top of the Tomboro moun-

* Raffles' Java. Vol. I. ch. I.

tain, (all of them apparently within the verge of the crater,) and after ascending separately to a very great height, their tops united in the air in a troubled, confused manner. In a short time, the whole mountain next Sang'ir appeared like a body of liquid fire, extending itself in every direction.

The fire and columns of flame continued to rage with unabated fury, until the darkness, caused by the quantity of falling matter, obscured it at about 8 p. m. Stones at this time fell very thick at Sang'ir, some of them as large as two fists, but generally not larger than walnuts. Between nine and ten p. m. ashes began to fall, and soon after, a violent whirlwind * ensued, which blew down nearly every house in the village of Sang'ir, carrying the alaps, or roofs, and light parts away with it. In the port of Sang'ir adjoining Sumbawa, its effects were much more violent, tearing up by the roots the largest trees, and carrying them into the air, together with men, horses, cattle, and whatever else came within its influence. (This will account for the immense number of floating trees seen at sea). The sea rose nearly twelve feet higher than it had ever been known to do before, and completely spoiled the only small spots of rice land in Sang'ir, sweeping away houses and every thing within its reach. The whirlwind lasted about an hour. No explosions were heard till the whirlwind had ceased, at about eleven a. m. From midnight till the evening of the 11th, they continued without intermission; after that time

* May not the occurrence of whirlwinds as a concomitant of volcanic eruptions, account for the fable of the giant Typhœus, who was at once considered the cause of both these effects. See Hesiod Θ 869, where Typhœus is described as producing those winds which are destructive to man; whereas Notus, Boreas and others, which are useful to him, are of divine origin.

Εκ δε Τυφωεος ες ανεμων μενος υγρον αεντων
Νοσφι Νοτυ Βορεω τε, και Αργεστεω Ζεφυροιο
Οι γε μεν εκ Θεοφιν γενεη. Θνητοις μεγ ονειαρ.
Αι δ'αλλαι μαψκυραι επιπνειυσι Θαλασσαν.

Ammianus Marcellinus, in describing the earthquake which destroyed Nicomedia, says that it was accompanied by hurricanes. Lib. 17. c. 7.

x

their violence moderatèd, and they were heard only at intervals, but the explosions did not cease entirely till the
15th of July. Of all the villages round Tomboro, Tempo,
containing about forty inhabitants, is the only one remaining. In Pekaté no vestige of a house is left: twenty-six
of the people, who were at Sumbawa at the time, are the
whole of the population who have escaped. From the best
enquiries there were certainly not fewer than 12,000 individuals in Tomboro and Pekaté at the time of the eruption, of whom five or six survive. The trees and herbage
of every description, along the whole of the north and west
of the peninsula, have been completely destroyed, with the
exception of a high point of land near the spot where the
village of Tomboro stood.* At Sang'ir it is added the
famine occasioned by this event was so extreme, that one of
the rajah's own daughters died of starvation.

Captain Bligh, in his voyage to the South Seas, mentions
the existence of a volcano on a high mountain in the Isle
of Flores: it seemed to him to have eruptions of so formidable a character, that the whole island was covered with
its products.

In the Island of Timor, the volcano of the *peak* served,
like that of Stromboli, as a sort of light-house, seen at more
than 300 miles distance. In 1637, this mountain, during a
violent eruption disappeared entirely: a lake at present takes
its place.

The Island of Daumer is also said to contain a volcano,
and Dampier in 1669 saw one burning between Timor and
Ceram.

Goonung-Api,† one of the Banda Islands, contains an
active volcano,‡ which had an eruption in 1820, and ejected
red-hot stones of prodigious size.

* Major Thorn (in his Memoir on the Conquest of Java, 1810. p. 320.)
speaks of another volcano, on the north coast of Sumbawa, which is considered to be in communication with that of Goonung-Api, three or four
miles off.

† It appears from Marsden's Sumatra, that this word signifies in the Málay
language volcano.

‡ Arago in the Annuaire for 1824.

On the southern side of the Island of Celebes, there are no volcanos known, but some are said to exist in the northern division.* This however does not appear to be ascertained.

In the archipelago of the Moluccas we hear of two in the Islands of Ternati and Tidore. Maltbrun (Annales des Voyages 24, p. 281) speaks of an eruption that happened in 1673 at Gilolo. An event which occurred in 1646 in the Island of Machian,† may also be attributed to this cause. It is stated that in that year a mountain was rent from top to bottom, emitted horrible streams of smoke and flame, and was divided into two parts, which now constitute two distinct eminences.

I have noticed a similar phænomenon when speaking of Mount Ararat, though the cause to which the fissure is attributable can in that case only be conjectured.

Dampier discovered in New Guinea in South Lat. 5. 33, an active volcano, which sent forth fits of flame at intervals.

The volcano in the Island of Sang'ir, one of the largest in the world, seems to connect those last mentioned with the burning mountains which I have before described as existing among the Philippines, thus appearing to establish a line of communication between those of Kamschatka and of the Indian Sea.

In the Great Pacific Ocean, the islands, according to Kotzebue,‡ may be referred to two classes, distinguished by their elevation into high and low. The latter class appear to be entirely of modern formation, the product of that accumulation of coral reefs, which Flinders and others have described in so interesting a manner.

The high islands on the contrary are chiefly volcanic, though in the Friendly and Marquesa Islands primitive rocks occur, and in Waohoo porphyry and amygdaloid.

The Mariana or Ladrone Islands constitute a sort of mountain chain, consisting of a line of active volcanos,

* See Appendix to Raffles's Java. Vol. 2. p. clxxviii.
† Ordinaire Hist. Natur. des Volcans. p. 177.
‡ Kotzebue's Voyage of Discovery. Vol. 2. p. 355.

especially towards their north, which is parallel to that of
the Philippine groupe, whereas the islands that lie detached
in the middle of the basin, of which these two groups are
the boundaries, seem for the most part to be extinguished.*

The Island of Ahrym, in the groupe of the New Hebrides,
contains an active volcano, and the same thing is stated by
Forster with regard to that of Tanna. A volcano is said
by Kotzebue to be burning in Tofua, one of the Friendly
Islands.

Mr. Ellis, a missionary, has given, in a narrative of a
Tour through the Sandwich Islands, (London, 1826, p. 199)
the most detailed account of the active volcano of Owhyhee,
that I have yet seen.

The plain, over which their way to the mountain lay, was a
vast waste of antient lava, which he thus describes :—" The
tract of lava resembled in appearance an inland sea, bounded by distant mountains Once it had certainly been in a
fluid state, but appeared as if it had become suddenly petri-
fied, or turned into a glassy stone, while its agitated billows
were rolling to and fro. Not only were the large swells
and hollows distinctly marked, but in many places the surface of these billows was covered by a smaller ripple, like
that observed on the surface of the sea at the springing up
of a breeze, or the passing currents of air, which produce
what the sailors call a cat's-pay. About two p.m. the crater
of Kirauea suddenly burst upon our view. We expected to
have seen a mountain with a broad base and rough indented
sides, composed of loose slags, or hardened streams of lava,
and whose summit would have presented a rugged wall of
scoria, forming the rim of a mighty cauldron. But, instead
of this, we found ourselves on the edge of a steep precipice,
with a vast plain before us, fifteen or sixteen miles in circumference, and sunk from two hundred to four hundred
feet below its original level. The surface of this plain was

---

* Chamisso in Kotzebue's Voyage of Discovery, vol. 2. p. 353.

uneven, and strewed over with huge stones, and volcanic rock, and in the centre of it was the great crater, at the distance of a mile and a half from the place where we were standing. We walked on to the north end of the ridge, where, the precipice being less steep, a descent to the plain below seemed practicable. With all our care, we did not reach the bottom without several falls and slight bruises. After walking some distance over the sunken plain, which in several places sounded hollow under our feet, we at length came to the edge of the great crater, where a spectacle sublime, and even appalling, presented itself before us. Immediately before us yawned an immense gulf, in the form of a crescent, about two miles in length, from N.E. to S.W. nearly a mile in width, and apparently eight hundred feet deep. The bottom was covered with lava, and the S.W. and northern parts of it were one vast flood of burning matter, in a state of terrific ebullition, rolling to and fro its ' fiery surge' and flaming billows. Fifty-one conical islands of varied form and size, containing so many craters, rose either round the edge, or from the surface of the burning lake ; twenty-two constantly emitted columns of grey smoke, or pyramids of brilliant flame ; and several of these at the same time vomited from their ignited mouths streams of lava, which rolled in blazing torrents down their black indented sides, into the boiling mass below. The existence of these conical craters led us to conclude that the boiling cauldron of lava before us did not form the focus of the volcano; that this mass of melted lava was comparatively shallow ; and that the basin in which it was contained was separated by a stratum of solid matter from the great volcanic abyss, which constantly poured out its melted contents through these numerous craters into this upper reservoir. The sides of the gulph before us, although composed of different strata of antient lava, were perpendicular for about four hundred feet, and rose from a wide horizontal ledge of

solid black lava of irregular breadth, but extending completely round, beneath this ledge, the sides sloped gradually
towards the burning lake, which was, as nearly as we could
judge, three hundred or four hundred feet lower. It was
evident that the large crater had been recently filled with
liquid lava up to this black ledge, and had, by some subterraneous canal, emptied itself into the sea or under the
low land on the shore. The grey, and in some places apparently calcined sides of the great crater before us—the
fissures which intersected the surface of the plain on which
we were standing—the long banks of sulphur on the opposite side of the abyss—the vigorous action of the numerous small craters on its borders—the dense columns of
vapour and smoke that rose at the N. and S. end of the
plain—together with the ridge of steep rocks by which it
was surrounded, rising probably in some places three or four
hundred feet in perpendicular height, presented an immense
volcanic panorama, the effect of which was greatly augmented by the constant roaring of the vast furnaces
below.''*

* Mr. Ellis's work, though not scientific, will prove of interest to the
naturalist from the notices of volcanic phænomena with which it is interspersed, and deserves on that account to be placed by the side of Henderson's
Iceland, to which I have had occasion to refer.

His account is enlivened by introducing a few of the legends to which the
striking natural phænomena of the island have given rise; for the natives of
Owhyhee, like the Greeks and Persians of old, have peopled the recesses of
the Mountain Kiraueca with a tribe of Deities, both male and female, the belief
in whose power, kept alive as it is by repeated volcanic explosions, was too
deeply rooted in their minds to be effaced, even by the authority of their sovereign, who in 1819 decreed the summary abolition of all idolatrous worship.

The natives still persist in believing, that the conical craters of the mountains are the houses of their gods, where they frequently amuse themselves by
playing at Konané (a game like draughts); that the roaring of the furnaces
and the crackling of the flames are the music of their dance, and that the red
flaming surge is the surf in which they play, sportively swimming on the rolling wave.

Some of their legends may remind us of those that prevailed among the
Greeks.

OWHYHEE.

Besides the volcano of Kirauea, which is in activity, there are several in an extinguished state. One of them, Mouna-Roa, is calculated by Captain King at 16,020 feet in height, estimating it according to the tropical line of snow. Another Mouna-Kaah, the peaks of which are entirely covered with snow, cannot be less, he thinks, than 18,400 feet. Mr. Ellis reckons the height at between 15,000 and 16,000 feet. The whole island of Owhyhee indeed, embracing a space of 4000 square miles, is, according to the observations of that missionary, one complete mass of lava, or other volcanic matter in different stages of decomposition. Perforated, he says, with ennumerable apertures in the shape of craters, the island forms a hollow cone over one vast furnace, situated in the heart of a stupendous submarine mountain rising from the bottom of the sea; or possibly the fires may range with augmented force beneath the bed of the ocean, rearing through the superincumbent weight of water the base of Hawaiah,* and at the same time, forming a pyramidal funnel from the furnace to the atmosphere."

Thus one of their kings, who had offended Pèlè the principal goddess of the volcano, is pursued by her to the shore, where leaping into a canoe he paddles out to sea. Pèlè, perceiving his escape, hurls after him huge stones and fragments of rocks, which fall thickly around, but do not strike the canoe. A number of rocks in the sea are shewn by the natives, which, like the Cyclopean Islands at the foot of Mount Etna, are said to have been those thrown by Pèlè to sink the boat.

I recommend the perusal of this legend (which may be seen in page 266 of Mr. Ellis's book,) as very characteristic of the manners and feelings of savage life. The king is represented as taking little pains to secure the escape of any one but himself, for his mother, wife, and children are all abandoned without compunction; his conduct to the friend who accompanies him, is the only trait which redeems his character from the charge of utter selfishness, nor among the natives who tell the story, is their praise of the adroitness with which he effected his escape, at all less commended on account of this desertion of his nearest relations.

* I adopt the more usual method of spelling. Ellis in his late Tour calls the island Hawaii.

OWHYHEE.

Other volcanos are stated to occur in different parts of that extensive tract, known by modern geographers under the name of Polynesia, as far as to New Caledonia and the New Hebrides. The separate mass of New Zealand, with which Norfolk Island is connected, may be viewed as the southern end of the bulwark;* its eastern can hardly be fixed at any nearer point than the coast of America, for I am assured that an active volcano at presents exists among the Galapagos, only 10 deg. W. of Quito.

Far therefore from believing that volcanos have been instrumental in the destruction of continents, or that their history lends any countenance to the fables respecting the Atlantis, we should rather be led to consider that they were more generally among the means which nature employs for increasing the extent of dry land, and for gradually converting an unprofitable tract of ocean into an abode for the higher classes of animals.

ON THE VOLCANOS OF AMERICA.

The Islands constituting the West-Indian Archipelago, have been divided by a French Geologist into four classes.†

The 1st of these includes such, as consist partly of primitive (or to speak more correctly of the more antient) rock-formations, and partly of those derived from volcanos.

The 2nd, those which are wholly volcanic.

The 3rd, those entirely calcareous. And

The 4th, such as are partly of volcanic origin, and partly composed of limestone with organic remains.

* Kotzebue's Voyage. Vol. II. p. 354.
† See Cortes Journal de Physique, tom. lxx.

TRINIDAD.

To the 1st of these classes are referred all the islands of the largest size, such as Trinidad, Jamaica,* Cuba, St. Domingo, and Portorico.

With regard to Trinidad however it may be observed, that it seems rather to make a part of the Continent, than to be a member of the system of mountains existing in the other West India Islands.

Dr. Nugent observes, that its rocks are either decidedly of primitive or of alluvial origin.† The great northern range of mountains, that runs from east to west, and is connected with the high land of Paria on the continent, by the islands of the Bocas, consists of gneiss; of mica slate, containing great masses of quartz, and in many places approaching so near to the nature of talc, as to render the soil quite unctuous by its decomposition; and of compact bluish limestone, with frequent veins of white crystallized carbonate of lime. From the foot of these mountains for many leagues to the northward, there extends a low and perfectly level tract of land, evidently formed by the detritus of the mountains, and by the copious tribute of the waters of the Ohio, deposited by the influence of currents.

The celebrated pitch lake exists in the midst of a clayey soil—it is a vast expanse of Asphaltum, perhaps three miles in circumference, which in the wet season is sufficiently solid to bear any weight, but in hot weather is often in a state approaching to fluidity. The manner in which it was originally formed, may perhaps admit of an explanation, by considering certain analogous phænomena that present themselves in its vicinity. Thus on the eastern coast there

---

* Jamaica, strictly speaking, does not belong to this class, since it appears from Mr. De la Beche's account, that besides transition rocks and some of the older secondary, it contains likewise an extensive formation of white limestone, referable to the tertiary period.—See his paper in the forthcoming Number of the Geol. Transactions.

† Geol. Transactions, vol. 1.

is a pit which throws up asphaltum, together, as it is said, with violent explosions, smoke and flames.

Almost in the same parallel, and also in the sea, but to the west of the island (near Punta de la Brea, south of the port of Naparaimo) is a similar vent.

At the south-west extremity of the island, between Point Icacos and the Rio Erin, are small cones, which appear to have some analogy with the volcanos of Air and Mud, which occur at Turbaco in New Grenada, and are described by Humboldt.

It is possible, that the whole of these phænomena may be analogous to those of Macaluba, and of the Lago Naftia in Sicily, which, though they appear to shew some kind of affinity to volcanos from their frequent occurrence in the vicinity of rocks derived from such operations, can hardly be viewed without more sufficient evidence as originating from the same deep-seated cause, to which the latter are commonly referred.

Respecting Jamaica, I am spared the necessity of collecting any details, by the promised appearance of a volume of the Geological Transactions, which will contain the memoir lately read by my friend Mr. De la Beche, on the physical structure of that part of the world.

Relative to a small extinct volcano which he discovered in one corner of the island, I have his permission for extracting the following particulars:

" In that part of Jamaica," says Mr. De la Beche, "which I examined, I observed rocks of a volcanic character only at the Black Hill, which is situated between Lennox, Low Layton, and the sea; this hill, when viewed from the neighbourhood of Buff Bay, has a somewhat conical appearance, and rises above the low hills that extend towards Savanna Point; the hill however, when approached, is seen to be no cone, notwithstanding its effect at a distance.

" The rock, of which the Black Hill is composed, is greyish green and hard, gives out an earthy smell when breathed upon, and may be described in general terms as a volcanic amygdaloid, the cells, with which it abounds, being mostly filled; those however which have been exposed to the weather are frequently empty, as are also those within a few inches from the surface. The interior of the empty cells is often incrusted with a little oxide of iron, while those that are full, and which convert the rock into an amygdaloid, contain chalcedony, calc-spar, &c."

Of the physical structure of Cuba, nothing appears to be known; so that it is rather from analogy than from any thing else, that I have ventured to refer this island to the same class as Jamaica.

Even to the remaining two a similar remark may in a degree apply; for St. Domingo has been too little explored by naturalists, to allow of our collecting any thing definite with regard to its constitution—we know only that it is extremely subject to earthquakes.

Cortes, in his Memoir on the Antilles, before referred to, speaks of Porto Rico as containing antient volcanic, as well as primitive rocks, but he does not enter into any details.

Of the second class of islands, which consist exclusively of volcanic rocks, the following is a summary, commencing with the most southern.

1. Grenada, an extinct crater filled with water; boiling springs; basalts between St. George and Goave.

2. St. Vincent, an active volcano, called Le Souffrier, the loftiest mountain in the chain which runs through the island. It first threw out lavas in 1718, but its most tremendous eruption was in 1812, when there issued from the mountain so dreadful a torrent of

lava, and such clouds of ashes, as nearly covered the island, and injured the soil in a manner which it has never yet recovered. The total ruin of the city of Caraccas preceded this explosion by thirty-five days, and violent oscillations of the ground were felt, both in the islands, and on the coasts of Terra Firma.*

3. St. Lucia contains a very active Solfatara, from 12 to 14 hundred feet in height. Besides a considerable condensation of sulphur given out from the crevices, jets of hot water likewise take place, which fill periodically certain small basins, like the Geysers of Iceland.

4. Martinique can hardly be said to belong to this class, for limestone is seen resting upon the volcanic products.

The latter however constitutes the fundamental rock throughout the whole island, and forms three principal hills called Vauclin, the Paps of Carbet consisting of felspathic lava, which are the most elevated summits in the whole of this series of islands, and Montagne Pelèe. Between the first and second of these is found in a neck of land a tract composed of ancient basalts, called La Roche Carrée. Hot springs at Prêcheur and Lameutin.

5. Dominica is completely composed of volcanic matter, but the action is extinct.

6. Guadeloupe may be divided into two parts, according to its physical structure.

The first, properly called Guadeloupe, consists entirely of volcanic rocks, and therefore belongs to this division of our subject; the second, named Grande Terre, is calcareous, consisting of a shelly limestone, covered by a bed of clay,

* Humboldt's Pers. Narr. vol. 4. p. 26.

and containing rolled masses of lava. The volcanic part of the island contains fourteen antient craters, and one in a state of present activity. The eruption of 1797 took place from an elevation of 4800 feet. Pumice, ashes, and clouds of sulphureous vapours were then ejected. The particulars are given in the report made to the French Government on the state of the volcano in 1797 by Mons. Amie.

7. Montserrat—a Solfatara; fine porphyritic lavas, with large crystals of felspar and hornblende, near Galloway, often much decomposed by the sulphureous exhalations.*

8. Nevis—a Solfatara.

9. St. Christopher's—a Solfatara at Mount Misery.

10. St. Eustachia—the crater of an extinguished volcano, surrounded by pumice.

The 3rd class comprehends the islands of Margarita, Desirade, Curaçoa, Bonaire, and in general all the islands of low elevation; they consist entirely of limestone of very recent formation.

The 4th class, partly composed of volcanic products, and partly of shelly limestone, comprises the Islands of Antigua, St. Barthelemi, St. Martin, and St. Thomas.

Antigua is stated by Dr. Nugent† to be composed, on the north and east, of a very recent calcareous formation, corresponding with that of Guadaloupe, in which an admixture of marine and freshwater shells is found; subordinate to this rock on its southern limit, we find extensive masses of coarse chert, full of casts of shells, chiefly cerithia. It is sometimes intermixed with marl.

* Dr. Nugent. Geological Transactions, vol. 1.
† Geological Transactions, vol. 1. new series.

Beneath these beds occurs an extensive series of stratified rocks, which Dr. Nugent has chosen to call claystone conglomerate, but which seems from his description to be a sort of trachytic breccia.

The rock has generally an argillaceous basis, with minute crystals of felspar imbedded, and a remarkably green tinge from the numerous spots of green earth intermixed.

Its brecciated character is derived from the fragments of silicified wood, chert, agate, jasper, porphyry, lava, and other substances imbedded. The silicified woods are particularly abundant.

Respecting the remaining islands, I possess no information that can be relied on, and it is much to be wished, that some Geologist, in imitation of what has been done by Humboldt on the American Continent, and by Von Buch in the Canaries, would present us with a detailed account of the physical structure generally of the Antilles.

The process, by which these islands, according to Moreau de Jonnes, are in many instances formed, is sufficiently curious; first a submarine eruption raises from the bottom of the sea masses of volcanic products, which, as they do not rise above the surface of the water, but form a shoal a short way below its surface, serve as a foundation on which the Madreporites and other marine animals can commence their superstructure. Hence those beds of recent coralline limestone, seen covering the volcanic matter in many of the islands.*

It may also be observed, that such as exhibit traces of the recent action of fire, are all situated in a line on the western boundary of the range, from N. latitude 12 to 18, and W. long. 61 to 63. Whatever indications of the kind occur farther to the west belong to eruptions of an older date.

Humboldt remarks,† that we must not suppose each island the product of a single volcano, but rather to be pro-

---

* Humboldt's Pers. Narr. vol. 4. p. 42.—See also Chamisso in Kotzebue's Voyage of Discovery.

† Personal Narrative, vol. 4.

duced by a series of eruptions proceeding from several orifices, and joined together by the masses subsequently ejected.

It is time however to proceed to consider the volcanos of the Continent of America, our acquaintance with which is exclusively due to the labours of a single individual, the celebrated Alexander de Humboldt.

It is much to be regretted that, owing to the surprising variety of subjects which he has undertaken to investigate, it has been impossible for him hitherto to do more, than present the world with fragments and glimpes as it were of that information with respect to the geology of the New World, which it is to be hoped he will one day unfold to us.

Beginning with the Peninsula of California, we may reckon three volcanos, supposed to be at present in activity: Mount Saint Elia, Mount del Buen Tiempo, and Volcano de las Virgines. Of these, however, little is known except their names, although the mountain of St. Elia ranks among the highest in the globe, being, according to Perouse, 12,672, and according to the Spanish navigators, who measured it in 1791 by precise means, no less than 17,875 feet above the level of the sea.

In Mexico appears to commence the great chain of volcanic mountains, which extends with little interruption from the 24th deg. of N. to the 2d deg. of S. lat.

The most northern point in this country in which any signs of volcanic action are known to occur, is near the town of Durango, in lat. 24. long. 104, where there is a groupe of rocks covered with scoriæ, called La Breña, which rise up in the midst of a level plain.

The latter consist of basaltic amygdaloid, and on the summit of one of the neighbouring mountains a regular crater was discovered.

No active volcanos however are to be met with until we

reach the parallel of the city of Mexico itself, and here almost in the same line occur five, so placed that they appear to de derived from a fissure traversing Mexico from west to east, in a direction perpendicular to that of the great mountain chain, which, extending from north-west to south-east, constitutes the great table land of the American Continent. It is interesting to remark, that if the same parallel line, which connects the active volcanos of Mexico, be prolonged in a westernly direction, it would traverse the groupe of islands called the Isles of Revillagigedo, which there may be reason to consider volcanic from the pumice found amongst them.*

The most eastern of these, that of Tuxtla, is situated a few miles to the north west of Veracruz. It had a considerable eruption in 1793, the ashes from which were carried as far as Perote, a distance of 57 leagues.

In the same province, but farther to the west occur, the volcano of Orizaba, the height of which is 17,300 feet, and the peak of Popocatepetl, 300 feet higher, the loftiest eminence in New Spain. The latter is continually burning,† though for several centuries it has thrown nothing up from the crater but smoke and ashes.

On the western side of the city of Mexico, are the volcanos of Jorullo and Colima   The elevation of the latter is estimated at about 9000 feet. It frequently throws up smoke and ashes, but has not been known to eject lava.

The volcano of Jorullo, situated between Colima and the town of Mexico, is of much more modern date than the rest, and the great catastrophe which attended its first appearance, is perhaps, (says Humboldt) one of the most extraordinary physical revolutions in the annals of the history of our planet.‡

---

* See the map of Mexico, which accompanies this volume.

† Mr. Bullock has called in question this statement of Humboldt's, but a still more recent traveller, Mr. Stapleton, has confirmed it. Bull. des Sci. for September, 1825.

‡ Nouvelle Espagne, y. 248. Fol. Ed.

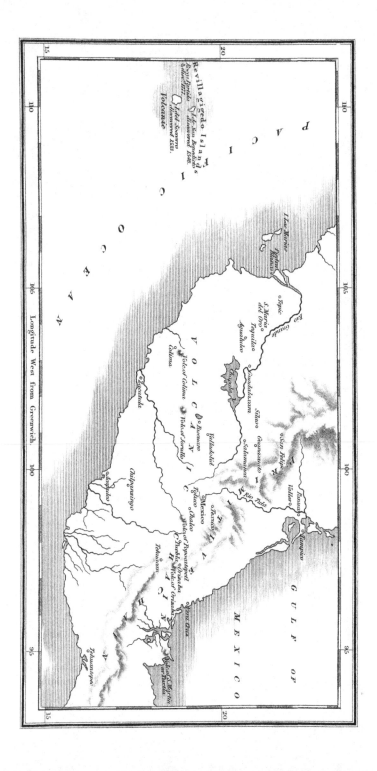

PACIFIC OCEAN

Revillagigedo Islands
Roca Partida
S. Benedicto
Island 1821.
Isle Socorro
discovered 1838.
Volcanic

I. las Marias
Iselan
Manzanilla

o Tepic
S. Maria
del Oro
Tequilao
Aguililao

Rio Grande

o San Felipe

VOLCAIN

Volc. of Colima
Colima

Valladolid

Zacatula

Volc. of Bacsan
Volc. of Jorullo

Silao
Guanaxuato
Salamanca

Chipansingo

Acapulco

Mexico
Ixtaco
Chalco
Volc. of Popocatepel
Puebla
Orizaba
Vera Cruz

Tezcuco

CORDILLERA

Tehuacan

Acapulco Velg nio
Sir Puebla

Tehuantepec

Rio Tula

Tula

Panuco
Tampico

GULF OF

MEXICO

15
20

110
105
100
95

Longitude West from Greenwich.

MEXICO.

Geology points out parts of the ocean near the Azores, in the Egean Sea, and to the south of Iceland, (?) where, at recent epoches, within the last 2000 years, small volcanic islands have risen above the surface of the water; but it gives us no example of the formation, from the centre of a thousand burning cones, of a mountain of scoriæ and ashes 1695 feet in height, comparing it only with the level of the adjoining plains, in the interior of a continent, 36 leagues distant from the coast, and more than 42 leagues from every other active volcano.

A vast plain extends from the hills of Aguasarco nearly to the villages of Teipa and Pelatlan, both equally celebrated for their fine plantations of Cotton. Between the Picachos del Mortero, the Cerros de las Cuevas, and de Cuiche, this plain is only from 2400 to 2600 feet above the level of the sea. In the middle of a tract of ground in which porphyry with a greenstone base predominates, basaltic cones appear, the summits of which are crowned with vegetation, and form a singular contrast with the aridity of the plain, which has been laid waste by volcanos.

Till the middle of the last century, fields covered with sugar-cane and indigo occupied the extent of ground between the two brooks called Cuitimba and San Pedro. They were bounded by basaltic mountains, the structure of which seems to indicate, that all this country, at a very remote period, had been already several times convulsed by volcanos. These fields, watered by artificial means, belonged to the farm of Don Pedro di Jorullo, and were among the most fertile in the country.

In the month of June, 1759, a subterraneous noise was heard. Hollow sounds of the most alarming nature were accompanied by frequent earthquakes, which succeeded each other for from fifty to sixty days, to the great consternation of the inhabitants of the farm. From the beginning of September every thing seemed to announce the complete

Y

re-establishment of tranquillity, when in the night of the 28th and 29th, the horrible subterraneous noise recommenced. The affrighted Indians fled to the mountains of Aguasarco. *A tract of ground from three to four square miles in extent rose up in the shape of a bladder.* The bounds of this convulsion are still distinguishable from the fractured strata.*

The Malpays near its edges is only 39 feet above the old level of the plain, called *Las Playas de Jorullo;* but the convexity of the ground thus thrown up increases progressively towards the centre to an elevation of 524 feet.

Those who witnessed this great catastrophe from the top of Aguasarco assert, that flames were seen to issue forth for an extent of more than half a square league, that fragments of burning rocks were thrown to prodigious heights, and that through a thick cloud of ashes, illumined by volcanic fire, the softened surface of the earth was seen to swell up like an agitated sea. The rivers of Cuitimba and San Pedro precipitated themselves into the burning chasms. The decomposition of the water contributed to invigorate the flames, which were distinguishable at the city of Pascuaro, though situated on a very extensive table land, 4592 feet above the plains of Las Playas de Jorullo. Eruptions of mud, and especially of strata of clay, enveloping balls of decomposed basalt in concentric layers, appear to indicate that subterraneous water had no small share in producing this extraordinary revolution. Thousands of small cones, from six to ten feet in height, called by the natives *ovens* (hornitos) issued forth from the Malpays. Although, according to the testimony of the Indians, the heat of these volcanic ovens has suffered a great diminution during the last fifteen years, I have seen the thermometer rise to 212° on being plunged into fissures which exhale an aqueous vapour. Each small cone is a *fumarole*, from which a thick vapour

---

* See the Frontispiece.

ascends to the height of from 22 to 32 feet. In many of them a subterraneous noise is heard, which appears to announce the proximity of a fluid in ebullition.

In the midst of the ovens, six large masses, elevated, from 300 to 1600 feet each, above the old level of the plains, sprung up from a chasm, of which the direction is from N.N.E. to S.S.W. This is the phænomenon of the Monte Nuovo of Naples, several times repeated in a range of volcanic hills. The most elevated of these enormous masses, which remind us of the Puys in Auvergne, is the great volcano of Jorullo. It is continually burning, and has thrown up from its north side an immense quantity of scorified and basaltic lavas, containing fragments of primitive rocks.— These great eruptions of the central volcano continued till the month of February, 1760. In the following years they became gradually less frequent.

The Indians, frightened at the horrible noises of the new volcano, abandoned at first all the villages situated within seven or eight leagues distance of the *Playas de Jorullo.* They became gradually however accustomed to this terrific spectacle; and having returned to their cottages, they advanced towards the mountains of Aguasarco and Santa Inĕs, to admire the streams of fire discharged from an infinity of small volcanic apertures of various sizes. The roofs of the houses at Queretaro, at a distance of more than 48 leagues in a straight line from the scene of the explosion, were at that time covered with ashes.

Although the subterraneous fire now appears far from violent, and the Malpays and the great Volcano begin to be covered with vegetables, we nevertheless found the ambient air heated to such a degree by the action of the small ovens (hornitos) that the thermometer at a great distance above the ground, and in the shade, rose as high as 109° of Fahrenheit. This fact proves, that there is no exaggeration in the accounts of several Indians, who affirm that for many years after the first eruption, the plains of Jorullo, even at a great distance from the ground which had been thrown up, were uninhabitable, from the excessive heat which prevailed in them.

The traveller is still shewn, near the Cerro de Santa Inĕs, the rivers of Cuitimba and San Pedro, the limpid waters of which formerly watered the sugar-cane plantations of Don Andre Pimentel. These streams disappeared in the night of the 29th September, 1759; but at a distance of 6500 feet farther west, in the tract which was the theatre of the convulsion, two rivers are now seen bursting through the argillaceous vault of the *Hornitos,* which make their appearance as warm springs, raising the thermometer to 126° of Fahrenheit.

The Indians continue to give them the names of San Pedro and Cuitimba, because in several parts of the Malpays great masses of water are heard to run in a direction from east to west, from the mountains of Santa Inĕs towards *l'Hacienda de la Presentacion.* Near this habitation there is a brook which disengages sulphuretted hydrogen. It is more than 23 feet in breadth, and is the most abundant hydro-sulphureous water which I have ever seen.*

---

* Mr. Poulett Scrope (whose views on the subject to which these Lectures relate I should have had frequent occasion to allude to, had not the present volume been almost completed before his " Considerations on Volcanos" came before the public) has questioned the soundness of the explanation which Humboldt has given of the phænomena presented by Jorullo.

I would observe however, that even those, who may be disposed to concur with this author in considering the evidence adduced by the Prussian traveller as insufficient to establish the case in point, supposing it to be, as his opponent represents, one altogether anomalous, and contradictory to the ordinary march of nature, may still see no reason for rejecting his view of the subject, if in conformity to that mode of formation which they may be led from other considerations to attribute to many detached and conical masses of trachyte and of basalt.

Now without meaning to enter generally into the consideration of this, or any other of the theories, which Mr. Scrope has often very ingeniously defended, I must remark, that the difficulty of supposing a line of eminences, perfectly detached from each other, composed of a material nowhere else found

MEXICO.

The five active volcanos just noticed appear to be connected by a chain of intermediate cones running in a parallel direction, and exhibiting evident indications of a similar origin.

in the vicinity, and apparently unaffected by diluvial action, to be merely the relics of one continuous stratum, the remainder of which has been swept away, inclines me to think, that all the conical hills near Clermont, composed of Domite, have been separately thrown up.—2dly. That having arrived at this conclusion with regard to these mountains in Auvergne, I am disposed from analogy to extend the same idea, not only to the trachytic rocks occurring near the Rhine, in Styria, and in many other localities, but likewise to certain basalts of similar form and position.—3dly. That having been thus brought to consider the elevation of separate dome-like masses of trachyte, and even of basaltic lava, by no means an uncommon occurrence, I am the more ready to credit the statements of Humboldt with regard to the modern case of Jorullo, although I consider my general views with regard to the older trachytes, independent of the accuracy of the report he has given us. I cannot however admit, that even if the case of Jorullo, as represented by Humboldt, were one of an unprecedented kind, there would be any sufficient reason for rejecting his statement; seeing that there are too few instances of Volcanos having broken out in new spots within the narrow limits of human observation, to warrant any general inferences on the subject. Indeed with respect to these few, so far as can be gathered from the imperfect statements handed down to us, there seems to be fully as much in the circumstances related, to favour Humboldt's views, as to contradict them. Witness the elevation of the promontory of Methone, or Trœzene in the Morea, and of the Islands in the Grecian Archipelago.

I make these remarks not with the desire of provoking controversy, but to acquit myself of any charge of disrespect—either to a fellow labourer in the same field of science as myself, in not noticing his objections—or to my readers in general, in imposing upon them in corroboration of my own views, a statement of facts, without informing them that that statement had been recently controverted.

I flatter myself however, that my disagreement with Mr. Scrope on this point will be found less wide in reality, than in appearance, for I am not aware that he means to deny the separate elevation of conical masses of trachyte, or that, admitting this, he would refuse to extend the same hypothesis even to basaltic lavas similarly circumstanced.

Now provided thus much be admitted, it is a matter of very little importance, so far as my own views are concerned, whether the *cause* of their holding the position they now occupy is to be sought for, in a swelling up of the materials caused by elastic vapours, or in a less perfect fluidity of the lava, which caused it to accumulate round a circumscribed space, so as to form a conical or dome-like mountain.

MEXICO.

Thus Orizaba is connected with Popocatepetl by the Coffre de Perote, and with Jorullo by the extinct volcano of Mexico, otherwise called Iztaccihuatl; and the geological structure of these and all the other lofty mountains, that rise above the table land of Mexico on the same parallel, appears to be the same, being composed of trachyte, from apertures in which the existing volcanos act.*

The same law prevails in the provinces of Guatimala and Nicaragua, which lie between Mexico and the Isthmus of Darien, but the volcanos here, instead of being placed nearly at right angles to the chain of the Cordilleras, run parallel to it.

In these provinces no less than 21 active volcanos are enumerated, all of them contained between 10 and 15 of N. latitude.

Those which have been most lately in a state of eruption, are Los Fuegos of Guatimala, Isalco, Momotombo, Talica, and Bombacho.

On the southern side of the Isthmus there are also several. Three occur in the province of Pasto, and three in Popayan, both I believe comprehended under the present Republic of Columbia, and there are five in Quito.

The connexion of that near the town of Pasto with those of the province of Quito, was shewn in a striking manner in 1797. A thick column of smoke had proceeded ever since the month of November, 1796, from the volcano of Pasto; but to the great surprise of the inhabitants of the city of that name, the smoke suddenly disappeared on the 4th of February, 1797. This was precisely the moment, at which, 65 leagues further south, the city of Riobamba, near Tunguragua, was destroyed by a tremendous earthquake.

Between Quito and Chili only one volcano is known to occur, and this is situated in Peru.

* Humboldt G. des Roches, 327.

QUITO.

Nevertheless the frequent occurrence of earthquakes in the intermediate country renders it probable, that no natural separation exists between the two provinces, but that the same operations are in fact proceeding throughout the whole intermediate tract.

It appears probable, says Humboldt, that the higher part of the kingdom of Quito, and the neighbouring Cordilleras, far from being a groupe of distinct volcanos, constitute a single swollen mass, an immense volcanic wall, stretching from south to north, the crest of which exhibits a surface of more than 600 square leagues. Cotopaxi, Tunguragua, Antisana, and Pichinca, are placed on this immense vault, and are to be considered rather as the different summits of one and the same volcanic mass, than as distinct mountains. The fire finds a vent sometimes from one, sometimes from another of these apertures. The obstructed craters appear to us to be extinguished volcanos; but we may presume, that, since Cotopaxi and Tunguragua have only one or two eruptions in the course of a century, the fire is not less continually active under the town of Quito, under Pichincha, and Imbaburu.*

In Chili no less than sixteen volcanos are said to be in a state of activity. They are all situated near the middle of the range of mountains, which runs in a direction nearly parallel to the coast. The lava and ashes discharged from them never extend beyond the limits of the Andes.

Only two volcanos are found among the maritime and midland mountains : one at the mouth of the River Rapel, which is small, and emits only a little smoke at intervals; the other, the great volcano of Villarica, distinguishable at the distance of 150 miles, and said to be connected at its base with the Andes.

* Personal Narrative, vol. 4. p. 29, Eng. Tr.

CHILI.

It continues burning without intermission, but its eruptions have seldom been violent. The base is covered with forests, and its sides with a lively verdure, but its summit reaches above the line of perpetual snow.

The most remarkable eruption of the Chilian volcanos was that of Peteroa, on the 3d of December, 1760, when the volcanic matter opened for itself a new crater, and a mountain in its vicinity experienced a rent several miles in extent. A large portion of the mountain fell into the Lontue, and having filled its bed, gave rise to a lake in consequence of the accumulation of the water.*

Thus a line of volcanic mountains may be traced at intervals from the 5th to the 40th degree of south latitude, running nearly parallel to each other; whilst the intervening spaces exhibit, in the frequent earthquakes that occur, phænomena of an analogous kind.

This apparent communication, or at least similarity of constitution, subsisting between the several parts of this tract, is the more remarkable, from the absence of all indications of volcanic action from the countries situated on the eastern side of the Andes, whether in Buenos Ayres, Brazil, Guyana, the coast of Venezuela, or the United States.

It is true there exist a little to the east of the Andes three small volcanos, situated near the sources of the Caqueta, the Napo, and the Morona, but these, in Humboldt's opinion, must be attributed to the lateral action of the volcanos of Columbia.

There is one remarkable phænomenon belonging to volcanos of the New World, which, though not altogether peculiar to them, is more frequent there than among those of Europe.

It often happens, that instead of ejections of lava proceeding from the volcano during its periods of activity,

---

* Molina. Hist. Natur. de Chili.

streams of boiling water mixed with mud alone are thrown out.

It was once imagined that the mud and water were genuine products of the volcano, derived from some spot in the interior of the mountain, equally deep-seated with that from which the lava itself proceeds; but a fact recorded by Humboldt has done much to dispel this illusion.

It seems, that with this mud are often thrown out multitudes of small fish (Pymelodes Cyclopum), sometimes indeed in numbers sufficient to taint the air. Now as there is no doubt that these fish proceeded from the mountain itself, we must conclude, that it contains in its interior large lakes suited for the abode of these animals, and therefore in ordinary seasons out of the immediate influence of the volcanic action.

Admitting the existence of these lakes, it is certainly most natural, to attribute the water thrown out to the bursting of one of them, and the mud to the intermixture of the water with the ashes at the same time ejected.

The conclusions to which Humboldt has arrived from his observations on the physical structure of America, some of which have appeared, but the greater part remain unpublished, are as follows:

All the most elevated points of the Cordilleras are of trachyte, which rock encircles in zones a large portion of the table land, but rarely extends towards the plains. When the trachyte does not exist in sufficient quantity to cover the entire soil, it is scattered in small distinct masses on the back and crest of the Andes, raised in the form of pointed rocks from the bosom of the primitive and transition formations.

Trachytes and basalt are rarely found together, and there seems to be a mutual repulsion between these two classes of

formations.* True basalts with olivine do not constitute
beds interstratified with the trachyte, but when they are
found near the latter, they are superposed.

These and other volcanic formations, such as clinkstone,
amygdaloid, and pumiceous conglomerates, are the principal
rocks met with above the trachyte ; sometimes, however,
small local formations of tertiary limestone and gypsum
occur.

The circumstance however which deserves to be considered with the greatest attention, as leading to the most
important consequences, is the apparent passage from the
trachyte into the porphyry beneath it.

This rock, which Humboldt considers as belonging to the
transition series, is distinguished from the porphyries which
are most common in the Old World, by the almost entire
absence of quartz, the presence of hornblende, of glassy as
well as common felspar, and sometimes of augite.   It would
therefore be difficult to fix upon any absolute line of separation between this kind of porphyry and trachyte; it is only
from the union of all these mineralogical characters with
the presence of obsidian, pearlstone, and scoriform masses,
and from the relative position of the rock, that we can determine it to be trachyte.

It is also equally difficult to pronounce where the trachyte
begins, and the porphyry which supports it terminates, for
glassy felspar gradually becomes more and more common as
we ascend to the upper strata of the transition porphyry,
and beds even occur in it, which are considered almost as
characteristic of the trachytic formation.

Thus between the porphyries which contain the rich silver
mines of Real del Monte, and the white trachytes with
pearlstone and obsidian which compose the Mountain of
Couteaux east of Mexico, an intermediate class of rocks

* This remark must be taken with very many exceptions, as Humboldt
seems to allow. Gissement des Roches, p. 349.

occurs, which partakes sometimes of the characters of the older, and sometimes of the newer formation.

In South America likewise, the same remark applies to the strata intervening between the transition porphyry covered with black granular limestone, and the pumiceous trachytes constituting the active volcano of Puracè.

In like manner, in the very midst of the Mexican porphyry, so rich in gold and silver, we observe beds, destitute of hornblende, but abounding in long narrow crystals of glassy felspar, which cannot be distinguished from the clinkstone porphyry of Bilin, in the trachytic district of Bohemia. In short Humboldt considers, that there is no more reason for admitting a *natural* line of separation between the transition and the volcanic formation of America, than there is between the former, and the secondary limestones which are found above it in the Old World.

Before however we proceed to draw any inferences from these observations, I would remind you of what I have said with respect to the parallel case of the porphyries of Hungary, and would observe moreover on the present subject, that the authority justly due in general to the opinions of an individual so distinguished as Humboldt is in some degree weakened here by the circumstance of their being founded on observations made more than twenty years ago, at a time when the very existence of several of the rocks, to which reference is repeatedly made, was unknown. It is therefore possible, not only that the observations of this philosopher, whilst in America, may have been deficient in that precision which the present improved state of the science would have imparted to them; but that his attempts to reconcile the facts themselves to systems since established, being made since his return, may not always have been successful.

When for instance Mons. de Humboldt speaks of the Muschelkalk, the Quadersandstein, and the Tertiary rocks

GENERAL REMARKS.

of America, it must be admitted that his decision on such points would have carried with it more authority, had he visited America since the characters of these rocks have been made out, since it is impossible for the most comprehensive mind to possess that full recollection of all the facts that bear upon a complicated subject, which should enable him to derive from this alone the same degree of certainty, as he might obtain upon the spot itself.

I hail therefore with the greatest satisfaction the intention which M. de Humboldt has held out to us, of again visiting the scenes, where in early life he has reaped so much distinction ; persuaded, that returning as he would do in full possession of all the facts that have since been accumulated in geology, he could not fail to clear up those doubts, which, in spite of his previous labours, still hang over the subject of the older formations of America.

The porphyry which has led to these remarks is distinguished by Humboldt into two kinds : the more antient, immediately covering the primitive rocks; the more modern, resting on slates and limestones belonging to the transition series. The former contains no metallic veins, the latter abounds in them. Both however possess the same characters, by which they become allied to the trachytes, just as the transition porphyries of Europe present many analogies to the red sandstone.

The older porphyry is found in South America immediately resting on the primitive rocks, and is covered by syenite, composed of hornblende and common felspar, with a little mica and quartz.

This syenite seems, in the province of Popayan, to pass into trachyte; the hornblende becoming more abundant, the mica more rare, and glassy taking the place of common felspar.

Sometimes, as at the foot of the volcano of Puracè, the syenite is separated by granite from the superimposed trachyte, which latter at top becomes glassy, and passes into obsidian.

The syenite is sometimes covered with a black limestone, so highly charged with carbon as to soil the fingers, and possessing all the characters of a transition rock, and yet the porphyry on which it rests has much resemblance to trachyte.

Humboldt also notices other circumstances which tend to establish a connexion between the transition porphyry and the trachyte; thus the former rock at Pisoja is columnar, and at Rio Guachicon contains, like the trachyte of the Siebengebirge, fragments of gneiss.

Notwithstanding this, he seems afterwards to admit, that it is questionable, whether these rocks are to be considered as belonging to the transition or to the trachytic series.

These porphyries in Quito alternate with gneiss and mica-slate, (rocks on which they usually repose) and are covered by trachyte, to which they present an approximation; hornblende and augite becoming more frequent towards the upper part of the formation.

The syenite associated with this porphyry also appears to contain salt springs, gypsum, and sulphur, and is covered by a greenstone, consisting of felspar united with hornblende, and not with augite. Mons. de Humboldt considers this an additional proof, that the particular porphyry alluded to is not a trachyte, for the latter rock has hitherto been met with associated only with augite rock, and not with genuine greenstone.

The porphyry of Equinoxial America is covered by a formation of transition clayslate passing into talcose slate, in which is situated the famous silver mine of Guanaxuato in New Spain. The thickness of this formation is no less than 3000 feet. It contains, as subordinate beds, not only syenite, but also serpentine and hornblende slate. No organic remains however, have as yet been discovered. It passes into a siliceous slate, which has sometimes the characters of Lydian stone.

This rock is covered in Mexico by a second formation of porphyry, distinguished by its extreme uniformity, and containing rarely any subordinate beds. Its thickness is calculated by Humboldt at 5000 feet at the least. Between Acapulco and Mexico, it appears to be covered by a formation of compact limestone of a bluish-black colour, full of large caverns, and a case of the same kind is more distinctly observed at Zumpango, where the order of superposition seems to be :

1st. The primitive granite.
2d. Alpine limestone (?)
3d. Granite.
4th. Porphyry with glassy felspar, which in one place supports an amygdaloid of a semivitreous character, and in others two different calcareous rocks, both cellular, which, though evidently less antient than that which rests on the granite, are certainly not so modern as the beds above the chalk.

The rocks too in the valley of Mexico, which I have before observed to pass in one place into trachyte, are covered in another by a calcareous rock, which Humboldt considers to be at least as old as the Alpine limestone.

On the other hand, the limit to the age that can be assigned to the metalliferous porphyry of New Spain, is best determined by the appearances presented at Guanaxuato, where it is seen resting immediately on a transition slate with Lydian stone, and both rocks are alike traversed by the same metallic veins.

This is the very rock which I have before alluded to as possessing in many cases all the characters of trachyte, and yet it presents so many analogies to the preceding rocks which are covered by the older limestones, that Humboldt ranks it among the transition porphyries.

GENERAL REMARKS.

Below the Equator, between the 5th and 8th degree of S. latitude, Humboldt observed porphyries resting upon clay-slate, which he has set down, although with doubt, as transition.

These porphyries have sometimes much the aspect of basalt, contain more augite than felspar, and alternate with beds of jasper, and of compact felspar having a black colour and uniform consistence.

At other times the porphyries have less affinity to trachyte, as at the Indian village of San Felipe, where they are covered by a black shelly limestone.

I have been induced to enter rather more fully than I should have otherwise done, into details respecting the older porphyries and trachytes of America, more particularly as to their mutual connexion, because the subject does not appear to have obtained the degree of attention here, which it has done on the continent, and which indeed its importance deserves.

In every nation, an influence is exerted upon the progress of science, partly by the physical constitution of the country itself, and partly by the direction which the inquiries of those most distinguished in each particular department may chance to have taken. The former circumstance naturally affords a *facility*, the latter a *popularity*, to certain subjects more than to others.

Thus it is easy to assign a reason for the *prominence* which has been given in Italy to the phænomena of volcanos, in Saxony to those connected with the operations of mining, and in England to the structure of the secondary formations, and the changes effected by diluvial action.

In like manner the igneous origin of the older rocks has been chiefly maintained in Scotland, by an appeal to the phænomena of granitic dykes, which Hutton and Playfair brought into notice; and in Germany, by reference to the analogies between the older and newer porphyries, which Humboldt has taken upon himself to investigate.

The individual however, who wishes to take a comprehensive view of this subject, ought not to neglect either branch of the enquiry, especially as both require much further elucidation, and as the contradictory statements with respect to them have not yet in all cases been explained away.

I hope therefore, that among the many individuals from this country who now resort to the New World with views either of emolument or information, some may be induced to investigate the volcanic phænomena which are there presented, and to such persons probably the above statements may not be without their use.

Should this be the case, I shall not repent having devoted a few pages to the above details, deficient as I am aware they must be considered in point of clearness as well as precision; and shall even consider myself as in some measure indemnified by the publication of these Lectures for the privation I have long experienced, in being cut off by my engagements at home from the possibility of personally investigating a portion of the globe, so full of interest to every lover of nature.

LECTURE IV.

GENERAL REMARKS ON VOLCANIC PHÆNOMENA.

———

1st. BRANCH OF THE INQUIRY, *viz.*

*A statement of the theories that have been proposed to ex-
plain the cause of Volcanos—Lemery's mode of accounting for them refuted, by the want of analogy between
their phænomena, and those arising from the cause he
assigned — Hypothesis suggested by Sir H. Davy's
discovery, with respect to the Metallic Basis of the
Earths and Alkalies.*

*Before considering this hypothesis, it becomes necessary
to define what phænomena are to be viewed as arising
from Volcanos—and therefore to inquire into the cause,
1st, of Earthquakes; 2dly, of Hot Springs; 3rdly, of
Exhalations of inflammable Gases, and "Salses" or
Mud Eruptions.*

*The hypothesis under examination implies the access of
Water to the seat of volcanic action—hence favoured by
the general situation of volcanos in the neighbourhood
of the Sea—Exceptions to this rule considered—in the
case of active Volcanos—and in that of extinct ones.*

*Phænomena of a Volcano examined in detail—so far as
regards the aeriform fluids given out—These consist of
Sulphuretted Hydrogen, Sulphurous Acid, Muriatic
Acid, Carbonic Acid, Nitrogen, and Steam—Explana-
tion of the presence of these bodies agreeably to the
Hypothesis proposed.*

*Structure of a volcanic mountain considered, 1st, in cases
where a conical mass of Trachyte has been thrown up—
2nd, in cases in which the whole appears to be composed
of alternating strata of Lava and Tuff.*

z

HAVING in my three preceding Lectures brought together such *facts* with respect to the phænomena of volcanos, as my own observations in the countries in which they occur, or the researches of others were capable of affording; I conceive it will not be altogether uninteresting, if I proceed to lay before you in my present such general inferences, as appear deducible from the foregoing details, either with respect to the cause of these *effects*, or their relation to the processes which were concerned in the production of certain of the older constituents of our globe.

With respect to the former of these enquiries (I mean the cause of volcanic phænomena,) it must be confessed, that our speculations, from the very nature of the subject, are involved in much uncertainty, since the processes we undertake to consider are placed beyond the scope of actual observation, and can be conjectured solely from certain of their remote consequences.

In the case before us too, the effects which we survey are often in themselves of such a nature as to preclude examination, so that indeed it is chiefly to a review of the minor operations, carried on by a volcano during the stages of its partial intermittence, and of its subsidence into a state of tranquillity, that we are indebted for any imperfect data we may possess, upon which to build an hypothesis.

I propose therefore to consider this part of the enquiry, with as much conciseness as is consistent with a clear explanation of my views; and passing over many hypotheses which have been advanced at various times to explain the nature of volcanos, to confine myself to the examination of one or two which possess some shew of probability, and are supported by the authority of illustrious names.

According to the first and most antient of these, volcanos were attributed to the combustion of certain inflammables, similar to those which exist near the surface of the earth, such, for instance, as sulphur, beds of coal, and the like; and, in order to account for the spontaneous inflammation of these substances, an appeal was often made to an experiment of Lemery's, which went to prove, that mixtures of sulphur and iron, sunk in the ground, and exposed to the influence of humidity, would give out sufficient heat to pass gradually into a state of combustion, and to set fire to any bodies that were near.

His process was, to mix as large a quantity, as could be conveniently had, of clean iron filings, with somewhat more of sulphur, and as much water as would make a firm paste, to bury this in the earth, (first wrapping it up in a cloth) and to ram the soil firmly above it. A mixture of this kind in a few hours grows warm, and swells so as to raise the ground. Sulphureous vapours make their way through

the crevices, and sometimes flames will appear. Rarely is
there any explosion; but when this happens, the fire is
vivid—and the heat and fire both continue for some time, if
the quantity of materials has been great.*

Breislac has proposed a theory not very different from the
above;† he supposes, that volcanos may arise from masses
of petroleum, collected in underground caverns, set on fire
by some third substance.

He imagines, that the presence of certain of the combinations of phosphorus or even of sulphuric acid may
originate the combustion, and he appeals to the conflagra-
tions that take place in coal mines, as proving that bituminous substances are in fact set on fire by the presence of some
body spontaneously combustible.

He shews, that petroleum is very abundantly distributed
in all parts of the globe, and that it is more particularly
found in the neighbourhood of volcanos, during the eruptions of which it is often exhaled in great quantities.

This hypothesis however seems to be supported on much
the same sort of evidence which has been adduced in favour
of the one before alluded to, and is saddled by objections of
a similar description; I shall therefore spare myself the
trouble of considering it apart, as the observations I shall
have to make will apply with equal force to both.

Another explanation of volcanic phænomena has been
suggested by the discoveries of Sir H. Davy, with respect to
the metallic bases of the earths and alkalies.

It being now ascertained, that the solid constituents of our
globe all contain some inflammable principle, and owe their
present condition to the union of this principle with oxygen;
it seems by no means an improbable supposition, that at a
certain depth beneath the surface, at which atmospheric air
is either wholly or partially excluded, these substances may
still exist in their pure unoxydized state.

* Mem. de l'Academie, 1700.
† Instit. Geol. Vol. 3.

Now if water were at any time admitted to them whilst in that condition, we know from the common principles of chemistry, that a great evolution of gaseous matter must take place, and that the combination of the oxygen of the water with these inflammables would give rise to heat, sufficient to account for the liquefaction of the surrounding rocks, and all the other phænomena attendant on an eruption.

Such being the opinions respecting volcanos, which appear, at first sight, to possess the greatest share of probability, let us next inquire somewhat more minutely into the arguments that may be advanced in their support.

And first with regard to the former hypothesis, I may remark, that it is chiefly favoured by the general occurrence of sulphur and of gases containing it in almost all volcanos; from whence it might seem a natural inference, that on the presence of this substance the phænomena themselves depended. Whether the same be the case with petroleum, which Breislac regards as the chief agent in the process, will appear hereafter; but whichever of these substances be the cause of the volcanic operations, it will at least be admitted, that there would be little difficulty in imagining means by which either of them might be brought into a state of combustion. When however we examine more narrowly into the analogies between the *effects* of volcanic fires, and those which we know to result from the combustion of either of these materials, we are soon brought to confess the inadequacy of such an hypothesis to account for the facts before us. What resemblance for example do the porcelain-jaspers and other pseudo-volcanic rocks, as they are improperly termed, which we observe in coal mines, that have been for centuries in a state of inflammation, bear to the lavas and the ejected masses of a genuine volcano; or where do we observe from them the same evolution of aeriform fluids, and of streams of melted materials which are so characteristic of the latter? The difference would appear still more striking, if I were to enter into other particulars with respect to the depth and geological position which must be assigned to the seat of the volcanic action; but these will be more in place at a later period of this Lecture.

The remaining hypothesis which I have to consider, has at least this advantage, that it supposes the agency of bodies which do not exist in nature on the surface of the globe; and this circumstance gives it at least a superiority over other explanations, in a case where the phænomena to be accounted for are of a description altogether different from any that result from other natural processes placed within the sphere of our observation. The individual therefore who maintains, that volcanos arise from the access of water to the metals of the earths and alkalies, is exempt from the necessity of pointing out, as the advocates of the contrary hypothesis ought to do, some process going on near the surface of the earth, resembling in kind at least, if not in degree, the phænomena which he attempts to explain.

Now if volcanos have arisen from this latter cause, the necessity of water to excite the combustion seems to imply, that they would be met with rather in the neighbourhood of the sea, than on the elevated table lands of extensive continents; and the existence of substances capable of decomposing that fluid indicates, that the process must have gone on at a depth sufficiently great to have precluded the access of air, which would have long ago imparted to them the very principle, to the absorption of which the volcanic action is attributed. Hence the rocks, which appear to proceed from the focus of a volcano, ought to be derived rather from granitic and other of the older formations, than from those of modern date; and the gases evolved during the process ought to consist, in part at least, of those which we know to be given out, when water is made to act upon the alkaline and earthy bases.

In order therefore to ascertain the degree of probability belonging to this theory, I shall consider, first, the geographical situation of volcanos; secondly, the character of the aeriform fluids evolved by them in different stages of activity; thirdly, the nature of their lavas and ejected masses; and fourthly, the depth from which they appear to have emanated.

Before however I enter upon these questions, it is right that I should define more accurately than has hitherto been done, what appearances are to be considered as establishing the existence of volcanic action, inasmuch as the ideas entertained on that head are singularly various and indefinite, and thus have occasioned much of the difference of opinion that exists on these questions. Some for instance are unwilling to admit earthquakes as any probable indication of subterraneous fire, whilst others not only include these, but go so far as to class hot-springs, gaseous exhalations like those of the Pietra Mala, and the eruptions of mud and petroleum commonly called " Salses " amongst volcanic phænomena.

With regard however to the first of these, I apprehend, that those who coolly examine the facts that have been collected on the subject, will scarcely entertain any other difference of opinion, than as to whether their connection with volcanos is universal; for in some instances earthquakes have occurred so immediately antecedent upon volcanic eruptions, and are so manifestly derived from the very same centre of action, that we want no better proof to establish an identity of origin.

In other cases the evidence, though not quite so direct, is perhaps as cogent as can be obtained in most questions of this description.

When for instance we observe two volcanic districts, both subject to earthquakes, which are ascertained to have a connection with the volcanic action going on, and find that an intermediate country, in which there are no traces of the operation of fire, is agitated by subterraneous convulsions, similar in kind, but stronger in degree than those which occur in the more immediate vicinity of the volcanos; have we not reason to conclude, that the same action extends throughout the whole of the above space, and that it is *this* which produces in the intermediate country the effects alluded to, which are only the more alarming from the absence of any natural outlet, from which elastic vapours might escape?

Now in proof of the former of these positions, it may be scarcely necessary to do more than appeal to the case of Etna or Vesuvius, which rarely return to a state of activity, after a long interval of comparative quiescence, without some antecedent earthquake, which ceases so soon as the mountain has established for itself a vent.* Such was the case before the celebrated eruption of 79 in Campania, and in that of Etna in 1537, where, says Fazzello, noises were heard, and shocks experienced, over the most distant parts of Sicily. In such cases no one would doubt the connection between the volcano and the earthquake.

* Humboldt gives us the following series of phænomena, which presented themselves on the American Hemisphere between the years 1796 and 97, as well as between 1811 and 1812.

1796.—September 27. Eruption in the West India Islands; volcano of Guadaloupe in activity.
...... November...... The volcano of Pasto begins to emit smoke.
...... December 14. Destruction of Cumana by earthquake.
1797.—February 4. Destruction of Riobamba by earthquake.
1811.—January 30. Appearance of Sabrina Island in the Azores. It increases particularly on the 15th of June.
...... May......... Beginning of the earthquakes in the Island of St. Vincent, which lasted till May, 1812.
...... December 16. Beginning of the commotions in the valley of the Missisippi and Ohio, which lasted till 1813.
...... December. .. Earthquake at Carracas.
1812.—March 26. .. Destruction of Caraccas; earthquakes which continued till 1813.
...... April 30. .... Eruption of the volcano in St. Vincents'; and the same day subterranean noises at Caraccas, and on the banks of the Apure.

Pers. Narr. Vol. IV.

See also Gemellaro on the Meteorological Phænomena of Mount Etna, extracted in the Journal of Science. Vol. 14. 1813.

The second point seems established, by considering the tremendous earthquakes which ravaged Calabria, and those mentioned by Humboldt as intervening between, and in the line of, the volcanos of Columbia, Quito, and Chili. Von Buch has well shewn, in his paper on Lancerote, the comparative immunity enjoyed by Teneriffe from those convulsions of nature which agitate the neighbouring islands, destitute of that great natural chimney or safety valve afforded by the Peak of Teyde.

If it be objected, that earthquakes are too general to be referred to this one cause, I may answer, that the preceeding details have shewn volcanos to be phænomena, which (taking in those extinct as well as recent) are seen in almost every part of the globe that borders upon the sea, and that earthquakes, like volcanos, though felt undoubtedly even in the centre of large continents, seem to produce their most frightful effects in countries not very far removed from the ocean.

Thus the most tremendous instances that we read of are those of Lisbon, where traces of antient lavas are discoverable, Asia Minor near the *Catacecaumene*, and the Caraccas situated between the volcanos of the Antilles and those of Columbia.

Dr. Stukeley,* who wrote on the causes of earthquakes, about the middle of the last century, at a time when the physical geography of the globe was much less generally explained, might have reason for believing that volcanos are too rare to account for a phænomenon so general as that which formed the subject of his enquiry; but at present this at least hardly can be said to furnish an objection, especially when we recollect the vast distances, to which sound and other vibratory motions may be propagated along the substance of solid bodies, and therefore are not obliged to place the seat of the action at that vast depth at which Dr. Stukeley imagines it must reside, considering the radius over which its effects are perceived.

* Phil. Trans. for 1750.

Gay Lussac* observes, that the shock produced by the head of a pin at one of the ends of a long beam, is distinctly transmitted to the other end, and Dr. Young has compared an earthquake to an undulation in the air, propagated along the solid mass of the earth.† Now if the shock caused in a substance like air is capable of shaking buildings, how much more tremendous must be that propagated through the compact strata of the earth, in consequence of the impulse occasioned by elastic vapours struggling to escape. Without enquiring therefore whether it be possible that earthquakes may in some cases be produced by other means, I think it may be fairly contended, that where we find them accompanying volcanic phænomena, or occurring in countries bordering upon, or placed between, mountains that indicate the action of fire, we have reason to conclude them to be dependant on the same cause, and that we may in such cases appeal to them as indications of the extent and direction of the processes themselves.

With regard to Hot Springs the case seems somewhat different; for whilst we know of no earthquake upon record, which may not with some shew of probability be referred to the action of a volcano, there are many, in which hot springs appear independent of such a cause.

I am aware indeed that even those of Bath and of Buxton, have been attributed to volcanic action;‡ but the fact, that they spring up in strata, which are very remote in geological position from the rocks amongst which volcanos have hitherto been known to occur, seems to shew, that we must look for other causes of a more local nature. I shall not anticipate the observations which I propose to make, when speaking of the temperature of mines, further than to remark, that the lias from which the Bath waters are thrown out, and the mountain or carboniferous limestone, in which

---

* Edinb. Phil. Journ. Vol. IX. p. 278.
† See Dr. Young's Natural Philosophy. Vol. I. p. 717.
‡ See Bakewell's Geology. p. 306.

those of Buxton, Matlock, and Clifton occur, both contain iron pyrites and other sulphurets, which give out during their decomposition a pretty considerable heat.

The steady temperature preserved by these, as well as all other hot springs, though somewhat perplexing, offers no argument in favour of the volcanic theory; it probably arises, not from the perfect uniformity of the process itself, but from the circumstance of the heat being produced at a considerable depth. Its cause, in short, is the same as that of the equable temperature of springs in general, which, notwithstanding the constant vicissitudes in the climate of the country in which they occur, always remain at the medium point, the excess of heat at one period of the year being balanced by the deficiency at the other.

Yet though I am disposed to be sceptical with regard to the volcanic origin of hot springs that appear in the newer rock formations, in parts far remote from any other indication of subterranean fire, I am fully aware, that in the great majority of instances, the geological position of these waters will be found to be either near volcanos active or extinct, or else amongst rocks of a granitic character.

In the former case, it is certainly most natural to attribute them to the volcanic action, which is either going on at the time, or which, though extinct, still continues to affect the temperature of the rocks which it pervaded; and there are not wanting reasons for believing, that in some instances, at least, the heat of springs arising in primitive districts may arise from the same cause. Thus the frequency of earthquakes among the Alps lends colour to the opinion of Von Hoff, that volcanic action may have been going on beneath that chain, though it has been prevented from manifesting itself by the vast mass of superincumbent rock, and thus has broken out only at some distance on either side, as in Auvergne, in the Vicentin, and near the Lake of Constance. Now if this be admitted, the hot springs which gush out among the Alps may be attributed to this volcanic action, prevented by the cause above assigned from exhibiting here more decided marks of its presence.

The same explanation may apply to the warm springs that arise at the foot of the Pyrenees and in some other localities, for there also we observe indications of volcanic action at some distance from the chain, as in the Vivarais and Cevennes on one side, and in Catalonia on the other.

I have before alluded * to the line of basaltic rocks which stretches across Germany, and stated at the same time, that the shocks of earthquakes are most common in the same direction as that of these masses, and round a certain distance on either side of the line in which they occur. It adds somewhat to the probabilities in favour of the volcanic origin of hot springs, that they are found to accompany these basalts throughout the whole of their extent. It is also a curious circumstance (for the knowledge of which I am indebted to another of the publications of the same author, who has supplied me with the facts just stated), that the existence of hot springs seems in some instances to be a preservative against earthquakes. Thus when in January and February 1824, earthquakes extended from the foot of the Saxon Mountains to the Circle of Elnbogen in Bohemia, the shocks were never felt nearer than two miles to the hot springs of Carlsbad.

The correspondence too in chemical composition among the ingredients of the hot springs met with in volcanic districts leads to the idea of a similarity of origin, and as the same constituents seem to exist in many of those found among granitic mountains, we might be induced to extend to them also the same inference. Thus it has been shewn in an interesting volume just published by Dr. Bischof, Professor of Chemistry at the University of Bonn,† that carbonate of soda saturated with carbonic acid gas is a common ingredient in all these waters, that it exists in those near the Lake of Laach and the Siebengebirge on the

* See my First Lecture. p. 89.

† Die Vulkanischen Mineralquellen Deutschlands und Frankreichs. Bonn. 1826.

Rhine; among the volcanic mountains of the Westerwald
and the Taunus; in those of the Habichtswald, Meissner,
Vogelsgebirge, and Rhongebirge; amongst similar rocks in
the Fichtelgebirge and Erzgebirge; in those of Bohemia;
and in the midst of the Giant Mountains of Silesia. He
further shews that similar springs exist in Auvergne, the
Vivarais, and in the Pyrenees, and conceives, that the same
remark applies to the mineral waters of Hungary, Spain,
and Great Britain.

I have before stated the exceptions that I should wish
to make to the admission of any such theory; but whilst I
avoid indulging in a hasty spirit of generalization with respect to the cause of hot springs, I cannot help regarding
the fact, that the mineral alkali is so commonly present in
lava, in basalt, and in all the products of volcanos, an argument in favour of the similar origin of such hot springs as
agree in possessing this as a principal ingredient.

Still more doubtful is the evidence of volcanic action
afforded by those gaseous exhalations, which occur in certain countries accompanied with eruptions of mud and
petroleum. That they are commonly connected with volcanos I am not disposed to deny, but they appear to be so
only in the relation of an effect, produced by the existence
of inflammable materials, brought together by the operations of a pre-existing volcano. That the process itself
is distinct from that which takes place in the latter, is
evident, from considering that in every case in which it
has been examined, the seat of the action is found to
be quite superficial, and to reside in a stratum of very
recent origin, known to be strongly impregnated with sulphur and petroleum. Such is the case at Macaluba in
Sicily, where the mud volcanos, as they are called, lie
quite detached from the true volcanic phænomena of Ætna
or of the Val di Noto, and seem manifestly dependent on
the combustion of the sulphur which exists there in such
large quantities. The same is likewise the case with the
" Salses " of Modena, which lie at a great distance from

every real volcano, and, if we believe Humboldt, with those of New Andalusia, and the Island of Trinidad.

Even therefore if we were disposed to admit the old theory, which attributes volcanic phænomena to the combustion of sulphur or beds of coal, we must still distinguish between the latter and the phænomena now under consideration, since it is evident that true volcanos are far more deeply seated, and productive of effects different, not only in degree but in kind, from those of Macaluba and similar places.

Nor need we be embarrassed by the occurrence of these appearances so generally in the neighbourhood of true volcanos, since it is well known that the latter are constantly giving out inflammable materials, which may readily become the cause of Air Volcanos. I have remarked in a former part of this essay, that the formation in which the processes going on at Macaluba reside contains the very materials, which are commonly found within the crater of a half-extinguished volcano; and hence it may be conjectured, that the latter have been accumulated at former periods of the earth in submarine solfataras, caused by that volcanic action which the lavas of the Val de Noto prove to have existed in Sicily, long before the earth acquired its present condition, or Mount Etna began its eruptions.

Restricting myself to the consideration of such phænomena as correspond with the above conditions, I shall proceed to consider, whether any thing can be collected, from the situation of those countries which afford indications of volcanic action, favourable to this or that hypothesis.

From the list of active volcanos given by Mons. Gay Lussac, it appears that there are no less than 163 in different parts of the globe, and amongst these there is this striking coincidence, that all, with a very few exceptions, are situated within a short distance of the sea.

The exceptions are, some of the American volcanos, and one or two, which can hardly be considered as fully ascertained, in the centre of Tartary.

With regard to the volcanos of the Asiatic Continent, it may be observed, that even granting their existence (which is by no means proved) we are not sure, that they may not be in the vicinity of some of those lakes of salt water which are scattered over the vast table-land of central Tartary; so that, with only a slight modification, the same general remark will hold good, namely that all groups of volcanic hills are in the neighbourhood of large masses of salt water in one part or another.

With respect to the American volcanos, it may be observed, that even those which lie inland make a part of hills similarly constituted, the extremities of which are close to the sea; and that, as we have reason to believe a subterranean communication to exist between all the members of the series, it is highly probable, that the water may find its way even to the remotest parts of the chain.

This is the case with Jorullo in Mexico, which, though itself no less than forty leagues from the nearest ocean, being in the midst of the table-land of Mexico, seems nevertheless connected with the volcano of Tuxtla on the one hand, and that of Colima on the other, the one bordering on the Atlantic, the other on the Pacific Ocean. This communication is rendered probable by the parallelism that exists between these and several volcanic hills intermediate.* The same, I believe, will be found to be the case with regard to other similarly constituted hills in the western hemisphere; so that, if we view the volcanos of Guatimala, Columbia, and Chili, as constituting so many distinct groups, connected with each other by subterranean communications, we shall find, I believe, that they are all in the vicinity of the ocean in one quarter or another.

Even those who are sceptical with respect to the supposed communication between the several members of the same chain, must admit, that it is a very remarkable fact,

* See the Map of Mexico attached to this volume.

and one that ought not to be overlooked in any theory whi ch professes to account for the phænomena of volcanos, that whilst the number of those in a state of activity is no less than 163, the greater part should exist in islands and maritime tracts, and that the whole extent of some of our continents should present not even a vestige of such appearances.*

It may indeed be objected that this remark does not extend to the class of extinguished volcanos, which have no such disposition, but are scattered indiscriminately over the central region of France, Silesia, Bohemia, Hungary, Transylvania, in parts, in short, the most remote from the access of the present ocean. But it will appear in the course of this Lecture, that at the period when these volcanos were in activity, the greater part were near the sea, if not underneath it, and that the rest were exposed to the access of water, derived from the lakes, which had been left in the low situations when the mass of the ocean had retired. Instead therefore of these being brought forward as exceptions to the generality of the rule laid down, the cessation of the action now that the water has left their neighbourhood seems to furnish a confirmation of it.

But as our hypothesis merely implies the presence of *water* as subservient to the volcanic operations, it may yet be asked why the existence of volcanos should be confined to the neighbourhood of the sea, whilst this fluid is so generally present on the face of our continents.— If the crust of the earth, it may be said, is so traversed by cracks and fissures beneath the bed of the ocean, as to

---

* The antients remarked the same thing with regard to volcanos. Thus Macrobius in Somn. Scip. lib. 2. c. 10. says, " Ignem æthereum physici tradiderunt humore nutriri;" and Justin lib. 4. sub initio. Accedunt vicini, et perpetui Ætnæ montis ignes, et insularum Æolidum, veluti ipsis undis alatur incendium. Neque enim in tam angustis terminis aliter durare tot sæculis tantus ignis potuisset, nisi humoris nutrimentis aliretur. He therefore supposes the waters to be sucked up by Charybdis, and thence transmitted to the volcano which they nourish. Eadem causa etiam Ætnæ montis perpetuos ignes facit. Nam aquarum ille concursus raptum secum spiritum in imum fundum trahit, atque ibi suffocatum tamdiu tenet, donec per spiramenta terræ diffusus, nutrimenta ignis incendat."

allow of water penetrating to a great depth below its surface, the same will hold good with respect to the land; and any of our freshwater lakes or rivers might therefore supply materials sufficient to feed the fires of a volcano.

But it may be replied, that in point of fact the fissures that penetrate the crust of the earth are too small, and too superficial, to allow of the descent of any considerable body of water to its nucleus, and that the same would probably be the case with respect to those underneath the bed of the ocean, were not the force of gravity assisted by the powerful influence of pressure derived from the vast column of superincumbent fluid. Owing to this, the water at the bottom of the ocean would be injected into the remotest pores and crevices of the subjacent rock, as quicksilver is made to pass into the finest vessels by a powerful syringe, and the enormous strain exerted laterally would have a tendency to enlarge and extend the fissures much beyond their original dimensions.

This joined to the fact, that the water at the bottom of the sea has a much smaller mass of rock to get through, before it reaches the inflammable materials upon which it exerts its action, may account for the occurrence of volcanos in its vicinity, without imagining that the salt it contains contributes in any degree to the effect.

Mons. Gay Lussac, in a short Essay which he has published on this subject,* has remarked, that it should seem according to this hypothesis, that the eruption ought to take place through the same aperture by which the water entered, rather than by a new one, and that jets of lava, as well as of gases, and scoriæ, ought therefore to take place at the bottom of the ocean, rather than on the adjacent coast.

But this illustrious chemist has surely forgotten, that as the specific gravity of lava can hardly be considered more than three times that of water, the pressure of an ocean

* Ann. de Chem. & de Phys. xxii. 415.

only two miles in depth would counterbalance that of a
column of lava sufficiently high to reach to the summit of
Vesuvius. When the volcanic action therefore took place,
either at a great distance from land, or where the incumbent strata opposed a resistance too great to be overcome,
the case supposed by Gay Lussac would actually occur,
and the products of the eruption would be thrown out by
the very aperture which admitted the water; but where,
as is more commonly the case, the pressure of so vast a
body of liquid proved superior to the resistance of the
rock above, joined to the weight of the lava itself, the phænomena would manifest themselves at the nearest point of
the coast which yielded to the force applied.

It must be added likewise that the original aperture would
be obstructed by the operations of the volcano itself, first by
the rise of the gases disengaged by the decomposition of the
water; secondly by the expansion in the rocks immediately
surrounding the place in which the action resided; and
thirdly by the injection of melted lava into the minutest
crevices of the rock.

Granting therefore the existence of the inflammable substances themselves at the spots in which the volcanic action
resides, it is not difficult to account for their being set on
fire in consequence of the water so constantly present.

It only therefore remains to be seen, whether the phænomena attendant upon an eruption are of such a nature,
as would result from the cause assigned.

The most constant and essential phænomenon of an
active volcano, is the evolution of certain aeriform fluids,
which, forcing themselves a passage through the incumbent
strata, carry up with them whatever comes within the sphere
of their violence, thus giving rise to ejections of stones, of
ashes, and even of water.

To determine the chemical nature of these gases, and to
ascertain by an extensive induction of particulars, which of
them are to be considered essential, and which as arising out

of the *accidents* of each particular volcano, would be a great step towards a knowledge of this subject, and enable us to speculate with some degree of confidence with respect to the cause of the phænomena themselves.

With a view to this object, I condensed a portion of the vapour given off round the crater of the Island of Volcano, as well as near that of Etna, and, by the kindness of my friend Mr. Herschel, assisted by Signor Covelli of Naples, became possessed of a similar quantity from the Solfatara of Puzzuoli.

These I have never indeed submitted to an accurate chemical examination, but I have satisfied myself as to the existence in that from the Solfatara of sulphuretted hydrogen, and in that from Volcano of sulphurous acid, whilst, as was to be expected, the sulphurous acid could not be detected in the former, nor the sulphuretted hydrogen in the latter.

These results were in strict conformity with what might be anticipated from the sensible properties of the vapours at the time I visited these spots.

I was disappointed however in finding, that although the interior of the crater of Mount Etna, which was at the time of my visit inaccessible, seemed fully charged with sulphurous acid, the vapour, which I condensed from the spiracles on its exterior, consisted of water with a trace merely of muriatic acid.

The latter, either in a free state, or combined with ammonia, was present also in the condensed vapours from Volcano and Solfatara, but in every case the proportion was so small as to be scarcely appreciable.

Such were the gases given out from these three volcanos during what may be termed their quiescent state, and it appears from Monticelli and Covelli's account, that the case may be considered as nearly the same at Vesuvius.

During an eruption however, it would seem that muriatic acid is given off in much larger proportion, visible in those white fumes which rise so copiously at the commence-

ment;* and various muriatic salts, particularly those of soda
and ammonia, are generally found in great abundance in the
cavities of the lava   Common salt indeed has been extracted by Monticelli in the proportion of 9 per cent. from lava
by simple washing, and sal ammoniac is sublimed in such
quantities as in some instances to become an article of commerce.

If we proceed to other volcanos, we shall find sulphur
combined either with oxygen or hydrogen, almost an universal occurrence.

Thus it is noticed at Guadaloupe, in the Azores, in the
Isle of Bourbon, in Java, in the Sandwich Islands, in
Kamschatka, in short the presence of sulphureous vapours
may be regarded almost as one of the characteristics of a
genuine volcano.

Humboldt has given an interesting account of a stream
derived from the extinct volcano of Puracé, near Popayan,
between Bogota and Quito, which, from its acid qualities,
is called Rio di Vinegro, or Vinegar River.   These waters
are found to destroy the fish in the river, into which they
empty themselves, for some distance, and it was found that
persons, who remain some time in the neighbourhood of the
cascades formed by the river, experience a smarting and
pain in the eyes, from the effect of the minute spray.   The
analysis made by M. Rivero of this water, gave per litre
(2.113 pints) sulphuric acid 16.68 grains; muriatic acid
2.84; alumine 3.7; lime 2.47; and some indications of
iron.

Sulphurous acid was exhaled in great quantities from the
crater itself, which was covered by a crust of very pure
sulphur, 18 inches in thickness, broken open on the northern
side, through which the vapours issued.

* Is not this what Livy refers to, when he speaks of clouds of wool being
seen to rise from the ground at Privernum (now Piperno), Priverni lana pulla
terrâ enata, l. xlii. c. 2.   Julius Obsequens notices a similar phænomenon at
Præneste (Palestrina) c. 140—and c. 89.   It is probable, that the evolution
either of muriatic acid or of muriate of ammonia, gave rise to this mistake.

Within the crater is a lake, the waters of which were found to be saturated with sulphuretted hydrogen. Humboldt thinks it probable, that the crust of sulphur arises from the mutual action of this gas with the sulphurous acid so largely present. The rock itself is a semivitreous trachyte, of a bluish grey colour, and conchoidal fracture.

Mons. Leschenault[+] has given an account of a similar phænomenon that occurs at Mount Idienne, in the eastern portion of Java. A stream, called the White River from the quantity of white clay which it carries along with it, is augmented by the addition of a little rivulet called Songi Pouti, which proceeds at once from the crater of this volcanic mountain, and which produces in it some extraordinary changes. When the latter stream is feeble, it is absorbed by the sandy soil through which it flows, and in that case the White River retains its colour and its other qualities to its very mouth; but when the tributary stream is sufficiently copious to empty itself into the White River, it changes its colour, and at the same time communicates to it such poisonous qualities, as cause the destruction of the fish, as well as that of the vegetation, in the country which it irrigates. This arises from the sulphuric acid contained in the waters that flow from the volcano. Vauquelin has analyzed a portion brought by Mons. Leschenault from a lake existing in the crater, and found it to contain sulphuric acid, muriatic acid, sulphate of alumina (simple), a small portion of alum, sulphate of lime, sulphate of iron, and a trace of sulphur. The sulphuric acid was the most abundant, next the muriatic, the sulphates of alumina, and iron; whilst the remaining substances existed in very small proportion.

The volcano of Mont Idienne is extinct, but the white and decomposed condition of the rocks which surround the

* Annales de Chemie, xxvii. 113.

† Annales du Musée, Tom. xviii. p. 425.

crater, and the sulphureous vapours given off, show that it
is in the state of a Solfatara.

It would be interesting to ascertain, whether the acid
earth said to be found in great quantities at a village called
Daulakie, in the south of Persia, between three and four
days journey from Bushire, on the Persian Gulf, has a connection with any thing volcanic. It is used by the natives
in making their sherbets, &c. and large quantities are thus
employed. A portion has been brought from thence by
Lieutenant-Colonel Wright, and examined by Mr. Pepys,
who finds that about a fifth of it is soluble in boiling water,
yielding an acid solution, which, when tested, gave proofs
of the presence of sulphuric acid and iron, and on evaporation yielded crystals of acidulous sulphate of iron.[*]

It would seem, that some mixture of this kind may have
given rise to the statements respecting solid sulphuric acid,
said to have been found in caverns among the volcanic rocks
of Radicofani in Tuscany.[†]

Carbonic acid is a common product of volcanos nearly
extinct; it is emitted, as we have seen, very abundantly from
fissures in the neighbourhood of Naples, as well as near
Rome, in the Vivarais, in the Eyfel, and in most of the localities specified in my former Lectures.

It is supposed that the Mofettes, which often succeed an
eruption of Vesuvius, consist of this gas; but it is remarkable, that during a state of vigorous action this volcano does
not appear to emit it.[‡] This is perhaps more accountable,
when we recollect, that Potassium, and probably the other
alkaline as well as the earthy bases when heated, decompose
carbonic acid, and indeed if we attribute its disengagement
to the action of heat upon calcareous strata, (which appears
to be its most probable source) it will perhaps appear, when
we come to consider the rocks amongst which volcanos are

---

[*] Phil. Mag. lxii. 75.

[†] Phillips' Mineralogy, p. 110.

[‡] Monticelli Storia de Fenomeni del Vesuvio, p. 149.

seated, that this gas proceeds not from the immediate focus
of the action, but from rocks to which the heat occasioned
by its continuance has gradually penetrated.

Nitrogen gas has been detected I believe at Vesuvius, and
in some other volcanos, and it is probable from the ammoniac salts which so abound, that it is present in all.

It is evident indeed that any atmospheric air which may
reach the spot at which the volcanic action resides, will be
deprived of its oxygen by the process, and that even at
the great distance from the volcano itself, the presence of
metallic and other sulphurets will more slowly, though not
less certainly, give rise to the same effect. Hence much
nitrogen gas must arise not only from the volcano, but likewise from the rocks for a certain distance round.

On this latter principle indeed we may account for its
presence, according to Humboldt,* among the exhalations
from the Vulcanitos of Turbaco, in New Andalusia, where a
similar formation seems to occur as that which I have described in Sicily, attended with phænomena similar to those
of Macaluba.

It would appear then, that, so far as we know at present,
the gases given out by volcanos are, muriatic acid, sulphur
combined with oxygen or hydrogen, carbonic acid, and
nitrogen, to which we must add aqueous vapour, of which
an enormous quantity is constantly exhaled, and this, being
afterwards condensed by the cold above the summit of the
volcano, may be the cause of those rains which frequently
succeed an eruption, as was the case in that of Vesuvius in
1794, which destroyed the town of Torre del Greco.

Now let us consider, how far this coincides with the
phænomena which might be expected to take place from the
sudden admission of salt water to the bases of the earths and
alkalies at a great depth below the surface of the ground.

The immediate effects of the chemical action thus excited

* Atlas Pittoresque, p. 239, Fol. Ed.

would be the decomposition of the water, the union of its oxygen with the metallic bases, and the disengagement of an enormous quantity of hydrogen gas.

But the latter substance, it may be said, has not yet been detected in a separate form among the gaseous products of a volcano, and Gay Lussac has even maintained that it *cannot* be present, since it would be inflamed in the air by the red-hot stones thrown out from the volcano, which has not been observed to happen. It might be answered perhaps, that this effect is prevented by the presence of large quantities of muriatic acid, for hydrogen gas is known to lose its inflammability when mixed with that substance; but the most probable solution of the difficulty seems to be, that of supposing this principle to have combined in its nascent state with sulphur, and the two bodies to have been evolved in the form of sulphuretted hydrogen gas.

In whatever degree indeed the presence of sulphur may contribute towards the processes taking place in the interior of a volcano, it is certain from the nature of the gases evolved, that it must exist either at or near the spot at which they are going on, and in either case the heat evolved by the inflammation of the earthy or alkaline metalloids would be amply sufficient to dissipate it in an elastic form, and to determine its combination with oxygen so soon as it came into contact with atmospheric air. In the former case it would be in a fit state to enter into union with any nascent hydrogen, in the latter it would constitute the sulphurous acid so common in all volcanos.

But it is well known, that sulphuretted hydrogen and sulphurous acid mutually decompose each other when in contact, so that the gas actually evolved from the crater only denotes the excess in which the one ingredient has been formed over the other. Where the process is carried on in places to which oxygen in a free state gains admittance, and causes heat enough to determine the ready combustion of the sulphur, it is probable that sulphurous acid

will be disengaged in sufficient quantity to decompose all
the sulphuretted hydrogen that may result. Hence water
will be reproduced by the union of the oxygen from one gas
with the hydrogen from the other, the sulphur from both
will be deposited, and the *excess* of sulphurous acid alone
escape into the atmosphere.

Such appears to be the case at Etna and at the crater of
Volcano, where the process is going on with some vigour—
it is only on the skirts of Etna near Jaci Reale, and in some
of the other Lipari Islands, where the volcanic action is
almost extinct, that sulphuretted hydrogen has been discovered.

The latter gas however is given off, as it would appear,
exclusively from the nearly extinct volcano of the Solfatara,
and has been met with at Vesuvius after an eruption.

The presence of nitrogen and of carbonic acid I have
already endeavoured to account for, and that of steam,
which is probably a principal agent in rending and heaving
up the contiguous rocks, is too obvious an effect of the heat
that would be excited by the action of water upon the metallic bases of the earths and alkalies, to require a particular
notice.

The muriatic acid disengaged from volcanos, is probably
derived from the muriate of soda contained in the sea-water,
which is disengaged in consequence of the chemical affinity
exerted at a high temperature between the alkali and the
clay or sand present; for it is found by experiment, that if
steam be passed through a mixture of either of these earths
with common salt, muriatic gas is disengaged.

So far as we have gone therefore, there is nothing to contradict, and much to confirm, the theory which derives volcanic phænomena from the access of water to the unoxydized nucleus of the earth. Let us now proceed to consider
whether the other phænomena are equally reconcileable
with this hypothesis.

A volcanic mountain is usually composed in one of two ways: it is either a conical mass of homogeneous rock, which appears to have been heaved up at once into its actual position by the agency of elastic fluids, or it is composed of a succession of beds, generally consisting of cellular lava and tuff, in alternating strata, which radiate in all directions away from a circular cavity, which occupies the centre of the mountain.

The former, if we believe Humboldt, is the case in many of the volcanos of Equinoxial America, where a huge conical mass, composed of trachyte, rises to an enormous height above the surrounding country, forced up, as he conceives, by elastic vapours in the manner described in the lines of Ovid already referred to, which appear so applicable to the recent case of Jorullo.

Of this mode of formation enough has been said in my third Lecture.

The latter kind of structure has been well illustrated by Von Buch in his Memoirs on Craters of Elevation, and on the Island of Lancerote,* and the same has been shewn by Mons. Neckar de Saussure to belong to the old crater of Monte Somma near Naples.

I have already stated my objections to the manner in which the latter geologist has attempted to account for this arrangement in the case of the Monte Somma, and I should even doubt its applicability to any mountain, composed of strata of lava divided by beds of tuff possessing any considerable degree of compactness; as I conceive, that wherever this is the case, the materials must have been cemented by the intervention of water.

In such cases I am more disposed to adopt an explanation, which, if not directly stated by Von Buch, seems to flow as a natural consequence from his remarks on Craters of Elevation.

That at a period antecedent to the commencement of the

* Memoirs of the Royal Academy of Berlin, 1818.

present order of things, the operations of a volcano would be carried on in a manner different from what they are at present, will appear from what I shall say in a future part of this Lecture; at present I need only remark, that there is abundant reason to believe, that the volcanic products were deposited more horizontally, and extended over a more considerable tract of country than modern lavas usually do, as likewise that they frequently exhibited a succession of tuff and lava beds, corresponding with the number of eruptions that had taken place.

Now let us suppose, that at any subsequent period, the volcanic action was renewed underneath these strata with an energy sufficient to upheave them round the point at which the expansive force was at its maximum, and we should then find a circular aperture formed in the midst of the raised mass, round which the successive beds would be seen dipping away in all directions from the centre, just in the manner which Von Buch and Neckar have described in the particular cases alluded to.

This then appears to me to be the true explanation of the kind of structure, exhibited in what Von Buch calls " Erhebungs Cratere"—craters which often have given out no lava whatsoever, being formed, like that particular one which I have described as existing in the Eyfel-country amongst rocks of transition slate, by the sole agency of elastic vapours.

Let us next proceed to consider the solid substances thrown out by a volcano during a state of eruption.

These may be divided into such as have undergone a complete change from the volcanic action, which are either lavas, or loose ejected masses; and such as are thrown out unaltered, or at least retain sufficient of their original characters to be recognized as belonging to some one of the known strata.

Beginning with the former class of substances, I shall lay before you: first an account of the phænomena exhibited by lavas which may be accounted chemical, and secondly of those connected with their mineralogical char cters.

Lava, when observed as near as possible to the point from whence it issues, is, for a most part, a semifluid mass of the consistence of honey, but sometimes so liquid as to penetrate the fibre of wood.* It soon cools externally, and therefore exhibits a rough unequal surface, but, as it is a bad conductor of heat, the internal mass remains liquid long after the portion exposed to the air has become solidified.† The temperature at which it continues fluid is considerable enough to melt glass and silver,‡ and has been found to render a mass of lead fluid in four minutes;‖ when the same mass, placed on red-hot iron, required double that time to enter into fusion.

Even stones are said to have been melted when thrown into the lava of Vesuvius and of Etna.§

On the other hand the temperature in some cases does not appear to have been sufficient to fuse copper, for when bell-metal was submitted to the action of that of 1794, the zinc was separated, but the copper remained unaffected.¶ This has led Dolomieu to believe that sulphur contained in lava acted as a flux, and caused its constituents to enter into fusion at a lower temperature than they would naturally do.

The presence of sulphur in lava, though long disputed, appears to be established by the recent experiments of Monticelli, who has shewn, that sulphurous acid is given out, when an heated mass of lava is exposed to the action of the air, but that when air is excluded, sulphur only is sublimed.**

---

* Fleurian de Bellevue sur l'action des Volcans, 441. Faujas St. Fond.

† That of 1822, some days after it had been emitted, raised the thermometer from 59 to 95 at a distance of 12 feet—three feet off the heat produced greatly exceeded that of boiling water. Monticelli Storia di Vesuv.

‡ Breislac, voy. dans la Camp, I. p. 279.

‖ Spallanzani, b. 4. p. 3.

§ Spallanzani, do. p. 8, and Fazzello as quoted by Spall. p. 11.

¶ Breislac in loco citato.

** See his last work, entitled, Storia dei Fenomeni dei Vesuvio.

Nevertheless it does not seem necessary to resort to this explanation of the ready fluidity of lavas, since their chemical composition, as ascertained by analysis, sufficiently accounts for this circumstance.

According to Dr. Kennedy, two specimens of the lava of Mount Etna contained each 4 per cent. of soda, and nearly 15 of oxyd of iron, to 51 of silex, 19 of alumine, and 10 of lime.* Now it is by no means inconsistent with chemical principles to suppose, that a mass so constituted would remain in a state of imperfect fluidity at the temperature between the melting point of silver and that of copper, and indeed some basalts are known to be reduced to fusion by a heat not more considerable than that of a common candle.

With regard to the mineralogical characters of lava, I shall appeal to the authority of Von Buch, who, whatever doubts may be entertained with respect to the soundness of his views in some other points, deserves to be listened to with respect on this, as a subject to which he has devoted his earliest attention, and on which he has thrown the most important light.

Almost all lavas he conceives to be a modification of trachyte, consisting essentially of felspar united with titaniferous iron, to which they owe their colour and their power of attracting iron.

True lavas, he conceives, which have flowed in streams from the sides of a volcano, agree in general in containing as a constituent, glassy felspar.

* Lava.

From Catania, west of Etna.	From Santa Vennera, Piedmont, west of Etna.
Silex .... 51.	.... 50.75
Alumina .... 19.	.... 18.5
Lime 9.5	.... 10.0
Ox. Iron............ 14.5	.... 14.25
Soda 4.0	.... 4.0
Muriatic Acid 1.0	.... 1.0
100.0 Kennedy.	100.0 Kennedy.

This felspar is derived immediately from trachyte, that being the rock which directly surrounds the focus of the volcanic action; for if we examine the strata that succes· sively present themselves on the sides of a crater, we are sure to find that the lowest in the series is trachyte, from which is derived by fusion the obsidian, as is the case at Teneriffe.

Agreeably with this view, I have myself remarked, that the oldest lavas of Mount Etna approached most nearly to the characters of trachyte, and that there even seemed a gradation, dependent upon the relative antiquity of the beds, down to the lavas of the present period, which possess the usual cellular and semivitreous aspect of such products.

But together with these, which may be viewed in the light of essential ingredients, lava often contains augite, leucite, hornblende, mica, olivine, specular iron, and other minerals, which appear not to have been present in the matrix, but to have resulted from the play of affinities induced amongst its ingredients, in consequence of the fusion which they had undergone.

Whether therefore we look to the original nucleus of a volcanic mountain, or to the products that have at successive periods been derived from it, we shall be led to the same conclusion, namely, that the substance on which the fire had acted was a felspathic rock allied to trachyte, and that the bodies ejected during an eruption may be regarded either as modifications of it, or as extraneous substances to which the action had accidently penetrated.

To the former class perhaps belong lavas in general, as well as pumice, obsidian, and the like; to the latter the marbles and other substances not volcanic, that lie scattered round the sides of many volcanos.

But considering the peculiar characters of trachyte, and the circumstance of its being limited to countries that appear to have undergone the action of heat, we can hardly regard it as a substance which makes a part of the original constitution of the globe, and shall be disposed to set it down as being itself a product, although a primary one, of the fusion of other descriptions of rock.

In order to determine this question, I propose to consider, 1st, amongst what formations volcanos commonly break out, and 2dly, what is the nature of those loose masses, which have been ejected without losing altogether their original characters, and therefore afford evidence of the kind of rock amongst which the volcanic action resides.

To begin with the Rhine, it appears that the formation on which the trachyte of the Siebengebirge rests, and among which the volcanos of the Eyfel have arisen, is a clay slate belonging to the transition series; in Auvergne, the rocks of Mont Dor and of Clermont rest immediately on granite, or are separated from it only by a tertiary deposit, whilst those in Cantal are incumbent on mica slate. In Hungary and Transylvania, the rock underneath is a porphyry, associated with syenite, clay-slate, &c. and referred by Beudant to the transition series; whilst in Styria, the rock most immediately surrounding the little trachytic formation of the Gleichenburg is gneiss.

In Italy the case is somewhat different—yet though the trachyte of the Euganean Hills rises from beneath chalk, we have reason to believe that primitive and transition rocks lie at no great depth beneath, as they are found near Schio, and support the alternations of Volcanic and Neptunian deposits in the Braganza.

To Vesuvius, the nearest rock formation is the limestone of the Appennines, but gneiss and clay slate seem to be the substratum through which the trachyte of Mount Etna has been protruded.

Humboldt has shewn, that the rock which supports the volcanos of the New World is generally a transition porphyry, and sometimes granite or syenite, and Von Buch reports, that the last mentioned rocks appeared as the lowest of those uplifted strata, which surrounded the crater of the Isle of Palma, and other of the Canaries.

Now, although the preceding enumeiation indicates such a variety with regard to the position of volcanic formations, as may seem at first sight to baffle all general conclusions, yet when we consider, that in the majority of instances, the rocks have been referred either to the primitive or transition series, and that in the remaining ones, the latter were at a depth far less considerable, than that at which we shall afterwards find reason to conclude the volcanic force itself to reside, I think it may not unfairly be presumed, that volcanos have universally broken out amongst the older formations, or those most near to the nucleus, whatever it may be, of the globe.

It is obvious indeed, that in those cases in which volcanos have appeared in the midst of primitive rocks, we cannot presume the seat of action to reside amongst those of a later date, but that the reverse does not hold good; so that if we only admit that any certain position is to be assigned to these *products,* a single case of their occurrence in the midst of older formations would overturn every inference, to be derived from their being observed to emanate from strata of a more recent date.

This presumption is farther strengthened, by considering the nature of the substances, found in the midst of lavas, which preserve any traces of their original characters, or the loose masses of unaltered rocks, that are occasionally thrown out.

Amongst the former, I have never seen or heard described any substance that bore the slightest resemblance to the constituents of secondary strata, but have often observed imbedded portions which present the appearance of altered granitic rocks.

Thus at the Puy Chopine in Auvergne, granite is found intermingled with the trachyte and greenstone, thrown together in such confusion, that we imagine the whole to have been elevated at one time, before the rock had been entirely acted upon by the heat.

In the lavas of the Vivarais, as well as in those of the

2 в

Rhine, I have met with masses imbedded that have much the appearance of an altered gneiss or granite, and the same thing is also seen in the tuff of Gleichenburg in Styria.

Humboldt mentions having found, in the midst of the new volcano of Jorullo in Mexico, white angular fragments of syenite composed of a small portion of hornblende and much lamellar felspar; and in the collection of Dr. Thomson, now in the Museum of Edinburgh, there is said to be a fragment of lava enclosing a real granite, which is composed of reddish felspar with a pearly lustre like adularia, of quartz, mica, hornblende, and lazulite.

I have likewise seen among the specimens from the Ponza Islands, presented to the Geological Society by Mr. Scrope, a piece of granite, or perhaps rather of a syenitic rock, which he states, in the annexed catalogue, to have been found in the midst of the trachyte from this locality.

But the most interesting fact perhaps of this description, is one related on the authority of a zealous cultivator of natural history, Signor Gemellaro of Catania,—I mean the presence of a mass of granite containing tin-stone, enveloped in the midst of a stream of lava from Mount Ætna. The specimen I have myself seen, and though I cannot pretend to have ascertained the presence of tin-stone, am able to vouch for the general accuracy of the account he has published of the substance.

It may be remarked, that these specimens of granitic rocks have, in general, a degree of brittleness, which accords very well with the notion of their exposure to fire.

The general character of the ejected masses, which are not imbedded in lava, is such, that it is difficult to refer them to any rock with which we are acquainted; it is true that M. Poli of Naples, has in his possession a fragment of rock thrown out by Vesuvius, which Humboldt pronounces to be a true mica slate, but in most cases they are mixtures of nepheline, mica, augite, leucite, and other minerals, which are rarely found associated together in the same manner in any of the original rock formations.

I have already alluded * to the probability, that the leucite crystals are cotemporaneous with the rocks in which they occur, and may add in addition to the fact before stated in support of this opinion, that Von Buch observed at Borghetto, in the midst of the crystals, a black substance analogous to the rock in which the whole was imbedded, and that in some cases, where the leucite did not completely envelope it, a connection was observable between the black substance and the rock. Sometimes in the place of the former a crystal of augite occupied the centre of the leucitic mass.

Admitting then these crystals to be cotemporaneous with the rock, and the whole to have been formed by the volcano out of the materials submitted to its action, it seems most probable that the latter, if derived from any known rocks, should have belonged to the primitive or at least the transition, rather than to any more recent order of formations.

Independently of these considerations, the general characters of trachyte favour the idea of its derivation from granite, or some analogous substance; it is from this resemblance that it has been called by Dolomieu granitoid lava,* and, although the two rocks may be distinguished by the presence of quartz in the one, and its absence in the other, yet the predominance of felspar in both seems to place them under the same genus, and to distinguish them from the constituents of secondary strata, where that mineral hardly can be said to occur, except where we have reason to suspect the agency of fire.

It would be interesting to possess a comparative analysis of a series of specimens of trachyte and granite, taken as nearly as possible from the same localities, since notwithstanding the mineralogical differences above pointed out, there is reason to believe that their chemical composition may vary but little.

* See page 124.

† Hence Strange mistook the trachyte of the Euganean Hills for granite. See his Paper in the Phil. Trans.

Thus though quartz is wanting in trachyte, and abundant
in granite, yet the siliceous earth contained in that mineral
may have united with the alumina present, in such proportions as would form felspar, and in this manner perhaps
the latter has become more abundant, at the expense of the
other two ingredients of granite.

The fusion effected by the volcanic operations would be
favourable to the play of affinities, and enable the particles
of the silica freely to combine with the other earths in the
requisite proportions.

In some cases on the contrary, where the material operated upon consisted chiefly of quartz, the result may have
been that variety called millstone trachyte, which, though
chiefly siliceous, betrays its igneous origin by the cells and
cavities it so abundantly contains.

We have thus arrived at the conclusion, that the characters of volcanic products in general are such as lead to a
fair presumption, that they are derived from some of the
older rock formations, a fact fully confirmed by a consideration of the phænomena attendant on an eruption, the
general tenor of which plainly denotes, that the focus of the
action is situated at a depth at least as great as that to which
granite extends.

I do not lay any stress on the remarks of Stukeley, who
calculates from the compass of country over which earthquakes have been felt, that the force must in some instances
be 200 miles beneath the surface,* because we have reason
to believe, that the vibrations may be propagated laterally
far beyond the immediate influence of the impelling force;
but I would argue from the immense mass of materials ejected by Vesuvius or Etna, without exhausting itself, or causing any diminution in its own dimensions; from the prodigious height to which the trachytic nucleus of a volcano

---

* As in the one that occurred in Asia Minor A. D. 17, which destroyed
thirteen cities, and extended over a diameter of 300 miles. See Stukeley on
the Cause of Earthquakes. Phil. Tranact. for 1750.

is raised, as at Teneriffe, and in Equinoxial America; and lastly from the immense violence of the eruptions, which would shiver to atoms any superficial covering of rock, that the elastic vapours must be disengaged at a depth at least as great as that, to which the crust of the earth can be supposed to extend.

These considerations will be viewed as more favourable to the hypothesis suggested by Sir H. Davy's discoveries, than to any other perhaps that has been proposed. Thus it has been shewn, that volcanos usually take place in situations, in which the element calculated to excite the combustion was largely present; that the aeriform fluids given out are such as would be generated by the chemical action superinduced by its presence; that the nucleus, as well as the products of a volcano, are of a nature likely to result from the action of heat upon the constituents of the nearest rocks we know of to the seat of the action; that the character of the unaltered masses ejected favour such an opinion; and finally, that the phænomena themselves indicate a cause at once deeply seated, and of wide extent.

All these circumstances on the contrary are opposed to the theory, which attributes volcanos to the combustion of beds of coal or sulphur, for though these substances may be often present in the neighbourhood of burning mountains, yet the rock, in which they are imbedded, belongs, as I have shewn, to a comparatively recent epoch in the history of our planet, and the phænomena, which they exhibit when in a state of inflammation, are such as denote a local and superficial origin.

Conceiving therefore that the former hypothesis affords the more plausible explanation of the facts detailed, it may be worth while to recapitulate the substance of the foregoing remarks, in such a manner as may enable me to point out the connection of the several phænomena one with another, and their dependence upon the cause assigned.

Let us suppose, that the nucleus of the earth at a depth of three or four miles, either consists of, or contains as a constituent part, combinations of the alkaline and earthy metalloids, as well as of iron and the more common metals, with sulphur and possibly with carbon. These sulphurets are gradually undergoing decomposition, wherever they come into contact with air and water, but, defended by the crust of the globe, just as a mass of potassium is by a coat of its own oxide when preserved in a dry place, the action goes on too slowly to produce any striking effect, unless the latter of these agents be present in sufficient quantity. Hence under our continents, the elastic fluids generated by this process are compressed by the superincumbent mass of rock, until they enter probably into new combinations, or diffuse themselves through the solid strata.*

But under the sea, where the pressure of an enormous column of water assists in forcing that fluid through the minutest crevices in the rock, the action must go on more rapidly, and the effects consequently be of a more striking nature.

These effects however will take place in the middle of the sea less generally than on the coast, because the pressure of the ocean itself opposes an impediment; and it will in general not be constant, but intermittent, because the heat generated by the process itself will have a tendency to close the aperture by which the water entered, first, by injecting the fluid lava into the fissure, and secondly, by causing a general expansion of the rock; nor will the water again find admission, until, owing to the cessation of the process, the rock becomes cool, and consequently again contracts to its original dimensions.

Now the first effect of the action of water upon the alkaline and earthy metalloids will be the production of a large

* Carbonic acid is known to be very commonly present in the waters of springs, and, as we are not aware that they attract it from the atmosphere, it seems most probable that it is the 'result of some process going on in the interior of the earth. (Bischoff, Vulkanischen Mineralquellen, p. 271.) The same remark perhaps may apply to the exhalations of nitrogen gas, which Dr. Davy detected in the warm springs of Ceylon, (see his Travels, p. 45.) and which have been also found in some parts of North America.

volume of hydrogen gas, which, if air be present, will combine with oxygen and return to the state of water, if it be *absent*, will probably combine with the sulphur, both being at the high temperature favourable to their union. In the former case nitrogen gas will be given off, in the latter sulphuretted hydrogen.

But in case of the presence of oxygen, the sulphur will also become inflamed, and give rise to the production of sulphurous acid, which will predominate among the gaseous exhalations emitted from the mouth of the volcano, provided sufficient quantity of air be present to combine with the hydrogen and re-convert it into water. So soon however as the oxygen is consumed, the hydrogen, no longer entering into combustion, unites with the heated sulphur, and escapes in the form of sulphuretted hydrogen, which, towards the latter period of the eruption, will predominate over the sulphurous acid, because it continues to be formed long after the want of oxygen has put a stop to the production of sulphurous acid. Now it is well known, that these two gases mutually decompose each other, and therefore cannot exist at the same time, so that the appearance of sulphuretted hydrogen from the mouth of the volcano may indicate, if not the entire absence of sulphurous acid at the place at which the process takes place, at least that its formation is stopped by the consumption of oxygen, or is going on with less energy than heretofore.

The very circumstance of the reproduction of water by the mutual decomposition of these two gases, might be the means of keeping up the action in a languid manner for an indefinite period. The slowness with which lava cools would cause it to give out for a considerable time sufficient heat to the adjoining strata, to place the sulphur at the temperature necessary to cause its combination with oxygen; hence a certain portion of sulphurous acid would be continually emitted, which however would be soon decomposed by the hepatic gas present. The water resulting from this process would percolate into the recesses of the

rock, act upon any portions of the alkaline and earthy
metalloids that might have escaped the original action, and
give birth to a fresh volume of hydrogen gas, ready in its
turn to dissolve a new portion of sulphur, and thereby to
contribute to the repetition of the same phænomena.

The separation of muriatic acid from the common salt
present in sea water is explained, on the common principles
of chemistry, by the superior affinity exerted by the base
for the siliceous or aluminous earth than for the acid, and
the sublimation of iron in the state of *fer oligiste,* rather
than of peroxide, may result from the deoxydizing property of the sulphuretted hydrogen at the same time disengaged.   The carbonic acid given off may be derived
either from the carbonaceous matters that have entered into
combustion, a view of the subject which is perhaps favoured
by the phænomena of the pietra mala, or from the action
of the high temperature upon the carbonates of lime and
magnesia, existing in the strata above the seat of the volcanic action.   I have already remarked, that this latter gas
is chiefly found in volcanos that have become extinct, or
have been long in activity, where time appears to have been
given for the heat to extend itself beyond the immediate
sphere of the volcanic action.

In short, on the supposition of salt water and air being
brought in contact with the sulphurets of the metals and
earthy metalloids, all the known phænomena of volcanos
may be deduced in the order in which they appear to occur:
in the first place, so long as air was present, an evolution of
large volumes of muriatic, sulphurous, and nitrogen gases,
together with aqueous vapour, would take place; at a later
period, when the oxygen was expended, sulphuretted hydrogen and carbonic acid, with a smaller quantity of muriatic acid, would appear; lastly, when all the other effects
had subsided, aqueous vapour and carbonic acid might
continue to be evolved.

If it be asked, how we can account for the presence of
atmospheric air in the interior of a volcano, I answer, that

as the first effect of the heat would be to produce a *softening* of the contiguous strata, it must necessarily happen, that the evolution of so large a portion of elastic matter would have the effect of bearing them up to a certain distance round the focus of the volcanic action.*

This aperture would undoubtedly be filled in the first instance by the gases given off by the volcano itself, but the slightest intermittence, or even inequality in the process would occasion a partial vacuum, which the air of the atmosphere would immediately fill.

After these remarks, which have for their object the nature and origin of volcanic rocks themselves, we are naturally led to inquire, what relation they may be supposed to bear to the other constituents of our globe.

It remains yet to be seen, up to what point we are justified in extending the operation of the same cause to the explanation of the phænomena of our globe, whether for example, there is sufficient reason from analogy to conclude, that the basalts and porphyries of older formation have resulted from a modification of the same process, or whether we can discover in them such differences of character, as imply something more than a mere alteration in circumstance, and baffle all attempts to refer them to a common origin.

So far as relates to the phænomena exhibited by the rocks themselves, the shifting, and disturbance they occasion in the surrounding strata, the hardening of the parts in contact, and the conversion of coal into coke by driving off the bituminous matter, I should despair of adding any thing to the luminous remarks of Professor Playfair; but

---

* Cornelius Severus, in his poem on Etna, seems fully sensible that gaseous exhalations are the cause of volcanic phænomena. See the lines beginning:

> Non propera est igni par et violentia semper,
> Ingenium velox illi, motusque perennis;
> Verum opus auxilio est, ut pellat corpora, nullus
> Impetus est ipsi; quà spiritus imperat, audit.
> Nunc princeps magnusque, sub hoc duce, militat ignis. &c.

it may not be altogether uninteresting to inquire, whether the direct inferences, to which he has been led, are borne out by the analogies subsisting between these formations, and the products of actual volcanos.

Now it is obvious in the first place, that no small degree of probability is attached to the igneous theory, when we discover every where among the oldest formations, rocks, whose mineralogical characters at least bear a manifest resemblance to those, which belong to recent and undisputed lavas.

The latter, as I began by observing, admit of a twofold division into trachyte and basalt, the former, including those rocks that consist essentially of no other ingredient than glassy felspar, the latter, those which contain in addition a portion of augite. Now, without entering into the disputed question as to whether the porphyries with glassy felspar, which occur in Hungary and the New World, are ever referable to the older formations, we can have little difficulty in pointing out cases, in which rocks of that age are found, having scarcely any other characters to distinguish them from the trachytes of existing volcanos, than such as relate to their mechanical texture.

Of this kind are the rocks of Sandy Brae in the county of Antrim, and of Drumodoon in the Island of Arran, which, if we overlook their want of cellularity, might be set down as belonging to the latest class of igneous products.

Still nearer is the resemblance subsisting between the augite rocks of these respective periods; for if we examine the lava which has flowed into the valley at the foot of the Graveneire of Montpezat in the Vivarais, we shall acknowledge that it possesses all the characters of an antient basalt, and even the rocks associated with trachyte on the borders of the Rhine, or resting upon it in Auvergne, differ from secondary traps only in the degree of their cellularity, or in some cases perhaps in a nearer approach to a glassy texture.

The same remark will apply to their chemical composition,

in which respect they have been shewn by Dr. Kennedy to
resemble each other very nearly, and in particular to correspond in the quantity of vegetable alkali they contain.

It may indeed be objected, that rocks so uniformly compact as the basalts and porphyries, which exist in the older
strata, are not common among the products of actual volcanos; that the cellular varieties found in the older rocks
differ from those existing in the more modern, in having
their pores for the most part filled with crystalline matter,
which is almost wanting in the other; and that vitreous
substances, which are rare in the former, constitute a considerable portion of the contents of the latter.

Considering nevertheless the remarkable resemblances
before pointed out, I think we should be disposed *à priori*
to view those minor differences, as indicating rather some
alteration of circumstances, than a cause altogether distinct; and this presumption would doubtless be strengthened, by considering, that the conditions, under which the
volcanic force must have been exerted at the earlier period
alluded to, were really different from those which operate
at the present.

But if it could farther be shewn, that the circumstances
of the former case were precisely such, as would be likely to
produce those characters which distinguish trap rocks from
the products of later volcanos, the notion of a similarity of
origin would be rendered more probable; and this probability would almost ripen into certainty, if it were proved,
that, as soon as the earth approached its actual condition,
the rocks resulting from the operation of fire presented a
nearer relation to the products of actual volcanos; insomuch that the characters of them, which date from a period
intermediate between the oldest and newest class of formations, appeared to constitute a connecting link between the
two former.

Dolomieu in France, and Strange in England, were I
believe the first who pointed out the occurrence of sub-

marine volcanos, which are not assumed to have existed for the mere convenience of explaining facts, but are fully established, by the repeated instances on record, of rocks consisting of volcanic matter suddenly elevated in the midst of the ocean.

The former explained in this manner that alternation of Neptunian and Volcanic deposits, which I have noticed in a former part of this inquiry as being found in Sicily and near Lisbon; and the latter referred to the same cause the analogous phænomena in the Vicentin; justly concluding, that these rocks, which in their general characters presented so close an analogy to the products of actual volcanos, were seen under circumstances, that excluded any other notion respecting them, than that of their having been ejected under water.

Availing himself of this distinction, though not agreeing with these writers in the precise view which they had taken of the phænomena, Dr. Hutton, in his celebrated Theory of the Earth, proceeded to point out, that the differences between the present effects of heat, and those attributed to it in earlier periods of the history of our globe, might be accounted for by the *pressure* exercised during the latter by the ocean, then incumbent upon the rocks so formed.

" The tendency of an increased pressure," to use the words of Professor Playfair,* " on the bodies to which heat is applied, is to restrain the volatility of those parts which otherwise would make their escape, and to force them to endure a more intense action of heat. At a certain depth under the surface of the sea, the power even of a very intense heat might therefore be unable to drive off the oily or bituminous parts from the inflammable matter there deposited; so that, when the heat was withdrawn, these principles might be found still united to the earthy and carbonic parts, forming a substance very unlike the residuum obtained after combustion under a pressure no greater than the pressure of the atmosphere. It is in like manner reason-

* Illustrations. In Playfair's Works. Vol. I. p. 38.

able to believe, that, on the application of heat to calcareous bodies under great compression, the carbonic acid
would be forced to remain; the generation of quicklime
would be prevented, and the whole might be softened, or
even completely melted; which last effect, though not deducible from any experiment yet made, is rendered very probable, from the analogy of certain phænomena."

These anticipations were fully realised by the masterly
experiments since undertaken by Sir James Hall, which
shewed, that the carbonic acid, usually driven off from limestone by the action of heat, may be retained in combination
with it by a pressure greatly inferior to that of the present
ocean, and that the calcareous matter under such circumstances enters into fusion, at a temperature which it completely resists when this elastic material is expelled.

Sir James Hall has applied this discovery with great success to explain the occurrence of calcareous matter in the
cavities of amygdaloidal traps, and that of water in those of
certain agates existing in the same class of rocks, whilst
such substances have never been met with amongst modern
lavas.

I do not find however, that he has alluded to the greater
cellularity, which the latter, taken in general, possess in comparison with basaltic formations, although this appears to be
an obvious consequence of the absence of that *quantum* of
pressure, which was necessary to control the volatility of the
gaseous ingredients; and may be justly attributed to such a
cause, from the fact, that the upper part of a stream of
modern lava, where the pressure is least, is generally most
vesicular.

The last distinction alluded to, namely the nearer approach to a vitreous aspect in the products of modern volcanos in general, may also, if I mistake not, be referred, in
great measure, to the same difference with respect to the
pressure exercised during their formation, and what took
place under the bed of the ocean.

Before however I attempt to account for this character

upon the above principle, it will be necessary for me to allude to some experiments of Sir James Hall and others, with respect to the cause of the vitreous texture assumed by many artificial products in the act of cooling.

The former, in a paper published in the 5th volume of the Edinburgh Philosophical Transactions, has the merit of first distinctly shewing, that the stony aspect, which distinguishes basalt from the usual products resulting from its fusion, may be imitated by allowing the substance to *cool* in a very slow and regular manner; and Mr. Gregory Watt,* who prosecuted these experiments on a more extended scale, shewed, that the various modifications of structure, exhibited from the most compact to the most vitreous lava, are successively produced, according to the different degrees of rapidity, with which the several portions of the same basaltic mass return to a solid condition.

Mons. Fleurian de Bellevue, who published a similar paper in 1805, pointed out, in addition to the preceeding results, an interesting analogy between the products of the slow cooling of glass and certain natural compounds, in the being similarly acted upon by chemical reagents, nitric acid reducing the former, as it does zeolites and other minerals, to a jelly, in consequence of the separation of the earthy and alkaline bodies soluble in that fluid from the silica previously combined with them. A similar property belongs to most felspathic lavas, especially to clinkstone; and even to certain bsaalts.

The analogies, between the products of artificial fusion and the substances found in trap rocks, have of late been extended by the experiments of Mitscherlich, who, it is said, has succeeded in producing crystals of augite in no respect different from those that occur in nature.

Hence it appears, that the vitreous structure, common to so many different substances, is assumed, in consequence, for the most part, of their rapid transition from a liquid to a

* Phil. Transactions for 1804.

solid state, owing to which the ingredients were prevented from exerting their mutual affinities, and uniting into those compounds which they have a natural tendency to form.

It is true, that something would appear to be attributable to the particular chemical composition of the substances themselves; for obsidians and pumice are found to be peculiar to certain volcanos, and even to certain stages of their action, being met with at Lipari, but not at Etna, among the antient Puzzolanas of Naples, but not among the modern products of Vesuvius; nor can their formation be in all cases attributed solely to rapid cooling, as we observe the former constituting entire currents, and the latter ejected in loose masses, intermixed with others of a more stony character.*

From the observations of Humboldt on Teneriffe, it may perhaps be concluded, that these products are derived in common from the direct fusion of trachyte; and thus we may account for their occurrence solely among the masses first ejected from the crater of the peak; nor does it seem altogether improbable, that this tendency to assume a *perfectly* vitreous condition may be connected with the presence of a larger quantity of alkali, and that the latter being in part dissipated by a longer continuance of the heat, the remaining ingredients may afterwards take upon themselves a form less perfectly vitreous, even where all the other circumstances remain the same.

These exceptions however do not affect the general truth of the proposition, that all lavas would assume a vitreous condition, if suddenly cooled; and a stony texture, if heat were abstracted with a certain degree of slowness; for though the rate at which the cooling process must proceed, in order to enable the ingredients to exert their mutual affinities, may be influenced by differences in chemical composition, still there is reason to believe, that every lava might be converted into glass, and that every kind of glass might by

* As in the Island of Lipari. See my Second Lecture.

proper management be made to develope a certain crystalline arrangement.

The question therefore occurs, to what are we to attribute the fact, that obsidians and pumice, so far from being
most common in submarine lavas, are never met with
amongst them, and that in general the products that have
resulted from the latter kind of eruptions do not present so
near an approach to the vitreous texture of the former substances, as is exhibited amongst those formed under existing circumstances.

Such an observation, seeming at first sight inconsistent
with the fact, that heat is commonly abstracted from bodies
with more celerity under water than in air, requires to be
considered somewhat in detail.

In order to understand this, it must be recollected, that
notwithstanding the rapidity with which a heated body
plunged into water has its temperature reduced, the power
of conducting caloric possessed by that fluid is exceedingly
small; and that it appears from the experiments of Count
Rumford and others, that, when heat is applied to a vessel
containing water, an equable temperature is established
among the several portions, chiefly in consequence of the
circulation induced in the strata of the liquid.

This circulation is effected in two ways: in some degree,
by the particles of water at bottom becoming specifically
lighter when heat is applied to that part of the vessel, and
consequently displacing those above; but in a still greater
degree, by the absorption and subsequent disengagement of
caloric, caused by the conversion of successive portions of
the water nearest the source of the heat into steam, and their
return to their original condition when they come into contact with the supernatant liquor.

Hence, if the heat be communicated to the top instead of
the bottom of the vessel, it is long before the water will
attain throughout an equable temperature.

Now it seems almost a *corollary* from the laws established
by Sir James Hall's experiments respecting compression,

that at the bottom of the ocean none of the water could be
converted into steam; for if, as this writer infers, the pressure
was sufficient to preserve the water existing in the midst of
the lava (where the heat must be supposed to be at its
maximum) in a liquid form, still more completely would it
prevent that, which was incumbent on the heated mass, from
assuming a gaseous condition in consequence of the heat
communicated from below.

I conceive therefore, that the agency of the water in
carrying off the heat of the lava would be limited to the
effect produced by the circulation occasioned among the
lower strata of that fluid; and this circulation would probably be less rapid than that occasioned in the strata of the
atmosphere by the bursting out of a similar stream of lava,
inasmuch as the expansion of water by equal increments of
heat is less considerable than that of air.

Perhaps likewise the superior density of submarine lavas,
the general absence of cells, and their not sending forth those
emanations of gaseous matter which appear to proceed from
modern currents, might contribute to the same effect, and
prevent the same vitreous appearance from manifesting itself,
which we should perceive, if the lava had come into contact
with shallow water, and which is observable, although in a
less degree, in that which has flowed in the open air.

It now only remains to be shewn, that the alteration of
circumstances, which took place as the earth approached its
actual condition, was such as, according to the principles
laid down, ought to have brought about that modification
of characters, which marks the volcanic products of this
intermediate age.

I assume as a position generally admitted among Geologists, that a large portion of the secondary and tertiary
strata have been deposited from aqueous solution, so that a
comparison of the height which they respectively attain
affords a standard, whereby to estimate the relative depth of
the ocean at the periods of their respective formation.

Now it was a favourite doctrine of the Wernerian school, that there existed an almost uniform relation, between the elevation which the strata attain, and their antiquity, so that it seemed to follow as a natural consequence, that the ocean, from which these several rock formations were deposited, gradually sunk from an elevation, perhaps nearly equal to that attained by the primæval granite, to the level which it occupies at the present moment.

In Mr. Conybeare's excellent Introduction to the " Geology of England and Wales,"* it is shewn, how far such an opinion is contradicted by the present state of our knowledge, and under what restrictions it may yet be received. Between the height of the same beds, deposited in different basins, little or no relation appears to be made out; but in those belonging to the same district, there seems, I think, little ground for scepticism as to the general truth of the proposition. Yet as these different basins can hardly be regarded as having been altogether insulated, it seems necessary to suppose, that the water, from which the strata were deposited, stood at an equal level in them all, and this obliges us to imagine great convulsions to have taken place, during, or subsequent to, the formation of secondary strata, which had the effect either of raising portions of them *above*, or depressing others *below*, the standard elevation.

Even those who are unwilling to admit this modified view of the Wernerian doctrine, will probably not deny, that at the period during which the tertiary rocks were deposited, the waters occupied an inferior level to that which they had previously attained; since, if we take a review of the principal localities in which these latter formations occur, we shall find, that they rarely, if ever, are found at so great a height, as the secondary formations on which they repose.

Thus the highest point attained by the tertiary rocks in Great Britain is at High Beech in Essex, composed of the

* Vide p. xx.

London clay, which rises to the elevation of 750 feet above
the sea—whereas the chalk is seen in Wiltshire at Inchpen
Beacon 1011 feet in height.*

In France the same rule appears to hold good; and if
there are any exceptions, they will be found, either in Switz-
erland, which exhibits in all its strata the most striking
proofs of derangement; or in America, where the operation
of volcanos affords a ready explanation of such apparent
anomalies.

Upon the most unprejudiced review therefore of the facts
before us, it will, I think, be concluded, that even if the
theory of distinct basins be considered inapplicable to the
secondary strata, it must be retained with regard to the
tertiary, since there seems every reason for believing the
latter to have been deposited at the bottom of lakes, either
of salt or fresh water, existing in the lower situations, after
the higher ground had been left dry.

Even in this case therefore, it seems necessary to imagine,
that the ocean, after having once covered the whole surface
of our present continents, underwent some diminution,
which had the effect of reducing it to its actual limits, and
that this reduction of volume was not sudden, but took
place gradually, seeing that we have evidence from the
phænomena of the tertiary strata of the existence of an
intermediate period, during which the hills were left dry,
but the low land was still occupied by the waters.

It is to this intermediate period that I am now alluding,
and if we apply the principles, which I have attempted to lay
down on the subject of submarine lavas, to the circumstances
of this particular case, it will, I think, be allowed, that any
influence that might be exerted by the pressure of an incumbent ocean in modifying the present effects of heat,
would at this time be more feeble than heretofore, and consequently that the products of any volcanic action which
might occur during its continuance, would possess less com-

---

* See Conybeare and Phillips's Geology of England.

pletely those characters which distinguish the two extremes of the series.

In the state of things now under consideration, the tendency to a vesicular structure, possessed by all kinds of matter which disengage elastic fluids whilst in a state of fusion, would be counteracted by a less efficient pressure, and the consequence of this struggle of opposite forces is therefore seen, in the occurrence of compact and cellular rocks intermixed, according as the resistance counter-balanced the expansive power, or the expansive power triumphed over the resistance.

The cells occurring in these latter rocks would likewise be rarely filled with crystalline matter, because the carbonic acid, having been retained less forcibly in combination with the alkaline earths present, would more rarely give rise to the formation of calcareous spar and similar products; and at the same time the other minerals, which require for their formation a very slow cooling of the ingredients, would be less abundantly formed, wherever the depth of water was insufficient to restrain the volatility of the elastic materials.

For the same reason, the lava itself would now assume a vitreous character more commonly than it did before, presenting often an appearance, intermediate between the effects of rapid cooling exhibited by the obsidian and the pumice which have been ejected, and the results obtained where a volcanic rock has cooled under the pressure of a larger body of water.

These differences will, however, be best pointed out by considering in detail the distinguishing characters of volcanic products formed at these respective epochs.

These I shall divide into three classes, marked, as I conceive, by their age, structure, position, and mode of origin, as distinct members of the same chain of formations; nor shall I scruple to include trap rocks under this general head, as, according to the view I have attempted to lay down, they have proceeded from the same cause as modern lavas, modified only by circumstances.

The 1st class then will comprehend those volcanic products, that have been formed since the earth obtained its present condition.

Being usually produced in the open air, and under a very limited pressure, they are usually more cellular and vitreous than the older formations, and their cells are more rarely filled with crystalline infiltrations. It may be observed, that although the number and size of the cavities are dependant on the mass of matter, the upper part of a current being always more vesicular than that nearer the centre, yet that it is rare to find among recent or post-diluvial lavas any portion of a stream that is altogether free from pores.

The volcanic products of this age may be divided into two kinds, namely, the loose fragments thrown into the air by the explosion of the gases generated, and the streams of melted lava that have flowed directly from the mountain.

The former consist of lapilli, having sometimes a compact, but more generally a scoriform character, of pumice, obsidian, and sand, which accumulate in heaps round the mountain, without any determinate order, so as to resemble a sort of breccia in all respects, except in the want of any cementing medium.

In some cases indeed the loose fragments cohere slightly by the intervention of the sand, but it is rare to meet with any bed possessing that degree of consistency which belongs in general to *the older* tuffs.

To this perhaps the nearest approach is exhibited in that congeries of volcanic matter which covers Herculaneum, where the sand and pumice appear to have been mixed up with water, and in consequence have united into an aggregate much more compact, than that resulting from the same shower of ashes which fell at Pompeii.

Even here however the degree of consistence can hardly be considered equal to that of tuff in general, so that it seems to follow, that something more than the mere presence of water is requisite, to bring about that intimate union of heterogeneous materials which we observe in the latter.

The lavas that have flowed from modern volcanos have been ejected usually through the medium of craters and not of dykes, for I believe it is a rare occurrence to meet with the latter phænomenon among rocks of this class.

Some indeed will appeal to the dykes of the Monte Somma, of Lanzerote, or the Isle of Bourbon, as instances; but I think it cannot be shewn, that in any one of these cases the dykes are co-eval with the modern eruptions; in all they appear to have been produced before the crater existed, and perhaps whilst the mountain was still under the surface of water.

The very existence indeed of a crater ought, it should seem, to be the means of preventing the injection of dykes; since the lava, finding a ready passage for its escape upwards, would be less likely to force its way through the solid substance of the walls that confined it.

Such an event may occasionally take place at a considerable depth below the crater, from the pressure of a vast column of lava causing the melted matter to be injected through the weakest portion of the containing walls; and something of this kind indeed occurs in every case, in which the lava, instead of rising to the summit of the mountain, escapes through its sides. Such dykes however, from the very cause assigned, would rarely come within the sphere of actual observation; since they would be nowhere so unlikely to occur as in the crater, that is, near the upper surface of the column of lava, to the pressure of which they owe their existence.

Another distinction between the rocks of this class and those belonging to the one I shall next mention, is, that they have been poured forth in narrow bands or streams, gradually widening indeed as they are removed from their source, but in all cases having a breadth in no degree proportionate to their length. More antient volcanic rocks on the con-

trary seem to form continuous strata, spreading more uniformly on every side over a large extent of country.

This is the case alike with the basalts of the Giant's
Causeway, the toadstones of Derbyshire, the porphyries of
Edinburgh, and the trachytes of Mont Dor. In none of
these cases do we discover any such tendency to a rectilinear
course, as may be traced in most of the streams that have
been derived from volcanos now in activity.

May not this remarkable difference be in like manner
attributed, to the greater resistance which would be opposed
in the one case to the escape of the melted matter, compared
to what takes place at present, and the consequent accumulation of the materials, until a passage having at length
been forced through the weakest part of the incumbent
stratum, a mass of lava considerable enough to cover an
extensive district, spread itself equably on all sides of the
aperture?

May not this be the reason, that in former periods of the
world long intervals appear to have elapsed between the
volcanic operations, sufficient indeed in some cases to admit
of the deposition of thick strata; whereas at present the
mountain is enabled by the absence of pressure to relieve
itself by a much slighter effort, and consequently gives rise
to less considerable, though more frequent eruptions,—to
rivers only instead of *seas*, of volcanic matter?

The case in a modern volcano which most nearly approaches to the one we have been considering, is where,
owing to the want of any permanent opening, such as a
volcanic crater presents, the lava is ejected from sundry
minute orifices on the surface of a level country. Such
appears to be the case in Lancerote, in St. Michael's, and
probably in Iceland.*

* See Sir G. Mackenzie's account of the cavernous lava of Iceland, and
compare it with Von Buch's description of the appearance of the stream at
Lancerote, beginning—Endlich hinter Tinguaton erschienen hohe Kegel zur
seite, von unten bis oben nur aus lockeren Rapillstucken gebildet; Kegel
über Kegel zeigten sich in der ferne, und von der Höhe sahe man die Lava-
masse einem schwartzen Gletscher ahnlich, sich herabsturzen. Eine stunde
weiter erreichte ich diese lava, wie ein meer von Verwüstung. &c. &c. p. 71.

In these instances, the lava, not having its direction determined by the slope of the country, flows equably towards
all points of the compass, and thus constitutes a bed, which,
if covered by a deposit of marl or sand, might be mistaken
for a mass of antient lava.

Cases indeed are mentioned, where a stream emitted from
the crater of a volcano appears to have spread itself equably
round the sides of the mountain.† But such an accident
can only arise from an uncommon regularity in the brim of
the orifice, and therefore must be of rare occurrence. The
regular and equable distribution of antient lavas could never
have happened generally, unless they had been ejected
through a narrower orifice than that of a crater.

I now proceed to consider the 2nd class of volcanic products, which, I believe, to be co-eval with the rocks denominated tertiary.

These, being ejected under a pressure less considerable
than what existed in the preceding period, naturally exhibit
some variation in characters.

They are composed of a mixture of vitreous and cellular,
with stony and compact rocks, the former connecting them
with the modern, the latter, as we shall find, with the more
antient products of fire. Crystalline infiltrations are more
common than in the former, but less general than in the
latter. Their subaqueous origin is proved by their repeated
alternations with Neptunian or freshwater deposits, often
containing shells which appear to have resided undisturbed
at the bottom of the water. They occur chiefly in beds, the
direction and origin of which it is often difficult to trace,
but which appear to be derived more commonly from dykes
than from craters.

They consist, either of lavas of an homogeneous texture

† Steininger (Erloschenen Vulkane in Sudfrankreich) has stated that the
crater of St. Nicholas near Cayres, between the towns of Puy en Velay and
Pradelles, in the south of France, is an instance of this kind. I did not visit it.

referable to the general heads of felspar and augite porphyry, or of tuffs consisting of an admixture of loose fragments of sundry volcanic products.

Of the former class, the one most abundantly distributed
during the above mentioned period, is some variety of that
felspathic porphyry, known under the general denomination
of trachyte.

If we take the trouble of considering in detail the spots in
which this substance has been principally met with, we shall
find it in general referable to a period, posterior to the date
of the secondary, and, so far as we can make out, co-eval with
the tertiary formations.

Thus it has been already shewn, that neither the trachyte
of Hungary, of Styria, of the Venetian states, of Germany,
nor of France, can be attributed to an æra more remote than
that of the latter class of deposits; whilst on the other hand
they exhibit marks of anteriority to the last general deluge
in the vallies which every where intersect them.

The first thing that strikes us in considering this class of
ignigenous products, is their vast thickness and extent, in
comparison with that of similar rocks thrown out either at
an earlier, or a later period; whence it would seem to follow, that during the formation of the tertiary class of deposits, the circumstances were peculiarly favourable to the
exertion of volcanic agency.

Possibly the retreat of so large a portion of that ocean,
which by its enormous pressure must have hitherto checked
the volatility of the gases generated by the processes we are
contemplating, may account for the difference above noticed;
since the elastic vapours, previously confined in the interior
of the earth, might in consequence of this change be enabled
to heave up the already softened materials of the incumbent
strata, and thus to form in the first instance those vast domes
of trachyte, which so often constitute the *nucleus* of a volcanic mountain, and the vent through which its lavas, &c.
are discharged.

The quantity of matter thrown out by this first operation

of the volcanic force, ought to have been more considerable than that discharged at any subsequent period, when, a permanent vent being established, a slight effort would be sufficient to effect a discharge, and might be expected even to exceed that emitted at earlier periods, when the action of the volcano was controlled by a greater pressure. It is true that in the older rocks, eruptions, where they took place, might be equally considerable ; but we have no proof that they occurred at once in so many parts of the globe, during any preceding epoch, as they appear to have done, during the period at which the tertiary rocks were being formed.

Under the circumstances above stated, I do not myself see any valid objection to the notion originally suggested by Humboldt or Von Buch, with respect to the heaving up of masses of softened rock in a dome-like form ; but those who cannot reconcile themselves to such an opinion are at liberty to form any other hypothesis with respect to those rocks, excepting that of considering them the relics of a single con-tinuous bed of lava; an idea quite irreconcileable, as it appears to me, with the linear direction in which they occur, the regularity of their form, their isolated condition, and the enormous height which they sometimes attain.

With regard to the origin of the tertiary volcanic pro-ducts which belong to the second department of our en-quiry, I mean the tuffs associated with the trachytes and basalts of this period, much difference of opinion appears to exist.*

By some they have been attributed to mud eruptions, analogous to those which are stated to have taken place from certain of the volcanos of America; by others they are deduced from showers of stones and ashes, agglutinated by the action of rain; whilst of late, since the attention of Geologists has been so much directed to the effects of what

* See some remarks on this subject in Mons. Neckar de Saussure's Paper on the Monte Somma. Memoire de la Societé d'Hist. Naturelle de Geneve.

is called diluvial action, it has been fashionable to attribute this class of products to the destruction of older volcanic rocks caused by the general deluge.

Before we proceed to consider the probability of these respective opinions, it will be necessary first to define the kind of tuff, of which we intend to speak, for it is certain, that under this vague general denomination a great variety of aggregates are included, possessing but little in common, except the presence of fragments of volcanic matter.

Thus we find the trass of the Rhine, the congeries of ashes and scoriæ which covers Herculaneum, and the decomposed trachyte of the Solfatara, all included under the name of tuff; although it is evident, that these formations belong neither to the same period, nor to the same class of deposits.

It is therefore not improbable, that each of the above hypotheses may be applicable to certain kinds of tuff, and that the discrepancy of opinion has arisen, from an attempt to extend to all an explanation suggested by the phænomena of one particular class.

Thus the hypothesis of mud eruptions may account for certain of those formations of trass, which, whilst they constitute a tolerably coherent mass, appear to be posterior to the formation of the vallies; to the lower part of which they are confined.

It would be interesting to obtain a detailed geological description of the mud eruptions of the American volcanos, as in the only case we know of in Europe of this kind, I mean in the tuff of Herculaneum, the aggregation of the parts is less considerable than we find it to be in the trass of the Rhine.

I was once disposed to consider this latter substance formed in the above manner, but the thickness and extensive distribution of the deposit in the neighbourhood of Andernach, and the want of any connexion between it and the neighbouring craters, obliged me to adandon such an idea.

Still less will an hypothesis of this kind account for the formation of so immense a mass of puzzolana as that, which occupies the neighbourhood of Naples, and which, from its division into distinct portions by interposed beds of loose scoriæ or pumice, indicates a series of ejections taking place at successive intervals, rather than one simultaneous operation.

In cases of this kind, the only probable opinion is to attribute the formation to vast masses of loose volcanic matter deposited at the bottom of water, through the agency of which, assisted by pressure, it became consolidated into its present form.

The action of this water, in the case of that near Naples, is evinced by the uniform manner in which the tuff has been washed into all the hollows existing in the older formations, as is evident at Sorrento, Vico, and other places on the coast, opposite to that from which the principal mass of puzzolana appears to have proceeded; and likewise in the valley of Maddelona, near Caserta, to the north of the city.

The origin of the substances of which this aggregate is composed is to be attributed, either to the detritus of the older rocks, or immediately to the ejections of the volcano, but the union of compact with scorified materials seems to shew, that both causes have operated. The former, I should imagine, are chiefly derived from older rocks, whilst the pumice, obsidian, and other analogous products, seem to have proceeded directly from the volcano itself.

To this class I would refer those beds of tuff, which occur in 'the Vicentin, in Styria, and in Auvergne,* containing shells, and alternating with beds of tertiary limestone.

---

* On the latter point I am happy to have the authority of Ramond, who remarks, that in the tuff of Mont Dor, there is no stratification in the arrangement of the integrant parts, the fragments are never disposed in parallel lines, nor accumulated with any respect to their bulk, but are heaped together without any order or arrangement whatsoever. There are no rounded masses, and when, as sometimes happens, the tuff contains fragments of prismatic trachyte that has been displaced, the prisms retain their angularity in a greater degree than they would do, if they had been subjected to the action of water. See his paper entitled, Nivellement Barometrique de Mont Dor, &c. Memoirs de l'Academie, 1813.

The pumiceous conglomerate of Hungary will belong, if we believe Beudant, to the same class; it is in part, he says, caused by ejections of loose fragments of rock, which took place under water; and in part the effect of this very water, which first detached portions of the existing rocks of the country, and afterwards cemented them into one coherent mass.

The latter view of the subject comes perhaps very near to the opinion of those, who attribute the whole to diluvial action; but it must be remarked, that, in several of these cases, we appear to have the best evidence, that the formation of the tuff dates from a period antecedent to the excavation of the vallies. Thus the puzzolana near Naples seems to be itself hollowed out much in the same manner as the older rocks which it reposes, and the pumiceous conglomerate of Hungary exhibits inequalities of surface nearly as great, as the older volcanic rocks, from whose destruction it is itself derived.

It may also be remarked, that the action of a temporary deluge, although it might have been the means of detaching the loose fragments of which the tuffs are composed, would not have united them into so compact and coherent a mass, nor have given rise to the formation of minerals, such as felspar, augite, mica, and leucite, which often appear, from their perfect condition, and intimate union with the paste, to have been produced subsequently to the formation of the aggregate in which they are found.

Such effects as these seem to require a much longer continuance of the action of water, than would be occasioned by a transient deluge, and may be best referred to that period of unknown duration during which the tertiary beds were deposited, when the boundaries of the land and sea appear to have been in a state of frequent oscillation, if we may judge from the proofs which the strata afford of the return of the ocean more than once to the spots from which it had retired.

It is perhaps introducing an unnecessary complexity into

the subject, to imagine, as some have done, an inroad of salt and fresh water alternately taking place during the period alluded to; for supposing a lake to have been formed in an inland situation by an irruption of the sea, it is evident, that provided an outlet existed whereby its redundant waters might be discharged, the mere influence of the rains and torrents would by degrees so diminish its saltness, as to render it at length unfit for the abode of any but freshwater molluscæ.

It is obvious indeed, that as every freshwater lake without an outlet must become more or less impregnated with salt, derived from the streams that flow into it; so every inland sea possessing one, will, if it maintains its level, become in process of time converted into a freshwater lake.*

The two seas, which may be considered as expansions of the River Jordan, sufficiently illustrate both these positions: the Sea of Galilee, though situated in the midst of a volcanic country, having an outlet, is filled with water equal in purity to the streams which flow into it; the Dead Sea, having none, is impregnated with salt to a greater degree even than the ocean, although the river, which passes through the former lake, is the very same which supplies it with water.

We have only therefore to imagine, that previous to the final retreat of the ocean within the limits at present prescribed to it, an intermediate period existed, during which the waters at several successive times gained possession of a portion of our continents; but that as these irruptions took place at distant intervals, the lakes of salt water left at each retreat became gradually converted into freshwater basins.

Such changes however could not have occurred, without giving rise to phænomena similar in some degree to those appealed to as evidences of diluvial action; but in the case before us, the existing rocks would not only be broken away and washed into the vallies, but time would have been

---

* My readers will recollect, that this was the principle, on which Dr. Halley accounted for the saltness of the ocean.

allowed during the continuance of the waters in contact
with them, for an union to take place between the fragments, and for the formation of a coherent and even compact mass.

Whilst therefore I fully acknowledge the probability of a
general and simultaneous irruption of water over the face of
our continents, at a period subsequent to the deposition of the
most modern strata, and consider the distinction which has
been established between the effects produced by it, and
those arising from the ordinary action of rivers and torrents,
as one of the points in which geology is most indebted to
the labours of English naturalists; I cannot consent to
attribute the generality at least of volcanic tuffs, to the
operation of a cause comparatively so recent.

The more fully indeed I am impressed with the soundness of the arguments adduced in proof of the reality of this
event; the more I am disposed to be cautious in attributing to it effects, the circumstances of which do not in all
respects correspond with the conditions of the case.

Now, independently of the foregoing considerations, it
seems highly improbable, that during the continuance of a
transient deluge those ejections of scoriæ and pumice, or
those depositions of Neptunian beds, should have taken
place, which are found intervening among many tuffs.
Neither would it be philosophical to apply to those formations of the same material which constitute a single
undivided mass an hypothesis, which must be abandoned
with respect to those which occur under the circumstances
above alluded to, since the similarity between them, in point
of situation, as well as in character, is in many cases such,
as should induce us to adopt an explanation which might
be applicable to both.

Such an explanation is in my opinion suggested by the
sudden inroads, and subsequent continuance of large masses
of water on the spots in which the tuff occurs, and by the
operation of subaqueous volcanos, during that period, in
ejecting loose masses of stone and ashes, which became con-

solidated, together with the fragments detached from the adjacent rocks, into an aggregate of this description.

The rocks belonging to the 3rd and last class of formations are distinguished by being formed exclusively through the medium of dykes. It is indeed difficult to prove by direct evidence that craters have never existed, because it may always be replied, that the revolutions which the earth has subsequently undergone have obliterated the traces of them. The general occurrence however of dykes, wherever these rocks are to be met with, renders it, to say the least, improbable that craters should have been produced, and is quite conformable to the principles laid down, with respect to the degree of pressure exercised upon such rocks at the period of their formation. If we consider indeed, that even at present, when the pressure of the ocean is removed, the lavas of Etna frequently make their escape through the flanks of the mountain instead of rising to its summit, we can easily conceive the greatness of the force, which would be required to elevate a mass of lava from the interior of the earth, in spite of the additional obstacle opposed by an incumbent ocean.

That the volcano in its greater efforts did nevertheless triumph over these obstacles, we know, from the beds of compact basalt that cover the face of many districts; but it seems more probable, that the latter originated from dykes, which traversed the strata intervening between the *seat* of the igneous action and the surface of the ground, than that they were poured out from a crater, as happens now that the pressure of the ocean no longer exists.

That this is really the fact in some instances, appears I think most satisfactorily from the knolls of basalt about Eisenach detailed in my first Lecture, where rocks, the volcanic nature of which cannot be questioned, are distinctly traced to a great depth below the surface of the sandstone rock which they are seen to cap. It is fortunate, that the individual, who had the care in that district of the roads, for which the basalt is extensively employed, was sufficiently aware of the important light which might be thrown upon the origin of such rocks by an examination of this spot, and should have caused the stone to be quarried in a manner which might enable him in time to display distinctly its connexion with the surrounding strata. In these indeed, as in many other instances, where the impelling force was less powerful, or the quantity of matter acted upon less considerable, the operations of the volcano were limited to the forcing of streams of lava through the strata, and have not given rise to beds extending far over their surface.

The basaltic masses met with in the country about Eisenach, and in various parts of Hessia, though agreeing in many particulars with the trap dykes of other countries, are distinguished from most of the latter, by their greater size and wedge-shaped structure, no less than by their forming isolated knolls often in the midst of a level country.

Some indeed might be disposed to consider the latter as the relics of a once continuous stratum that has been swept away, but I cannot reconcile this idea with their rare occurrence, and with the total absence of all traces of them in the intermediate spaces. It is most probable, that if excavations were made in the other rocks, similar to those at the Pflasterkaute and the Blaue Kuppe, they might all be found to proceed from a root or enormous dyke penetrating far into the earth.

The same arguments in short, which persuade us of the separate formation of the trachytic domes of Auvergne, seem to apply to the basalts of Hessia; for I do not see, that we should be justified in attributing in these cases any thing more to diluvial action, than the mere removal of a portion of the softer materials, by which these dykes were perhaps at first completely surrounded.

The dykes, which, from their connexion with beds of trap alternating with secondary strata, appear to be coæval with the latter class of formations, are said to be distinguished from those which appear to have been erupted during the deposition of the tertiary rocks, not only in their smaller dimensions, but likewise in the greater changes they have effected on the contiguous surfaces of the beds which they traverse.

Thus in the neighbourhood of basaltic dykes of this age, limestone is often converted into marble, claystone into flinty slate, and sandstone into jasper. Such changes would seem to require for their production something more than the common application of a heated body ; for it is by no means so usual to find a rock altered in the parts which immediately support a *bed* of trap, as it is in those which are traversed by the very same material in the form of a *dyke*.

Perhaps indeed the action of a continued stream of melted matter, fresh from the focus of the volcano, might be expected to produce more decided effects upon the walls of the fissure traversed by it, than would be occa- sioned by a mass of the same spread out over the cool and damp surface of a rock; since the heat in the former case must be at once more considerable and continued for a longer period. Even the pressure exerted upon the con- tiguous surfaces by the matter injected through the rock may contribute in some degree to the effect, and if there be really that distinction between dykes of older and younger formation, which has been above hinted at, it is possible that it may have arisen from the diminished force, with which the lava was propelled through the substance of the rock, when the resistance of the ocean above had been in great measure taken off.

In the case of the dykes of Monte Somma, which, Mons. Neckar remarks, have produced no change on the contigu- ous stratum, I should attribute more to the nature of the rock which they traverse, than to any difference in the dykes themselves. It is natural to expect, that the effect should be proportionate to the compactness of the former, and therefore that a stream of the same melted matter would pro- duce a more decided alteration, in passing, through a bed of granite or limestone, than one of tuff or gravel.

There is one species of alteration however, which, if we believe Von Buch, is effected by volcanic rocks of every age and kind. It consists in the conversion of common limestone into dolomite, or in other words in the impregnation of calcareous matter with magnesia, derived, as he supposes from the augite of the basaltic rocks contiguous.

I have before shewn the impossibility of applying this explanation, as Von Buch has attempted to do, to the dolomite found near the volcano of Gerolstein in the Eyfel.*

In the south of the Tyrol, where the magnesian limestone is greatly developed, Von Buch attributes its formation to the existence of augite porphyry, upon which it is always found incumbent.† He supposes this porphyry to have altered the original compact limestone, discoloured it, detroyed the organic remains which it before contained, changed it into granular dolomite, and thrown it up so as to form the lofty colossal precipices which it at present exhibits.

In the mountain of St. Agatha near Trent, the change is in a manner exposed to our view, in the numberless fissures which traverse the limestone summit. In this rock are situated the lead and calamine mines of Bleyberg, Carinthia, Schwartz in the Tyrol, &c. which Von Buch supposes to be injected by the same cause which elevated the entire mass, and introduced the magnesia. To the same agency he attributes the caverns which occur so frequently in magnesian limestone.

But it is clear, that many of these phænomena, which are appealed to as proofs of the action of fire upon the limestone, do in reality occur in situations where no such influence can be suspected; for there is no more probability, that the magnesian limestone of this country for example has been formed or altered by volcanos, than any other throughout the whole series of secondary formations.

* See my First Lecture, p. 52, 53.
† Über Dolomite als eine Gebirgsart. A Paper read before the Royal Academy of Berlin, and noticed in Ferussac's Bulletin des Sciences.

Even were we to admit, that in some cases the occurrence of dolomite is attributable to volcanic action, the manner in which Von Buch supposes it to have obtained magnesia seems equally incomprehensible; for it is difficult to conceive so uniform an admixture of this earth with the other ingredients to have taken place, without supposing the latter to have been reduced to fusion, and then how did the limestone chance to escape all admixture with the other ingredients, which the volcanic matter contained ?*

Upon the whole I believe most persons will agree with me, in considering the above hypothesis as unworthy of the distinguished geologist from whom it emanates, and in condemning it, as at once gratuitous in its assumptions, not very comprehensible in its details, totally inapplicable to most of the cases it is intended to explain, and unnecessary to be resorted to in any.

I have only indeed been induced to dwell upon it, from the respect I entertain for the services of this geologist in the particular branch of the study which forms the subject of these Lectures, services, which are so great, as almost to give a sanction to any error which he may have committed, and therefore render it the more necessary for us to ask ourselves, whether in adopting his opinions, we are governed by the cogency of his arguments, or the authority of his name.

The products of igneous action, from the earliest period down to the formation of the chalk, may be referred, if I mistake not, to this third class. Consistently with the fact of the greater depth of the ocean at that period, the characters that distinguish them from the more modern lavas are, greater freedom from cellularity, the more frequent occurrence of crystalline minerals imbedded, and the more stony aspect of the rocks themselves.

* The only circumstance which can be adduced in favour of this admixture, is that in some cases serpentine is seen to intervene between the trap rock and dolomite. Mr. Herschel has noticed a case of this kind in the Tyrol. See Edinb. Phil. Journ. Vol. 3.

With regard to the first of these distinctions I may remark, that the cells and cavities never seem to occur unoccupied by crystalline matter, insomuch that I was at first inclined to believe, that the occurrence of the latter may have
been the cause, instead of the effect, of the vacuities, the
carbonate of lime and other minerals having, as Sir James
Hall ingeniously suggests, when it entered into fusion with
the whinstone, kept separate from it, as oil separates from
water, thus giving rise to the spherical form which the
nodules of calcareous spar generally exhibit with more or
less regularity.* But the existence of hollows only partially
filled with zeolites and other crystals shews, that such an
explanation will not apply to all the cavities that occur
in basalt, and forces us to admit, that even under this vast
pressure the gaseous materials generated in the melted mass
would in some cases cause cells and vesicles in the midst of
it. These however had no sooner been formed, than they
were filled by infiltrations of crystalline matter, squeezed
into them by the pressure exerted from above; whilst in
more modern rocks, the cavities remain in general vacant,
the circumstances of the case being unfavourable to the
existence of such minerals.

With regard to the last distinction alluded to, I am inclined to doubt whether vitreous rocks are to be met with
in beds amongst the members of this class of formations;
for it seems not improbable, that the pitchstones, which
occur in stratiform masses pervading the sandstone of Arran,
may prove to be horizontal veins. This however will perhaps appear almost a dispute about words, after what I said
with respect to these rocks having in general originated from
dykes alone; for it seems a natural consequence from this
view of the subject, that the distinction between beds and
veins must be understood in a sense different from that, in
which those terms are usually employed. What I mean
however to convey is, that volcanic rocks of this age,
when they occur in extensive masses, do not usually

* Experiments on whinstone and basalt. Ed. Phil. Trans. Vol. 5,

put on a vitreous aspect, and that it is chiefly when they
have been thrown out in small detached masses, that
pitchstones have been formed, as it was only in such
cases that the cooling process went on with sufficient
rapidity.

There are many facts indeed to be collected from natural
as well as artificial processes, which prove, that the same
materials may form pitchstone, or basalt, according to the
rapidity with which they have cooled. Thus in the Isle of
Lamlash off the coast of Arran, I observed a dyke, the
centre of which was of basalt, the sides of pitchstone, and
Dr. Macculloch has noticed, that where veins of basalt ramify
into slender filaments, these gradually become pitchstone.

Now as I am not aware that pressure can have any influence upon the formation of vitreous products, except so
far as it tends to render the cooling process less rapid, it is
quite consistent with theory to find, that these portions,
which from the smallness of their dimensions, present a considerable surface from which heat could be abstracted,
should pass more frequently into pitchstone than masses of
greater bulk and thickness.

The more gradual cooling, which took place in submarine
lavas, may likewise account for the regular prismatic structure which distinguishes them from those of more modern
date. The latter indeed sometimes exhibit a cleavage at
right angles to the direction of their bed, which has been
compared to the columns of basalt; but is in reality distinct from it, as it arises merely from the shrinking produced by a diminution of temperature.* In basaltic rocks
on the contrary, the prismatic form arises, as has been well
explained by Mr. Gregory Watt in the paper alluded to,†
from the natural tendency which they possess to form
spherical concretions, which, pressing mutually upon each
other, will be converted into prisms, hexagonal where the

---

* This is well illustrated in the lava of Niedermennig on the Rhine. See my
First Lecture, page 49 and 50.

† Phil. Trans. Vol. 94.

texture, contraction, and adhesion to surrounding substances is every where uniform, but deviating into numerous irregularities in consequence of variations in the above
conditions. Hence the prisms in the last case so approximate, that often not even the blade of a knife can be passed
between them; whereas in prismatic concretions occasioned
by shrinking, the very cause assigned implies that a vacancy
is left. For the same reason the latter are never separated
longitudinally into joints, as is frequently the case with
columns of basalt.

Columnar basalt is however frequent among tertiary volcanic rocks, and is even found in a few instances amongst
the most modern, if those of the Vivarais and of the Eyfel
are to be placed in this class; but I am not aware of such
a rock being formed by the action of any existing volcano.
No genuine articulated columns at least occur either at
Vesuvius or Etna, for the rocks of Castello d'Aci and the
Cyclopean Islands are not derived from the present volcano.
It may be remarked, that in the cases alluded to, the basalt
is always seen covered by a bed of scoriæ or some other substance, which might have prevented the too sudden abstrac-
tion of heat.

There is also this further distinction.—In antient volcanic
products no difference of compactness or texture exists between the upper and lower portions of a bed, such as would
indicate a more sudden cooling or a greater evolution of
elastic matter in the superficial portions, as is generally the
case with those of the present day. This obviously may be
referred to the existence of a body of pressure in the former
case independent of the mass itself, whereas in the latter the
increasing weight of matter from above downwards alone
tended to check the formation of cavities, and the more
rapid cooling on the surface was calculated to render the
structure more vitreous.

In these points the characters of the preceding rocks present a difference, even with reference to those belonging to
the second class which are associated with tertiary *forma-*

tions, although a less striking one than with the most modern; but the most decided line of separation between the two former arises from the nature of the fragments, that occur in the tuffs with which they are respectively associated.

In the tertiary volcanic formations these consist, partly of compact, partly of scoriform materials, the former probably consisting of the debris of contiguous rocks; the latter, of the matters ejected by the volcano.

But the tuffs of secondary origin contain no other varieties of cellular lava but what are amygdaloidal, neither do we meet with those vitreous products so common in the more modern tuffs. I doubt indeed, whether a single fragment of pumice or obsidian is to be found amongst the tuffs of this age, and even pitchstone and pearlstone are rare accidents.

It now only remains to be inquired, up to what point it may be necessary, according to the principles laid down, to suppose the operation of fire to have extended; for if the secondary trap rocks, from the analogy they bear to modern lavas, are to be pronounced volcanic, it then becomes a question, whether the greenstones, syenites, and granites of primitive formation ought not, from their connexion with trap rocks, to be regarded as analogous.

In a Memoir on the Volcanos of Auvergne, which I inserted in the Edinburgh Phil. Journal for 1821, I took occasion to remark, that there seemed no absurdity in supposing that some trap rocks might be of aqueous origin, inasmuch as there existed an uninterrupted transition from them into greenstone, syenitic, and granitic rocks, respecting the formation of which the geological world is still divided. It is evident indeed, that the question, with respect to the hornblende rocks associated with granite, rests precisely on the same footing as that with respect to granite itself, and if primitive greenstone be of Neptunian origin, who will pre-

tend to deny, that even some of the trap rocks associated
with secondary formations may not be so also ?

Unless therefore the origin of granite be admitted to be
finally determined, I conceive it premature to dogmatize with
respect to the universal volcanic origin of flœtz trap formations ; all we can be warranted in doing in the mean time,
is to state from a careful examination of particulars, that
such and such members of the series appear to be so formed.

As so much therefore depends upon the conclusions to
which we may arrive with respect to the origin of granite
itself, let us proceed to consider briefly the arguments that
have been brought forward for and against the igneous formation of this rock.

The analogy already pointed out between granite and
trachyte may appear to some an argument in favour of the
igneous origin of the former, whilst others may be disposed
to view it in quite the contrary light. By the former it may
be said, that we have no instance of any rock composed
principally of felspar, which is known to be produced by
water, whereas trachyte and most lavas supply cases of its
formation by heat; and that the analogies between the two
rocks are such as seem to imply, that the process from which
they resulted was the same in kind, and differed only in some
subordinate circumstances.

The advocates for the contrary opinion may on the other
hand contend, that the presence of quartz in granite is ad-
verse to the idea of its having undergone fusion, since in a
state of liquefaction the silica of that mineral ought to have
attracted the alumina derived from the mica, and formed
felspar, as it has done in the case of trachyte.

The most cogent argument in favour of the igneous origin
of granite is derived from the analogy between the dykes it
sends out into the contiguous rocks, and those which proceed from basalt. But it must be confessed, that although
there are many appearances of the kind, which lead us to
suspect an igneous injection of the matter of the vein, yet,
on a calm survey of the phænomena, there will be found to

be a want of that direct and conclusive evidence afforded in the case of whin dykes.

I have myself examined the greater number of localities in Great Britain, appealed to by the Plutonists as the most triumphant proofs of their hypothesis, and I may state, that in no spot have I seen the phænomena of granitic dykes so well displayed as on the coast of Cornwall. Yet even there I came to the same conclusion as that expressed by my friend Professor Sedgwick,* whose exemption from any theoretical bias in favour of the Wernerian doctrines will hardly be disputed, and whose authority I had rather adduce than my own in support of an opinion, taken up in opposition to that of many distinguished naturalists of the present day.

After a detailed account of the phænomena of these veins, to the general accuracy of which I can speak, the conclusion to which he arrives is, that we have no other alternative, but that of considering them contemporaneous with the rocks through which they pass, as the position in which the beds of killas rest on the uneven surface of the granite, and their undisturbed direction, where most traversed by the veins above described, are, as he thinks, irreconciliable with the idea of subsequent injection.

In many cases too granite seems to pass by such regular gradations into gneiss, that it is impossible to fix the point, where the one rock begins, and the other terminates; so that if the igneous theory be maintained with respect to the one, that of all other primitive rocks seems to follow.

Dr. Boué, who has distinguished himself by several able and elaborate memoirs on Scotland, Germany, and various other parts of the continent, imagines, that this apparent transition arises from the action of the granite upon the rock contiguous, which it alters, and in a manner assimilates to itself.†

---

* See his paper in the Cambridge Philosophical Transactions. Vol. 1.

† Memoire Geologique sur le sud ouest de la France. In the Annales des Sciences Naturelles. August, 1824.

Thus in the Pyrenees, the granites, which, according to him, are allied with rocks of augite and serpentine in such a manner as to indicate a common origin, have been thrust through the midst of a formation of transition clay slate, and the latter, as it approaches these rocks, is converted, first into mica slate, and afterwards, when in closer contact, into gneiss.

This view of the subject will doubtless tend to do away with some of the difficulties, which beset the Huttonian theory as originally proposed, and I am only disposed to object to the admission that it is supported by the same degree of evidence, as that which may be adduced in favour of the igneous origin of whin dykes, and the rocks immediately dependent on them.

The question with regard to granite at present, seems to rest on about the same footing as that which concerned basalt did some years ago, when a very large and respectable body of geologists, with Werner at their head, saw no sufficient reason for admitting trap rocks to have been formed in a different manner from the strata with which they are associated. Werner, it is true, has been severely censured by some geologists of the present day; but I apprehend, that he is to be blamed, not for having withheld his assent to the propositions of the vulcanists, at the time when he first promulgated his views; but for having subsequently neglected to hold out to his pupils the question as one which required further investigation; and that the opinions which he had taken up in early life on this subject would in no degree have lowered his reputation, had they not been adhered to with so much pertinacity to the end of his career in spite of evidence subsequently brought together.*

* Some allowance nevertheless ought to be made for Werner, when we consider the advanced period of life to which he had attained, before the evidence in favour of the igneous origin of trap rocks had arrived at that degree of conclusiveness which would have justified a decided opinion on the subject. It was his misfortune indeed in some measure to have outlived his system, and to have remained stationary at the very time when geology was making its greatest progress; whence it has happened, that his services have been as much depreciated latterly, as they had been overrated before.

In like manner, I hope it will not argue any latent bias in favour of the exploded doctrines of the Wernerian system,

In order therefore to form a fair and candid estimate of his scientific merits, we ought to view him at the commencement of his career, or at least carry our ideas as far back as the period at which his school was resorted to by individuals of almost every nation, as the only then existing source of sound and practical information on the subjects which he taught.

Geology indeed, as it was studied at Freyburg, bore at that time about the same relation to its condition elsewhere, as History does to Mythology, or Chemistry to Alchemy; and, if it be objected, that even Werner did not altogether emancipate himself from the fables and chimæras that occupied his brethren elsewhere, it may be answered, that his defenders, at least at the present day, neither claim for him infallibility, nor an exemption from human infirmities.

I believe it may be laid down as a general law, though it is one no doubt sufficiently mortifying, not only that the groundwork of every thing great and good achieved by the human mind in whatever line is laid during the first thirty years of life, but that the opinions and researches, that *appear* to originate afterwards, have received, for the most part, all but their final development during the same period.

To say therefore, that Werner's geological system partook at first of the imperfections belonging to a new branch of knowledge, and that in his advanced years he felt reluctant to modify it, as a younger man might have done, in proportion to the new light that the science had received, is a reproach indeed, but one which applies too generally, to bear very heavily on his individual reputation.

It is also said, that he gave an undue prominence to theoretical opinions, and inculcated as dogmas, what, after all, ought only to have been brought forward as hypotheses.

But some excuse ought in candour to be made to a popular Lecturer, who, in the warmth of extemporaneous speaking, may sometimes indulge in speculations, which he himself would hardly deem admissible in his published writings; and if his disciples in some cases have chosen to build their faith on figments, which were introduced perhaps chiefly to enliven the dryness of practical details, they, and not their master, are to blame.

Without therefore professing that blind admiration for Werner, which his pupils at one time appear to have entertained, I cannot help considering, that the branch of natural history which he cultivated is greatly indebted to his exertions; and though the time, it must be confessed, is gone by, in which an addiction to the tenets of this or any other School of Geology can be defended, yet I am upon the whole inclined to think, that, up to a certain point in the progress of this science, even the exaggerated opinion entertained with respect to the merits of the Wernerian system may have had its use, as tending to inculcate more fully those principles of classification, and that method of discriminating rocks and minerals, which, with all their imperfections, must be allowed to possess no slight superiority over preceding ones, and to have facilitated upon the whole the advances that have been since made in this branch of knowledge.

if I suspend for the present my opinion with respect to the origin of granite; provided only, that I do not at the same time follow the example of Werner, in interesting myself so far on the opposite side of the question, as to receive with reluctance or incredulity the facts, which may perhaps hereafter serve to place the Plutonic theory on a less questionable foundation.

Indeed if we would avoid in future those oscillations of opinion on matters of geological theory, which, whenever they occur, serve, like vacillation in practical matters, to betray the infant state of the study itself, we must not adopt the opinions of the Plutonists, merely because they may be *more* probable than those opposed to them, but should be content to wait, until the evidence in support of them arrives at such a degree of force, as to place the theory on a level in point of credibility with those systems, which are received in other departments of science as established.

One of the questions which require to be more fully elucidated, before the origin of granite can be viewed as determined, is, whether this rock be ever found intruding itself into the midst of modern strata, as might be expected to be the case, if it were analogous in its origin to basalt. One instance has indeed been lately brought forwards, in which this rock is said to rest on a very recent limestone at Predazzo, in the Tyrol. The superposition is affirmed by Maraschini, Boué, and others; but it is questioned by Von Buch, who imagines, I believe, a fault to have taken place, which, by turning the strata completely over, has produced the delusive appearance described.

The same uncertainty seems to exist with regard to the formation of serpentine, as to that of granite. In the Lizard district of Cornwall, where rocks of the former kind extensively occur, the impression left upon my mind was, that the origin of the serpentine, greenstone, and clay-slate, was probably the same, and those who will peruse Professor

Sedgwick's paper on that country, will, I think, arrive at the same conclusion.*

Nevertheless in other situations serpentine seems to form dykes, possessing all the characters of igneous injection which distinguish those of trap. My friend, Mr. Lyell, late Secretary to the Geological Society, has described one of this kind in Forfarshire, and Dr. Boué has communicate to me other instances observed by Von Buch and himself, in various parts of the continent.

The existence however of a central heat, which some regard as demonstrated, may appear to lend considerable weight to the Plutonic theory, and accordingly deserves a short notice in this place.

In a late number of the Annals of Philosophy, is an ingenious paper by Sir Alexander Crichton,† in which the necessity of supposing the earth to be hotter in the interior, than it is at the surface, is inferred from the high temperature which he supposes to have prevailed in the Antediluvian World. This notion, which he entertains in common with Humboldt and other distinguished Naturalists, is chiefly derived from the existence of impressions of tropical plants in the coal strata even of the most northern regions, from which it seems fair to infer, that the climate enjoyed at those periods was equal to that in which these vegetables thrive at present.

Even at a much later epoch than this, during the formation of the beds above the chalk, a tropical climate seems to have prevailed in the latitude of London, as appears from the specimens of cocoa-nut and other analogous vegetable remains found in the Isle of Sheppey; so that, as Mr. Conybeare observes,‡ we may figure to ourselves the high mountain tracts, which at that time had raised their head

---

* See his paper " On the Physical Structure of the Lizard District," in the Cambridge Phil. Trans.

† Annals of Philosophy for November and December, 1825.

‡ Conybeare and Phillips' Geology of England and Wales, p. 30.

above the waters, as forming a groupe of spice islands, frequented by the turtle and the crocodile.

I feel therefore much less disposed to object to this part
of Sir A. Crichton's theory, than to that in which he refers
this high temperature to the fusion of the primitive rocks,
which he explains in the following manner.

The nucleus of the globe consists, he says, of the metallic
bases of the earths and alkalies, which, in the beginning of
things, took fire from the contact of air and water, and produced by their combustion granite. The latter, retaining its
temperature for a very long period, would impart to the earth
a source of heat independent of the solar rays, which must have
gone on progressively diminishing down to the present time.
But, as the earth must be supposed to have parted with
enough of its caloric, to allow of the existence of certain
animals as early as the epoch of the transition rocks, how
comes it, that it did not sink below the standard of tropical
heat by the time the coal formation was created, or, granting it an equinoxial temperature then, ought it not at least
to have become too cold for the existence of crocodiles, and
the growth of spices in this latitude, at a period so distant as
that of the tertiary formations?

Nor can it be contended, that the earth was receiving
from time to time fresh accessions of heat by the continuance of the same process which first gave rise to it, for we
have no proof of any general eruption of granitic matter
having taken place at a period subsequent to that of the
transition strata, and, even if we admit, that there are cases,
like the still disputed one of Predazzo, in which granite has
been thrown up at a comparatively recent period, such local
occurrences would have but little influence in modifying the
general temperature of the earth's surface.

Neither would it seem altogether satisfactory, if we were
to suppose that the heat in the interior of the globe has been
kept up by the volcanic action continued from the earliest
period down to the present; at least until it shall have been
proved, that the temperature is highest in those spots around

which these processes appear to go on with the greatest intensity, as in the neighbourhood of the sea.

The advocates however of a central heat call in to their assistance some observations recorded on the temperature of mines, which seem to shew that the substance of the earth is hotter in proportion to its distance from the surface.

It is one thing to admit the existence of a central heat, and another to decide upon the cause from which it may have arisen; nor am I prepared to deny the truth of the observations appealed to on this subject, which often appear to have been made by unprejudiced persons, and in particular those with respect to the progressive increase in that of the Cornish mines, in proportion to their depth.

Baron Fourrier has lately published a very elaborate treatise on this subject, in which however, if I am rightly informed, he has neglected to consider particularly the local causes of heat in mines, to which their temperature may perhaps be attributed.*

That these often interfere with the results, I feel disposed to believe from my own experience, such as it has been, on this subject; which leads me to the conclusion, that the temperature is often influenced by causes, less obvious, and less easily guarded against, than those arising out of the state of ventilation, the number of workmen employed, &c. &c.

In the course of my travels on the Continent I had repeated opportunities of examining into this subject, and in all the cases, where any remarkable degree of heat was discoverable, detected the presence of a large quantity of pyrites in a state of decomposition.

Such was the case in the mines of Hungary, where I was assured that in one instance, and that not the deepest part of the mine in which it occurred, the workmen were compelled from the heat to wear masks and gloves.

I was struck with observing the same thing in a differen

---

* I have only seen the Extract from his Memoir in the Annales de Chemie Tom. xxvii. Oct. 1824.

part of Europe, namely, at the Quicksilver Mines of Idria, where the metallic sulphurets likewise occur, and as it is well known that a considerable degree of heat is given out by such substances during decomposition, I cannot help believing that the temperature was, in those cases, affected by their presence.

I may add, that even respecting the Cornish mines, the statements given are very contradictory, for it is not long since there appeared an account by a Cornish gentleman of some trials, from which it would appear, that the heat below was in several instances no greater than that of the external air.

Now, as he justly observes, a single well-ascertained case of this kind ought to outweigh a hundred observations which shew an increase, since it is far more likely that the temperature of the earth should be raised above its natural standard by local causes, than that it should have been reduced below it.

Neither does it seem very consistent with the ordinary progress of nature to suppose, that the earth is gradually sinking in temperature, so that it will in time become unfit for the abode of the present race of animals, notwithstanding the warmth communicated to it by the sun.

It is true that the consideration of final causes ought not to be suffered to interfere with proofs of a more positive kind, but it may surely be introduced as an element into the calculation, when the utmost we can pretend to have arrived at, are probabilities.

Yet though there seems to be no reason for supposing the earth's temperature to be at present undergoing diminution, I am ready to allow that the presumption arising from a fair review of the phænomena is rather favourable than otherwise to the notion of a central heat;—all I object to is the bringing forwards, what ought only to rank in the light of an hypothesis, as the basis of elaborate mathematical investigations.

For my own part, however seductive it may be to the imagination to explain on some one broad principle the phænomena of our globe, and to lay down the great ends which volcanos are calculated to serve in the economy of nature, I think it more consistent with sound philosophy to limit myself, to those effects which have obviously been produced by their action, and to those final causes of their existence, which may be presumed from phænomena which we ourselves witness.

The former of these inquiries has already been insisted upon, and the occurrence of basalts in every class of rocks, under circumstances which establish igneous action, indi-cates that volcanos have existed almost from the commencement of our globe.

With respect to the latter point, I shall only remark, that whatever may have been the end, for the sake of which the accumulation of inflammable materials in the interior of our globe was ordained, their existence there, under circumstances which admitted of their undergoing from time to time inflammation, rendered the production of volcanos not only a natural consequence, but even an useful provision.

They are the chimneys, or rather the safety-valves, by which the elastic matters are permitted to discharge themselves, without causing too great a strain upon the superficial strata.

Where they do not exist, they give place to a visitation of a much more destructive nature; for those who have experienced a volcano and an earthquake will readily testify, that the consequences of the one are by no comparison lighter than those of the latter.

The same country is indeed often exposed to this double calamity, but that the existence of the volcano is even there a source of good, appears from the fact, that the most terrible effects are felt at a certain distance from the orifice, although the focus of the action is probably not far removed from the latter.

The agitations, which took place during six years at Lancerote, likewise shew, how much more destructive the effects of subterranean fire appear to be, where no permanent vent is established.

Thus far we have proceeded on solid grounds,—but if we are willing to push the enquiry farther, and to speculate on the other ends which volcanos may be intended to answer, it may perhaps not be too bold an hypothesis, when we consider their general distribution, to imagine that they are among the means which nature employs, for encreasing the extent of dry land in proportion to that of the ocean.

That such is the tendency of the processes daily taking place, appears from various considerations, and from none more remarkably than from the formation of coral reefs, a cause of increase to the quantity of dry land, with which the destroying agencies that are also at work have nothing to compete.

In speaking of the Canary Islands I observed, that volcanic processes seem much more frequently to have elevated, than to have submerged, tracts of country ; and if we consider, that coral reefs are mostly founded on shoals caused by volcanic matter that has been thrown up, a sort of consistency will appear in this instance to exist in the arrangements of nature, which leads to the belief, that fire and water are both working together to a common end, and that end, the preparation of a larger portion of the earth's surface for the reception of the higher classes of animals.

There may be something fanciful in what I am now going to suggest, with regard to another end which volcanos may be conjectured to fulfil ; yet if there be any truth in the idea, that the pressure of the ocean would be constantly forcing a certain portion of its waters through fissures into the interior of the earth, it would seem that there ought to be some compensating process, by which the ratio between the sea and land might be preserved unaltered.

This would perhaps be afforded by the action of volcanos, which restores to the surface just as much water as has been admitted to the spots at which the process is going on ; for though the first effect of the action is to decompose that fluid into its constituents, yet the immediate consequence is, as we have seen, the disengagement of a large volume of sulphuretted hydrogen and sulphurous acid gases ; so that by the action either of the latter fluid, or of atmospheric air upon the former, the whole of the hydrogen of the water, sooner or later, becomes re-united with oxygen. This indeed is *one* cause of the quantity of steam given out from the craters of all burning mountains.

The products of the volcanic action also, though, from the individual mischief they occasion, they can hardly be viewed by the inhabitants of the country overspread by them in any other light, than as serious present calamities, do not nevertheless deserve to be considered as permanent or unmixed evils.

It is true, that there is something gloomy and depressing in the contemplation of a volcanic mountain, when we consider the cities it has overwhelmed, the fields it has reduced to desolation.

Yet if we do not adopt the notion once so prevalent with respect to the speedy dissolution of the globe, if we take up the more pleasing, as well as, I conceive, the more probable opinion, that a world, which required so many ages to prepare it for the accommodation of its present inhabitants, is destined for many ages more to afford them a suitable abode; there is then something consolatory in the reflexion, that the very lava, which for so long a period has spread the most hopeless sterility over the ground it traverses, in process of time crumbles into the richest of soils ; and that, if we take the case of the neighbourhood of Naples as the volcanic district with which we are best acquainted, the experience of what has happened before justifies a belief, that the inflammable materials which supply the fires of Vesuvius will ultimately be expended, and that the mountain may at some future period return to the fertile condition, which Martial describes as belonging to it, when its heights were covered with vineyards, and the very spots surrounding the actual crater were considered the favourite resort of the Gods.

> Hic est pampineis viridis Vesuvius umbris,
>   Sparserat hic madidos nobilis uva lacus.
> Hæc loca, quam Nysæ colles, plus Bacchus amavit,
>   Hoc nuper Satyri monte dedere choros,
> Hæc Veneris sedes, Lacedæmone gratior illi,
>   Hic locus Herculeo nomine clarus erat.

ADDITIONAL NOTES.

Note to page 145.

On the Picture found at Herculaneum.

I find that the statement, given on Newspaper authority, respecting the antient picture of Vesuvius, lately dug up, is contradicted by the Editor of the Giornale delle due Sicilie.

Note to page 276.

On the neighbourhood of the Red Sea.

The Ichthyophagi of the environs of Ptolemais, in the Thebaid, preserved in the time of Agatharcides, the remembrance of an earthquake, during which the sea was left dry.—See Diod. Siculus, l. iii. c. 1.

That the following phænomenon also, the knowledge of which I owe to my friend Mr. Gray, of University College, Oxford, is connected with any thing volcanic, may be uncertain ; but as it is curious in itself, I shall insert his account of it, which I have extracted from the newspaper, called " L'Ermite du Mont Liban," published by Mons. Regnault, French Consul at Tripoli, in Syria, and is as follows :

No. 16, September, 1820.

M. François George Gray, Voyageur Anglais, qui a visité l'Egypte et l'Arabie Petrée, a bien voulu nous faire part, dans son passage au Mont Liban, d'un phenomène qu'il a observé avec la plus grande surprise dans un endroit appellé " Nakous," c'est à dire cloche, a trois lieues de Tor sur la Mer Rouge. Cet endroit recouvert de sable, environné de rochers bas en forme d'amphi-theatre, offre une pente rapide vers la mer dont il est eloigné d'un demi mille, et peut avoir trois cents pieds de hauteur sur quatre-vingts de largeur. On lui a donné la nom de Cloche, parcequ'il rend des sons, non comme faisait autrefois la statue de Memnon, au lever du soleil, mais à toute heure du jour et de la nuit et dans toutes les saisons. La premiere fois qu'y alla M. Gray, il entendit au bout d'un quart d'heure un son doux et continu sous ses pieds, son, qui en augmentant ressembla à celui d'une cloche qu'on frappe, et qui devient si fort en cinq minutes, qu'il fit detacher du sable, et effraya les chamaux jusqu' à les mettre en fureur.

M. Gray curieux de decouvrir la cause de ce phenomène, dont aucun voyageur n'a parlé, retourna au même endroit le lendemain, et resta une heure à attendre le son, qui vient en effet, mais beau-coup moins fort. Comme le ciel etait serein, l'air calme, il re-connut, qu'on ne pouvait attribuer ce son à l'introduction de l'air exterieur ; d'ailleurs il n'avait point remarqué de fente, par ou il put penetrer. Les Arabes du desert, dont il voulut connoitre l'opinion, l'assurerent, qu'il y avait sous terre un couvent de moines miraculeusement entretenue, et que le son, que l'on entendait, n'etait autre que celui de leur cloche. Des personnes moins portés pour les miracles, pourraient conjecturer, qu'il provient d'accidens volcaniques, à cause des eaux thermales qu'on trouve sur cette côte, notamment celles bien connues d'Hammam Pha-raoun (des Bains de Pharaon.)

Mr. Gray, in a letter to me, observes :—" The *camels* are an addition of M. Regnault's, as I had none with me, but I recollect repeating to him what the people of Tor told me respecting the effect of the sound upon these animals, and hence, I suppose, arose this little amplification. I beg leave to direct your attention particularly to the sound, which appeared to me so extraordinary—it *did* commence, as M. Regnault says, in a low continuous murmur, and then changed into *pulsations* as it became louder."

Seetzen (in Zach's Ephemerides, October, 1812) gives a similar account of this place.—It is worth inquiring, whether the noise may not proceed from the same cause as that of the Statue of Memnon, which is supposed to have stood on a sandstone rock. (See Keferstein Beitrage zur Geschichte und Kenntniss des Basaltes, 1te. Theil.) It is probable that the sound in that case proceeded, not from the Statue itself, but from the ground on which it was erected.

The French Naturalists, who accompanied Buonaparte, heard similar noises proceeding from the Temple of Carnac.

Note to page 283.

On the Waters of the Dead Sea.

Since the remarks on the Dead Sea, p. 283, were printed, I have seen a later analysis of its waters than that of Marcet, by Hermbstaedt., published in the Transactions of the Berlin Academy for 1820—21. This Chemist finds in it the following ingredients :

Free muriatic acid	0,507
Sulphate of lime......................	0,004
Sulphate of soda	1,597
Muriate of potass	0,275
Muriate of peroxide of iron............	0,333
Muriate of soda	4,859
Muriate of lime	4,250
Muriate of magnesia	15,755
	27,584
Water	72,416
	100,000

The water of the Jordan he finds to coniain—

Sulphate of lime........................	0.04
Muriate of soda	.35
Muriate of lime	.07
Muriate of magnesia	.03
Sulphuretted hydrogen	a trace
Loss	.01
Solid ingredients	0.50
Water	99.50
	100.00

Hence it is evident, that the Jordan, which supplies the Dead Sea, is quite as pure as most river waters, and consequently it is probable, that the salt in the Dead Sea arises only from the continued accumulation of the small quantity it is receiving from that stream. This renders it more unlikely, that this Lake should discharge itself, as some have imagined, by any underground channel.

Note to page 297.

There is a notice on the Geology of Ormus Island, in the last volume of the Transactions of the Geological Society, which mentions nothing of a volcanic nature.

Note to page 303.

On the Typhon or Typhœus of the Greeks.

As the fables of Grecian .Mythology generally appear to have had some foundation in fact, it seems.probable that those respecting Typhon or Typhœus, reported by Hesiod and others, may afford us a clue with respect to the existence of volcanic phæno-mena in countries at present of difficult access.

It is evident, I think, that the fable originally came from Egypt, where, as Dr. Pritchard observes, "Typhon stood opposed to

Osiris, just as Ahriman does to Orsmuzd, in the religion of Zoro-aster. The chief difference between the two schemes seems to consist in this circumstance, that the Egyptian fable is more entirely founded on physical principles. In the Persian doctrine Ahriman was not simply a personification of natural evil, his attributes comprehend also moral evil; but as we have seen that Osiris was physical good, or the productive or generative power, so Typhon seems to have represented all the destructive causes in nature."*

It is true that the Typhon or Typhœus of the Grecian Mytho-logy must, as Jablonski has observed, be distinguished from the Egyptian Dæmon of the same name; but as the former people seem to have derived their mythology from the latter, it will hardly be denied, that the original source of their notions respecting this evil genius, however they may have been afterwards modified, is to be sought for in Egypt.

The chief difference indeed between the two consisted in the more abstract sense in which the fable was understood by the Egyptians, than by the Greeks; the former regarding, as we have seen, the Typhon, as a general personification and cause of phy-sical evil; the latter confining it to certain particular objects of terror and aversion, such as earthquakes, volcanos, and whirl-winds.

It would perhaps not be difficult to explain why *these*, rather than other natural phænomena, were singled out as being imme-diately derived from the influence of the Dæmon.

In a climate like that of Egypt, the greatest physical evils, that were to be apprehended, would arise from the extreme heat, the absence of humidity, the suspension of the usual overflow of the Nile, the stifling winds of the desert, &c. &c.

Hence Plutarch (de Iside, p. 353) considers Typhon to be παν το αυχμηρον και πυρωδες και ζηραντικον ολως, και πολεμιον τη υγροτητι, and remarks that some regarded him as synonymous with the Sun, though this was not the *orthodox* opinion.

Now it was natural, that the populace, viewing the fable in a less abstract sense than the Priests, should consider the Dæmon solely

* Pritchard's Analysis of Egyptian Mythology, London, 1819, p. 79.

in this latter light, and that the Greek colonists, who probably were drawn from that class of society, should bring away with them the superstitions of the country from which they came, whilst they lost sight of the principles on which they were founded.

The considering Typhon the particular cause of earthquakes and volcanos may have happened in two ways,—1st, from the fiery nature and origin of these phænomena, and 2dly, from the general opinion entertained by the antients, that they were caused by winds pent up within the bowels of the earth.*

This is the idea expressed by Aristotle, Meteor. lib. 2, Cornelius Severus in his interesting and philosophical poem on Etna, and various other writers; and the Egyptian word *Bebin*, which answers to Typhon, signifies, according to Jablonski, *something pent up*, or an *underground cavern*.

This opinion was indeed suggested, not only by the phænomena themselves, which seem to indicate the action of elastic vapours struggling to escape, but likewise by the simultaneous occurrence of hurricanes with earthquakes and volcanos.

Thus in the recent eruption in the Island of Sumbawa, near Java, the greatest damage was done, not by the volcano itself, but by the whirlwind that accompanied it.†

That Typhon was considered in an especial manner the cause of whirlwinds (originally indeed of the scorching wind of the desert, but afterwards among the Greeks of all pernicious blasts),‡ is evident from the lines in Hesiod's Theogenia, quoted at length at the bottom of page 321.

It may be remarked, that the Egyptian, as well as the Grecian Mythology, agree in a remarkable manner with the systems of certain modern Geologists, who imagine volcanos to be the expiring efforts of that force, to the more general action of which they ascribe the elevation of mountains, and the formation of the older strata. They also correspond with the *fact*, that volcanos

---

* See note to page 321. It is alledged also by Tilesius, who accompanied Krusenstern, that the cause of the Typhon of the Chinese Sea is to be sought for in the bowels of the earth, and depends on agitations at the bottom of the sea.—Edinb. Journ. of Science, vol. 9. p. 204.

† See page 321.

‡ Hence Locusts are called " the children of Typhon."

at present are less extensively distributed than they were at former periods of the earth, and consequently that their influence may be inferred to be on the decline.

Conformably with this, the Egyptians imagined that Mercury has cut the sinews of Typhon, and that he has since lost the power that originally belonged to him, having sunk from his "high estate" as the general worker of evil, to that of the framer of earthquakes, thunder, and hurricanes.*

So also Hesiod, after describing the manner in which Jupiter overcame the Titans, and other monsters, which Mother Earth had successively produced to contend with him, says, that Typhœus, the author of whirlwinds and earthquakes, was the youngest of her children.

It may be remarked, that this poet notices the Typhon in an earlier part of the same poem, and therefore appears to distinguish him from Typhœus; but it will hardly be supposed that the attributes of the two Dæmons, both resembling each other in name, and derived in all probability from the same Egyptian source, would be very accurately distinguished.

Let us therefore, without attending to this distinction, proceed to consider how far the descriptions given of either, or both of these monsters, correspond with the hypothesis as to their being more especially intended as personifications of volcanic action.

The account given by Apollodorus is perhaps the most circumstantial, and agrees very well with that of Hesiod, the earliest writer, except Homer, by whom Typhon or Typhœus is mentioned.

This mythologist describes him as surpassing in size and force all the children of Earth; he was taller than all the mountains, his head often touched the stars; his arms stretched, the one to the setting, the other to the rising of the sun. The serpents, which were twisted round his thighs, rose to his head, and sent forth an horrible hissing; fire gleamed from his eyes; he hurled stones to heaven with a loud and hollow noise; (μετα συριγμων ομ8 και 6ons) surges of fire boiled up from his mouth (πολλη εκ τ8 ςοματος εξεβρασε ζαλη).

How well this corresponds both with the structure and phænomena of a volcanic mountain, I consider it needless to point out,

---

* See Plutarch de Iside et Osiride.

and shall therefore proceed to mention the spots in which the scene of the monster's adventures are laid ; for, as in the many instances where these have been explored, their nature is found to be volcanic, there seems a reasonable presumption, that the same may be the condition of such as have not been examined.

Now Apollodorus mentions, that Typhon was born in Cilicia, where we know of the existence of an extensive volcanic district called the Catacecaumene. When pursued by Jupiter, he fled to the neighbourhood of Mount Casius on the borders of the Lake Serbonis, near the Pelusian branch of the Nile. What the nature of this lake and of the mountain near it may be, I have not been able to ascertain. Jablonski however says, that the lake has a great affinity to the Dead Sea, which I have shewn to owe its existence to a volcanic eruption. Maillet in his Description de l'Egypte, p. 129, says, that the Egyptians got their bitumen for embalming from thence, and not, as is generally supposed, from the Dead Sea. The word (Ser) in Coptic, it is said, means to sprinkle, and (Bon) fœtid, and Manetho says, that the lake in his time emitted hot exhalations.

One of the cities, called Typhonia, formerly existed there, and the Egyptians called the lake Τυφωνος εκπνοαι. Plutarch, Vit. Ant. p. 917.

It must however be confessed, that the pestilential vapours which arose from the water may alone have caused it to be considered the abode of Typhon, and that with regard to the particular passage referred to commentators are not agreed, for Heyne proposes to substitute, for Casius, Caucasus, where there was a rock which went by the name of Typhonian. It is indeed not improbable, that the latter may be the true reading, as it appears that volcanos actually exist there,* and if we suppose that the mountain alluded to was Demavend, which stands near the famous Caucasiæ portae, and therefore may perhaps have been viewed by the antients as belonging to that chain, the Typhœus of the Greeks would then be the Zohag of the Zend-avesta, confined, according to the Persian mythology, under their volcano, as the Grecian monster was under Etna, or Cumæ.†

* See page 297, et seq.
† See page 303.

The other places mentioned by Apollodorus as the scene of these adventures are, Mount Hæmus in Thrace, so called from his blood which was there spilt; the peninsula of Pallene; and Mount Etna; but others have mentioned Lydia, Phrygia, and Bæotia, as the spots where he was finally vanquished (See Szetzes Scholia in Lycophron).

Homer, as I have already stated, makes the Arimean mountains (which perhaps may have been those near the Dead Sea,*) the bed of Typhœus, and it is worth remarking, that neither that poet, nor Hesiod, allude to Mount Etna as the abode of the monster, as Pindar and other later writers have done. This may be considered as an additional proof, that this volcano was not in action about the period at which they lived. Now we have abundant proofs of volcanic action in most parts of Asia Minor, and particularly in the provinces of Lydia, Phrygia and Cilicia;† such are the extinct volcanos near Smyrna and Scandaroon, the Plutonium, or Corycian Cave, noticed by Strabo and re-discovered by Chandler, and the destructive earthquakes so common throughout that country; we have accounts likewise of a mud eruption in the Lelantic fields near Chalcis in Eubœa,‡ which, if it was not itself of a volcanic nature, indicates, like the phænomena of Macaluba in Sicily, the accumulation of materials brought together by previous volcanic agency. It remains to be seen, whether the same holds good with regard to the other spots alluded to, viz. the Lake Serbonis, the Peninsula of Pallene, and Mount Hæmus in Thrace. I believe the Ceraunian mountains are also mentioned as the seat of Typhon, but this has arisen from a lambent flame which still plays on the summit of some of them, arising from the escape of an inflammable gas, as on the top of the Appennines between Florence and Bologna.‖ The same seems to be the case with regard to the δικορυφον σιλας mentioned as occuring on the top of Parnassus.

I cannot close this note, without pointing out the curious coincidence of names between those places in Asia and in Europe, noticed either as the abode of Typhon, or as the site of igneous phænomena.

* See page 292.   † See page 295.
† See page 239.   ‖ See page 337.

EUROPE.

Nysa, one of the peaks on Mount Parnassus noted for the flame that emanated from its summit.

Corycian Cave, on Mount Parnassus.

Hæmus in Thrace, where the blood of Typhon was spilt by Jupiter.

Inarime, now called Ischia, under which Island Typhœus is said by Virgil to be oppressed. Its other name was *Pithecusa* from the Apes which abounded there, (πιθηκοι) and the name Inarime was probably derived from the same cause, as we are told that *Arimi* in the Etruscan language signified *Apes.*

ASIA.

Nysa in Cilicia, the spot where Typhœus was struck with lightning.

Corycian cave in Cilicia, where Jupiter was confined when vanquished by the monster. It was noted for its Plutonium, or a Grotto del Cane like that near Naples.

Hæmus near the Lake Serbonis in Egypt, where were the exhalations or breathing-holes of Typhon, Τυφωνος εκπνοαι.

The Arimæan mountains, mentioned by Homer as the bed of Typhon, the Τα Αριμα, are placed by some in Cilicia, by others in Syria. There is no doubt however as to the existence in the former country of mountains that went by that name, and though they probably derived it from being peopled by Syrians, the descendants of Aram, yet it is a singular coincidence, that *Apes* were sacrificed in a temple of Diana that stood in this country. (See Strabo.)

Note to Page 408.

On the Tertiary Lavas.

Mons. Menard de Groye, in his account of the volcano of Beaulieu, seems to have adopted in some degree the same ideas with those expressed above, with respect to the cause of the differences between basalt and lava. The differences, he says, are precisely such as we might expect, from the one being submarine, the other produced in the open air.

L'air de vetusté, qui se voit empreint, si l'on peut dire, sur toutes ces terrains trappeens ; la destruction de tout cratere, s'il y en eut parmi eux ; cette stratification, qu'on leur assigne pour ordinaire, mais que je n'ai presque jamais bien reconnue ; leur alternation, observee en Saxe, dans le Vicentiu, dans le Derbyshire et ailleurs, avec des couches de sable, de pierre calcaire, et ces corps heterogenès coupees mème quelquefois dans le basalt et dans la wacke ; la fragilité qui se fait remarquer generalement dans les matieres trappéens ; leur etat plus crystallin ou du moins plus grenu ; la stratification très rare parme elles ; toutes ces singularitès si inexplicables dans d'autres hypotheses devrennent faciles à concevoir, et à expliquer, dans celle que nous proposons. —Menard de Groye, Journ. de Physique, Vol. 82.

He then goes on to consider the *third case*, namely, that of incomplete immersion, when a volcano is bathed in water at its base, whilst its summit is elevated above the waters, in which class he places those which I have called Tertiary volcanic rocks, such as Beaulieu, Vicentin, Meisner, &c.

But I cannot agree with this author in attributing the formation of columnar basalts to their being elevated above the waters, nor can I admit the fact to which he appeals, namely that the basalt at Beaulieu passes into greenstone in the lower part, as a proof, that the former was ejected above, the latter below, the level of the then existing water. Such an idea accords indeed very well with what I have elsewhere said *with regard to the dependence of crystalline arrangement upon pressure*, but unfortunately it happens, that the basalt is seen quite as frequently passing into greenstone in its upper as in its under part, as is the case in the very instance of the Meisner, to which Menard alludes.

I believe I have stated fairly the theory proposed by Monsr. Menard de Groye, to explain the distinction between basalt and lava, and have enabled my readers to judge, how far my ideas have been anticipated by this author; for, although I did not read his memoir, until after my treatise was composed, he has of course a fair claim to *priority,* where the opinions are the same. It will be seen however, that though Menard finds himself obliged (as I conceive every geologist will be, who examines with attention the Tertiary volcanic rocks) to make a distinct class of them, yet he does not explain the cause of their differences on the same principles as myself. According to him, the differences arise from their being formed partly above, partly below the surface of water; whereas my hypothesis assumes them to arise from their being formed under a body of water less considerable than the ocean. With respect to basalt his views are diametrically opposite to mine, as he considers it to be so formed in consequence of the absence of water, whereas I have explained its compactness from its being produced underneath that fluid. Indeed Monsr. Menard does not appear in any part of his memoir to allude to the difference in the state of compression produced by the presence or absence of water.

Dr. Boue in his memoir on Germany, has also alluded to a similar distinction (Journ. de Phys. Vol. 95) of volcanic products, which he divides into those caused by volcanos burning in the open air, and by the same more or less submarine, or burning under water.

Under the head of those partially submarine he includes trachytes and many basalts, and this division obviously corresponds with my second class of Tertiary volcanic products. The rocks included by Dr. Boue and by myself are nearly the same.

Dr. Boué has also noticed in common with myself the following distinctions between submarine and subaerial volcanos: viz. that they originate from dykes, form mountains of inferior height, and are associated with tuffs possessing a strong degree of aggregation. He also remarks the greater frequency of crystalline infiltrations, and the more decided changes effected upon the surface of the ambiguous rocks. It is satisfactory to find my observations thus confirmed, or rather anticipated, as it proves that they have not been *imagined* for the sake of propping up an hypothesis, but that the principles laid down by Sir J. Hall admit of a more extended application, than appears, so far as we can collect from his writings, to have been anticipated by their author.

———

Note to Page 429.

That I am warranted in speaking as I have done of the oscillations of opinion which have prevailed respecting the origin of basalt, will be evident from the following passage in Daubuisson's account of the basalts of Saxony, translated by Neill. " It appears, says the celebrated chemist of Berlin,* that naturalists are recovering by degrees from the *volcanic illusion.* It is about fifty years since a French naturalist revived the opinion concerning the volcanic origin of basalt, and he lived to see almost all Europe adopt his sentiments. Bergman, the first of the chemists who employed himself with diligence and success in examining mineral substances, and who, to an intimate acquaintance with the effects of heat, joined an extensive knowledge of mineralogy, could not bring himself to consider basalt as a product of volcanic eruptions. The Swedes adopted his view of the question. *It is scarce forty years since every body in Germany considered basaltic mountains as antient volcanos. Werner lifted the Neptunian standard; and now, among all the German mineralogists of any reputation, I know but one (Voigt) who still retains the old doctrine.* We have already seen in how decisive a manner Klaproth has pronounced on the subject : he, of all the German chemists, has had most opportunities of observing the effects of fire on mineral substances, and he has besides studied the history of basaltic mountains with that correctness for which he is remarkable. In Ireland, Mr. Kirwan was a supporter of the volcanic doctrine ; but the numerous chemical experiments which he made on minerals, and other considerations, led him to a change. Dr. Mitchell, one of the very best mineralogists, and Mr. Jameson, the author of the Mineralogical Travels in Scotland, and the greater part of British naturalists, consider basalt as having been produced in the humid way.

The geologist who of all others possessed the greatest experi-

———
* Klaproth, Journ. des Mines, No. 74.

ence, —— Saussure, the illustrious mineralogist of the Alps—
found it necessary, in the latter part of his life, greatly to limit his
notions as to basalt being of volcanic origin. In speaking of the
extinguished volcanos of Brisgau, he says, " I acknowledge, that
before studying the writings of Werner, I felt no hesitation ; but
that philosopher has taught me to doubt." Dolomieu, who was
at the head of the Vulcanists, but in whom the love of truth was
paramount to the spirit of party, admitted that some basalts had
been produced in the humid way. He observes,* " I have circumscribed the volcanic empire more than any other mineralogist,
French, English, or Italian, having withdrawn from its dominion
many mineral substances formerly placed under it. I hold that the
basalts of Saxony, of Scotland, and of Sweden, may claim Neptunian origin." When treating of the basalt of Ethiopia, he adds,
" I may affirm with certainty, that it is not of volcanic production."†

I shall not extract the remarks that follow, in which Daubuisson,
not very logically, I think, anticipates the complete overthrow of
the volcanic theory, from the change that he represents as having
taken place in the opinions of naturalists on this subject. I confess, I should have rather have been disposed to argue from the
foregoing statements, that the question had not been sufficiently
sifted; for, on speculative subjects at least, such oscillations of
opinion as are described show a defect of evidence, even more
than the influence of authority. If the *former* did not exist, it is
hardly possible that the *latter* should prevail, on a question of pure
science, to such a degree as to overpower reason.

Experience has since shewn such to have been the case, for the
preponderance of opinion at present in favour of the volcanic
theory is fully as great as it ever was in opposition to it, even
when Werner's reputation was at its height. It is for the Geologists of the present day to take care, that they also have not been
misled by the influence of great names, and that the evidence on
which they proceed be such, as to afford a reasonable security
against any similar change of opinion.

These considerations rendered me cautious in making up my

---

* Journal de Physique, tom 37.
· † Daubuisson on Basalt. Neill's Transl. p. 163.

mind with regard to the universal igneous origin of basalt, even after visiting Auvergne; and at present they lead me to regard a detailed examination of every basaltic country a necessary preliminary to a decision with respect to its origin.

It is only thus that we can proceed on sure grounds, and ought to flatter ourselves that we are laying a stable foundation for a future theory of the earth.

I amused myself some years ago with drawing up a Table, intended to represent the gradation of opinion on subjects of this nature, entertained at that period by numerous Geologists, and though I am aware, that it does not express in many instances exactly the notions of these individuals at the present time, any more than it does of myself, yet, as it has already found its way into a Journal, it cannot be improper to give it a place in this volume.

GEOLOGICAL THERMOMETER.

Shewing the Opinions attributed to various Geologists with respect to the Origin of Rocks.

———◆◆◆———

WHISTON, Theory of the Earth, 1725. — 100   *Plutonic*   All Rocks affected by
BUFFON, Théorie de la Terre.   *Region.*   heat. The Earth struck
LEIBNITZ, Protogæa, 1768.   off from the Sun by a
DESCARTES. — 95   Comet.

BOUÉ, Essai sur l'Ecosse, 1822.   All Rocks of Chemical
  — 90   origin igneous.

HUTTON, Theory of the Earth, (Ed.Trans. v.i.)   All the older rocks either
PLAYFAIR, Illustrations, 1820.   fused or softened by
Sir J. HALL, Edin. Trans. vol. vi. 1806. — 85   heat. Metallic Veins
Sir G. MACKENZIE, Travels in Iceland, 1810.   injected from below.

Sir H. DAVY, on Cavities in Rock Crystal,1822. — 80   Granitic Rocks igneous.
MAC CULLOCH, Various Papers in Geol.   Granitic Veins inject-
  Trans. from 1814 to 1817.   ed from below.
KNIGHT, Theory of the Earth, 1820. — 75
BAKEWELL, Elements.

                           — 70

BRIESLAC, Journal de Physique, vol. xciii.   Some Granites and Sie-
                                  nites igneous.
FAUJAS ST. FOND, Essais Géologiques. — 65   *Volcanic*   All Trap Rocks igneous.
HUMBOLDT, Travels and Memoirs.   *Region.*
SPALLANZANI, Sur les Isles Ponces.
Sir W. HAMILTON, Memoirs, &c.   — 60
DOLOMIEU, Voyage aux Isles de Lipare, 1783.
SAUSSURE, Voyages dans les Alpes, 1787.
WHITEHURST, Theory of the Earth, 1786. — 55

CORDIER, Sur les Substances Minérales dites — 50   Augite Rocks igneous.
  en masse, 1815.
VON BUCH, Travels, Memoirs, &c.
                           — 45
BUCKLAND, Memoirs.   Flœtz-trap Rocks igne-
CONYBEARE, Geology of England, 1522.   ous. Whin-dikes in-
SEDGWICK, Memoirs. — 40   jected in a fluid state
                                   from below.
DOLOMIEU, Journ. de Phys. vol. xxxvii. 1790.   Some Flœtz-traps igne-
SAUSSURE, Journal de Phys. (an. 2.) 1794. — 35   ous; others aqueous.
DAUBUISSON, Ib. 1804, Sur les Volcans d'Au-
  vergne.
DAUBENY, Edin. Philos. Journ. 1821, On the — 30
  Volcanos of Auvergne.

DAUBUISSON, on the Basalts of Saxony, 1803. — 25   *Neptunian*   Igneous origin of any
DELUC, Treatise on Geology, 1809.   *Region.*   Trap Rocks question-
KLAPROTH, Beiträge, vol. iii.   ed. Whin-dikes co-
JAMESON, Edinburgh Philos. Journal, 1819. — 20   temporaneous with the
RICHARDSON, On the Giant's Causeway,   rocks they traverse.
  Ph. Tr.
                           — 15
MACKNIGHT, Wernerian Memoirs, 1811.   All Rocks (except the
JAMESON, Geognosy, 1808.   Volcanic) deposited
MURRAY, Comparative View, 1802. — 10   from aqueous solution.
MOHS, Memoirs, &c.   Metallic Veins poured
KIRWAN, Geological Essays, 1795.   in from above.
WALKER, Lectures, 1794. — 5
WERNER, Theory of Veins, 1791.

LAMARCK, Hydrogeologie.   Secondary Rocks secret-
DEMAILLET, Telliamed.   ed by animals and ve-
                                  getables, and formed
                                   out of Water.

LIST OF WORKS

Relative to Volcanos.

ON THE VOLCANOS OF FRANCE.

Guettard ...... Memoire sur quelques Montagnes de la France,
qui ont été des Volcans. Mem. de l'Acad.
des Sciences. 1752.
Memoire sur la Mineralogie de l'Auvergne, ditto
1759.

Desmarest ..... Memoire sur le Basalt.
Ditto 1771.

Faujas St. Fond. Volcans du Vivarais. Fol. 1778.
[One of the best of Faujas' works, containing much information, from
which Geologists even of the present day may profit.]

Giraud Soulavie. Histoire Naturelle de la France Meridionale.
1780.

Legrand d'Aussy Voyage en Anvergne, 1st and 2nd Vol. 1788.
3rd Vol. 1794.

Montlosier ..... Essai sur la Theorie des Volcans d'Auvergne.
1789.
[A concise, but interesting description of the country about Clermont.]

Buchoz .... Sur le Cantal. 1792.

Dolomieu ...... Rapport. Journal des Mines. Ann. 5 & 6.
Description des courses Lithologiques. Journ.
de Pharm. Tom. 1. p. 127.
Observations. Soc. Philom. Ann. 6.

Lacoste .... Sur les Volcans de l'Auvergne. Ann. 11.
Lettres, &c. 1805.

Marzari Pencati. Orittographia del Monte Coiron. 1806.

Cocq .... Journal des Mines. Tom. 19. 1806.

De Laizer . . . . . Sur le Puy Chopine. 1808.

De Laizer, Weiss, and Gillet Laumont, sur les Laves avec parties
bleues. Journ. de Mines. Tom. 23.

De Laizer . . . . . :Sol d'Auvergne. Journ. des Mines. Tom. 23.

Ramond . . . . Hauteurs Mesurees Barometriquement. Journ.
des Mines. Tom. 24.

The same Memoir in a more complete form.
Acad. des Sciences. 1813—1815.

[This Memoir is more than it professes to be, including, together with
barometrical measurements, judicious remarks on the Geology of the
country.]

Cordier . . . . Sur le Mont Mezen. Journ. des Mines. Tom. 26.

Sur le Breche de Mont Dor. Tom. 4. 1819.

Von Buch . . . . . Mineralogische Briefe aus Auvergne. In the 2nd
Vol. of his Geognostichen Beobachtungen.
1809.

Journal des Mines. Tom. 13.

Vital Bertrand . Essai sur l'Histoire Naturelle du Puy. 1811.

Bertrand Roux. Description du Puy en Velay. 1823.

[An excellent account of the physical structure of that part of France.]

Steininger . . . . . Erloschenen Vulkane in Sudfrankreich. Mainz,
1823.

Bemerkungen über die Eifel und Auvergne.
1824.

Nöggerath . . . . . Vorkommen von Fossiler Saugethier-resten in der
Auvergne. Karsten Archiv. 1824.

Menard de la Groye. Sur Beaulieu. Journ. de Phys. Vol. 82.

[There is also a Memoir on the same subject by Saussure, in the same
Journal for the year 1787.]

Daubeny . . . . . . Letters on Auvergne addressed to Professor
Jameson. Edinb. Phil. Journ. for 1820—21.

Letters on Auvergne, translated into German by
Professor Nöggerath, with Notes and Correc-
tions. Bonn. 1825.

Bakewell's . . . . . Travels in the Tarentaise. 2 Vols. 8vo. London. 1823.

VOLCANOS OF GERMANY.

Near the Rhine.

[I have been so much indebted to Dr. Boue in preparing this list, especially in what relates to German works on Geology, that I thought proper to affix an Asterisk to such Books mentioned below, as would have been probably omitted, had it not been for his kind assistance.]

* Collini ...... Observations sur les Monts Volcaniques. Manheim. 1781.

[Gives an account of the Volcanic country near the Rhine.]

* The same Author in 1775 had published Observations sur les Agates et Basaltes.

Sir W. Hamilton. Phil. Transactions for 1778. part 1.

[Gives a short account of the volcanic rocks of the Rhine.]

Nose .... Orographische Briefe über das Siebengebirge. 2 vol. quarto. 1789.

[The most full, though a tedious and desultory account of the neighbourhood of the Rhine.

* Humboldt.... Beobacht. über einigen Basalten am Rhein. 1700.

* Forster ..... Ansichten des Niederrheins. 1791.

* Dettier ...... sur les Volcans de la Kyll. Paris. 1804.

* Wurzer...... Taschenbuch des Siebengebirge. Köln. 1815.

* Camper...... Reise nach den Vulkanen des Niederrheins, edited by Von Mark.

Keferstein ..... Beiträge zur Geschichte und Kenntniss des Basaltes. In Schriften der Naturforsh Gesellsaft zu Halle. 1819.

[Gives an account of the Eyfel Volcanos. See also the Numbers of his "Teutchland geologish-geognostich dargestelt."]

Steininger ..... Geognostiche Studien. 1819.

Erloschenen Vulkane in der Eifel. 1820.

Neue Beiträge. 1821.

Gebirgskarte. 1822.

[Various Memoirs in the work entitled "Das Gebirge in Rheinland Westphalen," edited by Professor Nöggerath of Bonn, 4 Volumes of which are now published.]

SWABIA.

Leonhard......Taschenbuch for 1823, gives an account of the Rocks of Brisgaw, and those near Constance.

Boué ....in the Annales des Sciences Naturelles for August 1824, gives an account of the same rocks.

Saussure ....in the Journal de Physique Vol. 44, had also treated of them.

Pictet.........has inserted, in the Transactions of the " Societe d'Histoire Naturelle of Geneva," a short notice respecting the same country.

* Razoumofsky in the Bergmannisches Journal. Vol. V. p. 188. has spoken of them.

* Wiedenman ..(Vol. VI. of the same work) has commented upon the foregoing paper.

* Heller in Moll's Annal. der Berg. und Huttenkunde. Vol. 1.
[On the Rhongebirge]

* Voigt ....Beschreibung des Hochstifts Fuld. 1783.
[Relates to the Rhongebirge.]

* Voigt ....Mineralogischen Reisen von Weimar über den Thuringerwald, Rhongebirge, &c. 2 Vol. 8vo. 1802.

Voigt ....Kleine Miner. Schriften. Weimar. 1802.
[Relates to the Basalts of Hessia,]

* Jacob and Hoff. Der Thuringerwald für Reisende geschildert. 2 Vols. Gotha. 1807—1812.
[Gives an account of the Basalts on the S. W. side of Thuringerwald.]

Sartorius ......Die Basalte von Eisenach. 1802.
[Contains the first notice of the curious Dykes, mentioned p. 71—72.]
Geognostiche Beobacht. Eisenach. 1823.
Nachtrag. 1823.
[All these Memoirs relate to the Basalts about Eisenach.]

* Hoff.........Magazin der Berlin. Gesellschaft Naturfors-chender freunde. 5th year. p. 347. on the Blaué Kuppe near Eschwege.—with a plate.

See also De la Beche's Geological Memoirs.

Ditto on the Dyke of Hörsel near Eisenach. Same work 7th year.

* Leonhard ....Taschenbuch for 1822.
[Gives an account of the Rocks about Heidelburg.]

* Scubler......in the Wirtemb. Jahrbucher for 1822.
[Gives an account of the Basalts of Wirtemburg.]

HESSIA *and the neighbourhood Parts.*

Raspe.........on some German Volcanos.   London. 1776.
[Relates to the Habichwald near Cassel.]

* Schaub ......Beschreibung des Meissners in Hessen. 1799.

* Catalogue des Min. du pays de Cassel. 1808.

* Reiss........Beobacht. über einige Hessische Gebirsgegende.
Berlin. 1790.

* Becher ......über die Nassauische Gegende.   1786.
[Gives an account of the singular Porphyries associated with Clay Slate
about Dillenburg.]

* Klipstein ....Miner. Briefen. Giessen.   1779—1784.
[Relates to the Vogelsgebirge.]

FICHTELGEBIRGE.

Flurl ....Beschreibung von Baiern.   1792.

* Goldfuss and Bischof.Beschreibung der Fichtelgebirge.

SAXONY.

Charpentier ....Min. Geographie der Chursachsischen landes.
Leipsic. 1778.

Daubuisson ....on the Basalts of Saxony.
[Translated by Neill. Edinburgh. 1814.]

SILESIA.

Von Buch .....on the Environs of Landeck.
[Translated by Dr. Anderson. 1810. Pronounced to be the best Essay
in Mineralogical Geography that had appeared in Germany.]

* Oehnhausen ..Versuch einer geognostichen Beschreibung von
Obersilesien. 8vo. Essen. 1822.

BOHEMIA.

* Reuss.,......Sammlung Naturhist. Aufsatze. 1796.
Min. & Bergmann. Bemerk. über Bohmen. Berlin. 1801.

* Reuss.......Orographie des Nordw. Mittelgeb. Dresden.
1790.

Miuer. Geographie von Bohmen. 2 vols. 4to.
1723—4.

* Lindacher ...in Sammlung. Phys. Aufsatze von Mayer. 1791.
[Shews the Volcanic origin of the Basalt of Wolfsberg in the Circle of
Pilsen.]

. . . . .

HUNGARY, TRANSYLVANIA, STYRIA, BANNAT.

Fridwalozky ...Mineralogia magni Principatus Transylvaniæ
Claudiop. 1767.

Scopoli ....Crystallographia Hungarica. 1776.

Born..........Briefe. Dresden. 1774.
[Also translated into French. Relate to the Bannat of Temeswar.]

Fichtel........Min. Beiträge von Siebenburgen. 1780.
Bemerkungen über den Karpathen. 1791.

Ferber ..,.....Abhandlungen über die Gebirge in Ungarn.
1780.

Bredetzki......Beiträge zur Topographie von Ungarn. 3 vols.
8vo.

Buchholz ......Reise auf die Karpatischen Gebirge. Ungrisches
Magazin. tom. 4. 1787.
[The same Author published anonymously another Description of the
Carpathians, with some remarks on Hungary. 1783.

Townson ......Travels in Hungary. 4to. 1797.

Esmark ....Bescreibung einer Reise durch Ungarn. 1798.
[Represents the Trachytes of Hungary as of aqueous origin.]

Asboth........Reise von Keszthely nach Vesprim. Wien. 1803.
[Relates to the Basalts of Lake Balaton.]

Leonhard's ....Taschenbuch 1816. p. 413. Memoir on some
parts of Hungary by Jonas.

Ditto ....1813—1815—1816 — 1817 — 1819 — 1820, for
different Memoirs by Zipser.

Zipser ....has also published " a Manuel" of the Mineralogy of Hungary. Œdenburg. 1817.

Bright.........Travels in Hungary. 4to. 1818.
[Chiefly Agricultural.]
On the Hills of Badacson, &c. Geological
Transactions for 1819.

Beudant.......Voyage en Hongrie. 3 Vols. 4to. Paris. 1822.
[An excellent work, but would have been benefitted by compression.]
Von Buch.....Transactions of the Berlin Academy for 1818—21.
[On the Volcanos of Styria.]

ON THE VOLCANOS OF ITALY.
Venetian States.

Arduini ....His observations on the Euganean Hills and
Vicentin, in which their volcanic origin is
asserted, were first published in 1765.
Other publications of his appear in the Saggi
Scientifici, &c. dell. Academia del Padova,
and in the Atti dell. Academia di Sienna.
1760—61.
He also published 1774 " Saggio Mineralogico
di Lithogonia and Orognosia." Padova. 4to.
A German Translation of some of his works ap-
peared at Dresden in 1778, entitled " Samm-
lung Mineral. Abhandlungen."
Salmon .....;...on the Euganean Hills. Journal de Physique.
Vol. 53.
Strange........Phil. Trans. Vol. 65. for 1775.   ]
Fortis.........Memoires pour servir a l'Histoire Naturelle de
l'Italie. Paris. 1802.
[Relates to the Vicentin, which he shews to be of Volcanic formation.]
Brongniart.....Sur les Terrains calcareo-trappiens. du Vicentin.
1823.
[Relates chiefly to the Shells contained in the Tertiary Rocks.]
Maraschini.....Saggio Geologico sulle Formazione delle Rocce
del Vicentino. Padova. 1824.
[The most complete account that has appeared of this country, since
Geology has assumed its present form.]
Fleurian de Bellevue. Journ. de Phys. 1790.
[Maintains the igneous origin of the Rocks of Grantola.]
Pini ....sul alcuni Fossili singolari della Lombardia. 1790.
[On the Rocks near Grantola.]
Gautieri ....Confutatione dell. opin. di alcuni Mineralogiste
sulla Volc. di Mont. di Grantola. 1807.
Follini ....Descrittione del Monte Baldo.

CENTRAL ITALY.

Targioni Tozzetti Relatione dei Viaggi in Toscana. Firenze.
2 vols. 8vo.
[The same translated into French. Paris. 1792.
Ferber ....Histoire Naturelle d'Italie. Traduit par Die-
trich. Strasbourg. 1778.
[Translated into English, 1776.]
Santi ......Viaggi d'Historia Naturale. 3 Vols. Pisa. 1795—98.
[The last Volume relates to the Monte Amiata. The same in French.
Lyons. 1802.]
Gmelin ....Dissert. di Hauyne. Heidelburg. 1816.
[Gives a Map of the Craters about Albano.]
Kephalides.....Reise durch Italien & Sicilien. 2 Vols. with
Maps. 1818. 8vo. Leipsic.
Brocchi ....Conch. Subappeninna. 4to. Vol. 2. Milano.
Catalogo Ragionato di una Raccolta di Rocce.
Milano. 1817.
Suolo di Roma. 1820.
Menard di Groye sur le Feux di Baragazzo. Journ. de Phys.
Vol. 85.
Prystanowsky ..über die ursprung der Volcanen in Italien.
Berlin. 1823.
[There are also several Memoirs by Brocchi and others, in the Biblioteca
Italiana, on central Italy.]

SOUTHERN ITALY.

Sorrentino .....Istoria del Vesuvio. Napoli. 1734.
Duca del Torre Istoria e Fenomeni del Vesuvio. 1755—1768.
4to. Napoli.
Gioeni.........Litologia Vesuviana.
Sir W. Hamilton. Campi Phlegræi. Naples. 1776. Fol.
Supplement to Ditto. 1779.
[First published in the Philosophical Transactions.]
Dolomieu ......sur les Iles Ponces. 1788.
Breislac ....Voyages en Campanie. 2 Vol. 8vo. 1801.
Von Buch .....Geognostiche Reise. Berlin. 1809.
[Vol. 2nd relates to Rome and Naples.]
Menard de Groye. Journ. de Phys. Vol. 80. on the Eruption of
1813, &c.

Lippi ....Sotteraneo di Pompeo e di Ercolano. Opera delle alluvioni, è non dell Erurzione del Vesuvio. 1819.

Odesleben .....Reise in Italien. 2 vols. 1821.
[Relates to Naples, Rome, and the Euganean Hills.]

Necker ....sur le Monte Somma. In the Transactions of the Natural History Society of Geneva. Vol. 2.

Geological Transactions. Vol. 3. On the Dykes of ditto.

Monticelli & Covelli Storia di Fenomeni del Vesuvio. Napoli. 1823.

Göthe ....zur Naturwissenschaft. 2nd vol. contains an account of the Geological Phænomena presented at the temple of Puzzuoli.

Canonico de Jorio—A pamphlet on the same subject.
[See also sundry Papers by Brocchi and others on this part of Italy— especially on Mount Vultur and the Lago di Ansanti.]

SICILY.

Brydone ......Travels in the Two Sicilies. 1774.

Borch.........Mineralogia Siciliana. 1780.
Lettres sur la Sicile & sur Malte. Turin. 1782. 2 vols. 8vo. with a volume of plates.

Spallanzani ....Voyages dans les deux Siciles. Traduits de l'Italien. 6 vols. 8vo. Paris. An. 8. (1800.)

Recupero ......Storia Naturale dell' Etna. 2 vols. 4to. Catania. 1815.

Ferrara ....Campi Flegrei della Sicilia. Messina. 1810.
Descrizione dell' Etna. 1818.
Memorie sopra il Lago Naftia, e sopra l'Ambra Siciliana. 1805.
sopra i. Tremuoti della Sicilia in Marzo. 1823.

Gemellaro .....Giornali dell' Eruzione dell' Etna avvenuta 1809. Catania. 1816.
sopra alcuni pezzi di Granito, &c. trovati presso ai a cima dell' Etna.

* Moricand ....Observ. Geognostiques. Journ. Britann. 1819.
Translated in Gilbert's Annals. 1820.

462 *List of Works.*

Brocchi ....in Bibl. Italiana. Vol. 20. on the Cyclopean
          Rocks.
Gourbillon.....Voyage à l'Etna. Paris. 1820.
Sayve ....Voyage en Sicile. 1822.
Smith ....Tour in Sicily. 4to. 1824.
    [Little on Geology.]
Daubeny ......Sketch of the Geology of Sicily in the Edin-
          burgh Philosophical Journal. 1825; and in
          Silliman's American Journal. 1826.

LIPARI ISLANDS.

Dolomieu......sur les Iles di Lipari. 1783.
          See also Spallanzani, Voyages dans les deux
          Siciles. Ferrara Campi Flegrei, &c.

ICELAND.

Von Troil......Letters. London, 1770.
    [Giving an account of Sir Joseph Banks's Journey.]
Olafsen ....Reise durch Island (Teutsche Uebersetz.)
          Kopenhagen. 1774.
Sir G.Mackenzie.Travels. Edinburgh. 4to. 1811.
Hook er.......Recollections, 2 vols. 8vo. 1813.
Henderson .....Residence in Iceland. 2 vols. 8vo. 1819.
Garlieb ....Iceland rucksichlich seiner Vulkane. Freiburg.
          1819, 8vo.
Menge ....Edinburgh Philosophical Journal, vol. 2.

GREECE, TURKEY, AND ARCHIPELAGO.

Tournefort.....Voyage in the Levant. 2 Vols. 4to. London.
          1718.
Philosophical Transactions for 1707. Island of Santorino.
Choiseul Gouffier,Voyage Pittoresque de la Grece. 2 Vols. folio.
Clarke ....Travels in various parts of Europe, 1st volume.
Dodwell.......Travels in Greece. 4to. 1815.
Andreossi......Voyage a l'embouchure du Mer Noire. Paris.
          1818.

AFRICA.

Von Buch . . . . . in the Transactions of the Academy of Berlin has
given remarks on the Canary Islands, espe‑
cially Lancerote, (Ann. 1818—19.)—Tene‑
riffe (Ann. 1820—21.)*

Bory St. Vincent Voyage aux Iles d'Afrique. 3 Vols. 4to. Paris.
1804.

[On the Isles of France and Bourbon.]

Dr. Webster . . . Boston, 1820. Account of the Azores.

Ehrman . . . . Weimar, 1807. On St. Helena.

Beatson's . . . . . . Tracts on St. Helena. 4to. London. 1816.

Annales des Mines, 1824, gives some account of basalts in Sene‑
gal, from the observations of a Frenchman who
died there

Bowdich . . . . in his Posthumous Work, London, 1825, has
given the most recent geological description
of Madeira and Porto Santo.

See also Humboldt's Personal Narrative, Vol. 1.
for Teneriffe.

---

* Whilst this sheet was going through the press, I saw for the first time a
later and more complete publication of Von Buch's, entitled " Physicalische
Beschreibung der Canarischen Inseln," published at Berlin in 1825 in 1 Vol.
4to, with a Folio Atlas accompanying it. This work appears to contain the
substance of all his Memoirs above referred to, together with a great deal of
new matter. I regret that I was ignorant of the existence of such a work
whilst engaged in drawing up the materials of my Third Lecture, as this
Geologist has in the latter part of his publication given many interesting
details with regard to Volcanos in other parts of the world. I may instance
particularly his account of the Volcanic Islands in the Grecian Archipelago,
as containing some facts to which I could not have access, as they were com‑
municated by an intelligent naturalist, (Signor Parolini of Venice) who has
never published any account of his observations in those parts of Europe.

ASIA.

No works strictly Geological. The Travels which contain most information on the physical structure of this quarter of the Globe, are,

ARABIA, SYRIA, ASIA MINOR, PERSIA.

Niebuhr . . . . Beschreibung von Arabien. Copenhagen. 1772.
4to.
[Translated into French 1773.]

Sestini . , ... Voyage de Constantinople à Bassora, en 1781.
[Translated from the Italian.]

*Sestini . . . . Briefen über die Vulkane von Syria und Meso-
potamia, in the Deutschen Mercur.

Volney . . . . Voyage en Syrie and Egypte. 3d edition. Paris.
1800.

Olivier . . . . Voyage dans l'Empire Othoman. 3 Vols. 4to.
1807.

Morier . . . . Journey through Persia. 2 Vols. 4to. 1808—9.

Seetzen . . . . Researches, published under the title of, A Brief
Account of the countries adjoining the Lake
Tiberias, the Jordan, and the Dead Sea.
London, 1810.
Also in Zach's Correspondence, Vol. 13, p. 551.
Vol. 18, p. 425.

Buckingham . . . Travels in Palestine. 1821.

Burckhardt . . . . Travels in Syria. 1822.

CAUCASUS.

Pallas . . . . Travels in the Russian Empire. 5 Vols. 4to.
1788.

Reineggs . . . . . . Beschreibung des Kaukasus. Gotha und St. Pe-
tersburg, 1796.

Klaproth . . . . . . Reise in dem Kaukasus. Halle und Berlin. 1812.

Engelhardt & Parrot, Reise in dem Krym und Caucasus. Berlin,
1825.

Sir Ker Porter . . Travels in Georgia, 4to. 1822.

See also Tournefort, Clarke, and others.

ISLANDS OF ASIA.

Kraskeninicoff ,. History of Kamscatka and the Kurule Isles.
[Translated from the Russian into French 1767.]
Marsden . . . . Sumatra. 3d edition. 1811.
Sir S. Raffles . . . Java. 2 Vols. 4to. London.
Kotzebue . . . . . . Voyage of Discovery during the years 1815—18.
[Translated into English 1821.]

SOUTH SEA ISLANDS.

Forster . . . . Observations during a voyage round the world.
London, 1778.
Ellis . . . . Tour in the Sandwich Islands. 1826.
Von Buch . . . . . On Van-Diemen's Land, in the Magazin der
Naturforschender Freunde zu Berlin.
See also Capt. Cook's, Bligh's, Dampier's, Vancouver's, and La
Billardiere's Voyages.

AMERICA.

Molina . . . . Histoire Naturelle de Chili.
Humboldt . . . . . Political Essay on New Spain.
Personal Narrative.
Essai sur le Gisement des Roches.

ANTILLES.

Torrubia . . . . Historia Naturale d'Espanola, Madrid. 1754.
Dauxion Lavaysse Voyage a l'Ile de Trinitè.
Le Blond . . . . . . Voyage aux Antilles.
Nugent . . . . in the Geological Transactions, vol. 1. old series,
and vol. 1. new series.
Cortes . . . . Journal de Physique, tom. lxx.
Moreau de Jonnes, Statistique des Antilles. Paris. 1822.

WORKS ON EARTHQUAKES.

Beuther .... Compendium Terræmotuum. Strasburg. 1601.
Bernhertz ..... Terræmotus (a Register of Earthquakes).
       Nurnberg. 1616.
Dr. Vincenzo Magnati.. Notizie istoriche de' Terremoti.
       Napoli. 1688.
A Chronological and Historical account of Earthquakes.
Seyfarth .... Allgemeine Geschichte der Erdbeben.
       Leipsic. 1756.
Bertrand ...... Memoires Historiques sur les Tremblemens de
       Terre. La Haye. 1757.
Bertholon...... Journal de Physique. Tom. 14.
Vivenzio ...... Istoria e teoria de Tremuoti avvenuti nella Pro-
       vincia della Calabria, &c. di 1783—87.
       Napoli. 1788.
Cotte .... Tableau Chronologique des principaux pheno-
       menes meteorologiques observes en differens
       pays depuis 33 ans. &c. " Journal de Phy-
       sique. Tom. 65."
The London Philosophical Transactions contain several Memoirs
       on Earthquakes by Dr. Stukeley and others,
       especially about the middle of the last
       century.

FINIS.

ERRATA.

P. 60, line 3 from the bottom, insert " may" before " be."

81, line 15, *for* " contain," *read* " contains."

107, line 14, *for* " Remebyel," *read* " Remetzel."

110, line 8 from the bottom, *for* " Gräbz," *read* Grätz."

146, note, *for* " Reishe," *read* " Reiske."

155, line 9 from the bottom, leave out " their."

160, line 7 from the bottom, *for* " sulphurous and muriatic acid gases, the former chiefly produced during a period of calm, the latter at, and immediately subsequent to, an eruption," *read* " muriatic and sulphurous acid gases, the former produced chiefly at the commencement, the latter during the later periods of an eruption."

212, line 15, leave out " Persian Gulph."

221, line 10, *for* " one," *read* " our."

260, line 5, *for* " last year," *read* " the year before last."

264, line 15, *for* " phonolite," *read* " clinkstone."

do. line 3 from the bottom, dele " long."

272, note, *for* " ause," *read* " aux."

314, line 5 from the bottom, *for* " composed in," *read* " composing."

327, line last but one, *for* " Hawaiah," *read* " Owhyhee."

362, line 10, *for* " preceeding," *read* " preceding."

367, line 8 from the bottom, *for* " Gay Lussac," *read* " Arago."

DIRECTIONS TO THE BINDER.

Plate of Jorullo to front the Title Page.
Map of the Dead Sea to front page 286.
Map of Mexico to front page 336.

PRINTED BY W. PHILLIPS,
GEORGE-YARD, LOMBARD-STREET, LONDON.

Printed in the United States
By Bookmasters